# 朗文英漢化學圖解詞典

## LONGMAN ENGLISH-CHINESE
## ILLUSTRATED DICTIONARY OF
# CHEMISTRY

Longman 朗文　YORK PRESS

English edition © Librairie du Liban 1982
(Original title: Longman Illustrated Dictionary of Chemistry)
Bilingual edition © Librairie du Liban & Longman Asia Limited 1991, 1992, 1995

Published by
Longman Asia Limited
18/F., Cornwall House
Taikoo Place
979 King's Road
Quarry Bay
Hong Kong
Tel: 2811 8168
Fax: 2565 7440
Telex: 73051 LGHK HX

朗文出版亞洲有限公司
香港鰂魚涌英皇道 979 號
太古坊康和大廈十八樓
電話：2811 8168
圖文傳真：2565 7440
電傳：73051 LGHK HX

Third impression 1995
一九九一年初版
一九九五年第三次印刷

ISBN 962 359 053 9

Produced by Longman Asia Limited
Printed in Hong Kong
SWT/03

The
publisher's
policy is to use
**paper manufactured
from sustainable forests**

# Contents 目錄

# How to use the dictionary　本詞典的用法

This dictionary contains some 1500 words used in chemistry. These are arranged in groups under the main headings listed on pp.3–4. The entries are grouped according to the meaning of the words to help the reader to obtain a broad understanding of the subject.

At the top of each page the subject is shown in bold type and the part of the subject in lighter type. For example, on pp.12 and 13:

**12 · PROPERTIES OF SUBSTANCE**/CHANGE OF STATE

**PROPERTIES OF SUBSTANCE**/PHYSICAL PROPERTIES · 13

In the definitions the words used have been limited so far as possible to about 2000 words in common use. These words are those listed in the 'defining vocabulary' in the *New Method English Dictionary* (fifth edition) by M. West and J.G. Endicott (Longman 1976). Words closely related to these words are also used: for example, *characteristic*, defined under *character* in West's *Dictionary*. For some definitions other words have been needed. Some of these are everyday words that will be familiar to most readers; others are scientific words that are not central to chemistry. These are contained on pp.211–33 and will be found listed in the alphabetical index.

## 1. To find the meaning of a word

Look for the word in the alphabetical index at the end of the book, then turn to the page number listed.

The description of the word may contain some words with arrows in brackets (parentheses) after them. This shows that the words with arrows are defined near by.

(↑) means that the related word appears above or on the facing page;
(↓) means that the related word appears below or on the facing page.

A word with a page number in brackets (parentheses) after it is defined elsewhere in the dictionary on the page indicated. Looking up the words referred to in either of these two ways may help in understanding the meaning of the word that is being defined.

The explanation of each word usually depends on knowing the meaning of a word or words above it. For example, on pp.91–2 the meaning of *crystalloid*, *crystallization* and *polymorphism*, and the words that follow

---

本詞典共收集化學用詞約 1,500 個，這些詞按照第 3－4 頁所列的主要標題分類編排。所有詞條均按詞義歸類，旨在幫助讀者對所查找科目獲得概括的瞭解。

每一頁的上方用黑體字印出有關科目名稱，並以秀麗體印出該科目下的分段名稱。例如在第 12 頁和第 13 頁：

**12 · 物質的性質／物質變化**

**物質的性質／物理性質 · 13**

釋義部分所用的詞盡可能限於常用的 2,000 個詞左右；這些詞列於 M · 韋斯特和 J · G · 恩迪科特合編的《新法英語詞典》（第五版，朗文公司 1976 年版）中的釋義詞彙內。一些和這些詞密切相關的詞，例如：使用韋斯特詞典中在 "character（特徵）" 條下釋解的 "characteristic（特徵）" 這個詞。對於某些釋義要用另外一些詞。其中有些是多數讀者都熟悉的日常用詞，還有一些是非偏重於化學方面的科學用詞。這些詞都收於第 211 至 233 頁中，並編列於字母順序索引中。

## 1. 查明詞的意義

在詞典末尾的字母順序索引中找出欲查的詞，然後翻到該詞所示明的頁數。

詞的釋義中遇有一些詞後面帶有箭號括在圓括弧（圓括弧）內，表示該詞的解釋就在附近。

(↑) 表示和這個相關的詞出現在本詞條之前或成後一頁上；
(↓) 表示和這個相關的詞出現在本詞條之後或後一頁上。

遇到詞後用括號註明的頁碼上，表示該詞的解釋在所註明的頁碼上。參照這兩種方式查出這些詞並閱讀其解釋，可幫助您更好地理解原先所查那個詞的詞義。

對每一個詞的闡釋通常都依賴於理解在該詞前面出現的一個詞或幾個詞的意義。例如，在第 91 至 92 頁、"晶質"、"結晶作用"、"同質多晶現象" 以及接着出現的幾個詞的詞義，

## 2. To find related words

depends on the meaning of the word *crystal*, which appears above them. Once the earlier words are understood those that follow become easier to understand.

Look in the index for the word you are starting from and turn to the page number shown. Because this dictionary is arranged by ideas, related words will be found in a set on that page or one near by. The illustrations will also help here.

For example, words relating to crystal systems are on pp. 96–7. On p. 96 *crystal systems* is followed by words used to describe various kinds of systems and types of structures; pp. 98–100 give words for the related subject of colloids; p. 101 lists words for the properties of different types of colloidal dispersions.

## 3. As an aid to studying or revising

There are two methods of using the dictionary in studying or revising a topic. You may wish to see if you know the words used in that topic or you may wish to revise your knowledge of a topic.

(a) To find the words used in connection with crystals look up *crystal* in the alphabetical index. Turning to the page indicated, p.91, you would find *crystal, crystalloid, crystallization, recrystallization,* and so on. Turning over to p.93 you would find *pattern, symmetry,* etc.

(b) Suppose that you wished to revise your knowledge of a topic; e.g. *isotopes.* If, say, the only term you could remember was *relative molecular mass* you could look it up in the alphabetical index. The page reference is to p.114. There you would find words relating to *isotope, isotopic ratio* and *relative isotopic mass,* etc. If you next looked at p.113, you would find words relating to the structure of the atom which would help you to understand the descriptions given for the words connected with isotopes.

## 4. To find a word to fit a required meaning

It is almost impossible to find a word to fit a meaning in most dictionaries, but it is easy with this book. For example, if you had forgotten the word for the substance from which metals are obtained, but you knew such substances came from a mine, all you would have to do would be to look up the word *mine* and turn to the page indicated, p.154. There you would find the word you wanted, which is *ore,* and also related words such as *deposit, seam* and *lode.*

## 2. 查找相關的詞

幫依賴於在上述這些詞前面出現的" 晶體 "這個詞的意義。理解了在前面出現的那些詞的意義之後就比較容易理解接着出現的那些詞的意義。

在索引中查找出作爲思想源頭的那個詞,然後翻到所註明的頁碼。由於本詞典是按照概念編排的,因此近似在同一頁或後的一頁上可找出一組相關的詞。再參看插圖也可幫助自己。

例如:在第96至97頁上列有一些晶系有關的詞;在第96頁,在" 晶系 "這一詞條後面編排的各詞條用於描述各種晶系和結構型式;在第98至100頁編排的是一些用於有關膠體方面的詞;在第101頁編列一些用於不同類膠體性質的詞。

## 3. 學習和複習的輔助工具

本詞典可用作學習和複習某一個課題時,有兩種用法。您可以利用本詞典查核自己是否認識該課題中所用的詞,也可以用以複習某一項課題的知識。

(a) 引查找晶體方面用到的詞時可在字母順序索引中查出" 晶體 "(crystal)這個詞,翻到所註明的第91頁,即可找到" 晶體 "、" 結晶作用 "、" 再結晶作用 "等詞。翻到第93頁,可查到" 圖像 "、" 對稱 "等詞。

(b) 假設您希望複習某一項課題的知識,例如有關同位素的知識。如果您能記住的唯一名詞是" 相對分子質量 ",您只需在字母順序索引中查出這個詞條,翻到標註的第114頁,就可查到" 同位素 "、" 同位素比 "、" 相對同位素質量 "等詞。再翻到第113頁,可查到與原子結構有關的一些詞,這可幫您理解與同位素有關的那些詞的解釋。

## 4. 查找適當的詞,以表達確切的意義

在大多數詞典中,您要查找一個適當的詞來表達某一意義,這幾乎是不可能的,但用本詞典就可輕易做到這一點。例如,您忘了用於表達可獲得各種金屬的那種物質的名詞,但您知道這種物質來源於一個礦(mine)這個詞,翻到標註的第154頁,您就可查出這個詞,即" 礦石 "(ore)這個詞,再查到相關的詞,如" 礦床 "、" 礦層 "、" 礦脈 "。

# THE DICTIONARY 詞典正文

**material** (n) a material has some general properties (↓) by which it can be recognized. Other properties can vary between different kinds of the same material. Examples of materials are wood, leather, rubber and brass. Different kinds of wood have slightly different properties: their colours vary, their densities (p.12) vary, their hardness varies. The variation of the properties, from one kind of the same material to another, is small. The chemical composition (p.82) of a material can also vary, but the variation is small. **material** (adj).

**substance** (n) a substance has properties (↓) by which it can be recognized. These properties do not vary from one piece of the substance to another. The chemical composition (p.82) of a substance does not vary. Examples of substances are: iron, cane-sugar, salt. Many substances are compounds (↓); some substances are elements (↓).

**compound** (n) a substance (↑) that can be decomposed by chemical action (p.19) into simpler substances. The chemical constitution (p.82) of the substance is known, and a chemical formula (p.78) can be given for it. For example, lime is a compound of calcium and oxygen; one atom (p.110) of calcium combines with one atom of oxygen to form one molecule (p.77) of lime (calcium oxide); the chemical formula is CaO. To contrast material (↑), and compound: a material (e.g. wood) has a chemical composition (p.82) and properties (↓) which may vary between limits; a substance (e.g. a protein) has a definite chemical composition, but its constitution may be too complex to be described; a compound (e.g. sulphuric acid) has a definite chemical composition, a known chemical constitution, and can be given an exact chemical formula.

**element**¹ (n) a substance (↑) that cannot be decomposed by normal chemical action (p.19) into simpler substances. All substances and materials (↑) are composed of elements which are chemically combined. See page 116 for a modern definition of element. **elementary** (adj).

材料：物料 (名) —— 一種材料具有某些可辨認的共同性質(↓)。同一種材料的不同類別之間可有所不同的其他性質。例如，木材、皮革、橡膠和黃銅是不同種材料的例子。各種木材的不同類別之間，其性質略有不同，即各種木的顏色、密度(第12頁)、硬度各異，但同一種材料的不同類別之間，性質差別很小。同一種材料的不同化學組成(第82頁)雖也可不同，但差別其小。(形容詞為 material)

物質：實物 (名) 物質具有可辨認之性質(↓)。這些性質並不因同一種物質中之不同部分而有所不同(同一種物質的化學組成(第82頁)也不會改變。例如：鐵、蔗糖、鹽均為物質。有許多物質是化合物(↓)；還有一些物質則是元素(↓)。

化合物 (名) 可藉化學作用(第19頁)分解為較簡單物質的一種物質(↑)。物質的化學構造(第82頁)為人們所知，可寫出其化學式(第78頁)。例如，石灰是鈣和氧的化合物，由一個鈣原子(第110頁)和一個氧原子(第77頁)化合而成，組成一個石灰(即氧化鈣)分子(第77頁)，其化學式為 CaO。材料(↑)、物質(↑)和化合物間的差別在於：材料(如木材)有某一種化學組成(第82頁)及可在一定範圍內改變的性質(↓)；物質(如蛋白質)有明確的化學組成，但其構造可能過於複雜以描述之；化合物(如硫酸)有明確的化學組成，人們已知其化學構造，因而可寫出正確的化學式。

元素 (名) 不能藉一般化學作用(第19頁)再分解成更簡單物質的一種物質(↑)。一切物質和材料(↑)都是由一些元素化學結合而成。有關元素的"現代定義"可參見第116頁。(形容詞為 elementary)

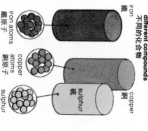
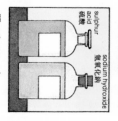

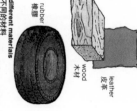

**different materials** 不同的材料
rubber 橡膠
wood 木材
leather 皮革
plastic 塑膠

**different substances** 不同的物質
iron 鐵
sugar 糖

**different compounds** 不同的化合物
sulphur acid 硫酸
sodium hydroxide 氫氧化鈉

**different elements** 不同的元素
iron atoms 鐵原子
copper atoms 銅原子
sulphur atoms 硫原子
iron 鐵
copper 銅
sulphur 硫

**property** (n) a property can be seen, heard, smelt, or felt by the senses and it allows one material (↑) or substance (↑) to be recognized as different from another material or substance. All materials and substances have physical properties (↓) and chemical properties (p.19).

**physical property** a property (↑) which does not depend on the effect of other materials or substances. Examples of physical properties are: shape, colour, odour (p.15), solubility (p.87), melting point (p.12), density (p.12). See chemical property (p.19).

**intensive property** a property which does not depend upon the amount of material or substance; such properties are used to identify different specimens (p.43) of the same material or substance, e.g. colour, odour, density, boiling point (p.12), concentration (p.81).

**extensive property** a property which depends upon the amount of material or substance; such properties are used to identify different specimens (p.43) of the same material or substance, e.g. mass, volume.

**characteristic** (adj) describes a property which readily distinguishes (p.224) an object, material, substance, pattern (p.93) from all other similar things. A characteristic property provides an easy means of recognition, e.g. copper has a characteristic reddish-brown colour, which readily distinguishes it from other metals. **characteristic** (n).

**feature** (n) a distinctive property common to a group of materials or substance.

**description** (n) a list of the properties of an object, material, substance, pattern, form of energy, or collection, or a list of the events in a process. **describe** (v), **descriptive** (adj).

**state of matter** solid, liquid, and gas are the three states of matter. Any material or substance is a solid, a liquid or a gas.

**change of state** a physical change (p.13) of a material or substance from one state of matter (↑) to another e.g. from solid to liquid; from liquid to gas. A change of state is commonly caused by heating or by cooling.

性質：屬性 (名)　性質係可由感覺器官視、聽、嗅或感知的，亦可作爲辨別一種材料或物質的區別。一切材料和物質都具有物理性質 (↓) 和化學性質 (第 19 頁)。

物理性質　不受其他材料或物質影響的一種屬性。例如：形狀、顏色、氣味 (第 15 頁)、溶解度 (第 87 頁)、熔點 (第 12 頁)、密度 (第 12 頁) 皆爲物理性質。參見 "化學性質" (第 19 頁)。

強度性質　指和材料或物質(的)量無關的一種性質。強度性質可用於鑑別 (第 225 頁) 材料或物質的不同樣品 (第 12 頁) 爲示量性質。

示量性質：廣延性質　指和材料或物質的(總)量有關的一種性質。它使一種物體、材料相似物品中區別(第 224 頁)爲人們提供一種簡便的辨認方法。(名詞爲 characteristic)

特徵的 (形)　描述一種性質，它使一種物體、材料、物質、圖像(第 93 頁)易於從所有其他相似物品中區別(第 224 頁)出來。特徵性質爲人們提供一種簡便的辨認方法。例如，銅有特徵的紅棕色，使之易和其他金屬相區別。(名詞爲 characteristic)

特性 (名)　指爲一類材料或物質所共有的一種特殊性質。

描述 (名)　敍述一種物體、材料、物質、樣品、能量形式或蒐集的一系列性質，或一個過程中的一系列事件。(動詞爲 describe，形容詞爲 descriptive)

物態　物質具有固體、液體和氣體三種存在形態。任何一種材料或物質都或者是一種固體、或液體、或氣體。

物態變化　指材料或物質從一種物態 (↑) 變化爲另一種物態的物理變化 (第 13 頁)。由固體變爲液體；由液體變爲氣體。物態變化通常是受熱或冷卻引起的。

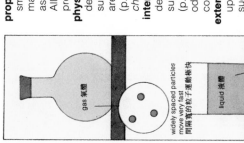
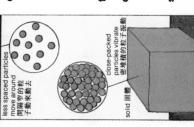

gas 氣體

widely spaced particles move very fast
間隔寬的粒子運動極快

liquid 液體

less spaced particles move around
間隔窄的粒子來動去

close-packed particles vibrate
密堆積的粒子振動

solid 固體

**three states of matter**
物質的三態

**solid** (n) one of the states of matter (p.9). A solid has a definite (p.226) mass, a definite volume and a definite shape. For example, iron is a solid at room temperature. **solid** (adj), **solidity** (n).

**melt** (v) to change a solid (↑) to a liquid (↓) by heating; a solid to change to a liquid when heated. For example, heat melts ice; ice melts when heated. Only one material or substance is concerned in the action. Compare dissolve (p.30) which concerns two or more substances. **molten** (adj).

**molten** (adj) describes a material or substance in the liquid state. The material or substance is solid at room temperature.

**solidify** (v) to change a liquid (↓) to a solid (↑) on cooling. Solidifying is the opposite action to melting. Only one material or substance is concerned in the action. The word is used for materials and substances which are normally solid at room temperature, e.g. molten iron solidifies at about 1500°C.

**set** (v.i.) of suspensions (p.86) in liquids, to form a solid as the liquid evaporates.

**freeze** (v) to change a liquid (↓) to a solid (↑) on cooling below room temperature. The word is used for substances which are normally liquid at room temperature, e.g. water freezes to form ice. Freezing is the opposite action to melting. **freezing** (adj), **frozen** (adj).

**liquid** (n) one of the states of matter. A liquid has a definite (p.226) mass, a definite volume, but no definite shape, e.g. water and kerosene are liquids at room temperature. A liquid takes the shape of its containing vessel (p.25). **liquefy** (v), **liquefaction** (n), **liquid** (adj).

**boil** (v) to change a liquid (↑) into a gas (↓) by heating. Bubbles (p.40) of gas are formed in the liquid and the gas is given off (p.41). The temperature of the liquid remains constant (p.106) during boiling. **boiling** (adj), **boiled** (adj), **boil** (n).

**boiled** (adj) describes water that has been boiled for some time. The water no longer contains dissolved (p.30) air.

固體（名） 物質狀態（第 9 頁）之一。固體有一定的（第 226 頁）質量、一定的體積和一定的形狀。例如，水在室溫過是固體。（形容詞爲 solid，動詞爲 solidity，名詞爲 solidification）

熔化（動） 加熱使固體（↑）變爲液體（↓）；固體受熱變爲液體。例如：熱使冰熔化；冰受熱熔化。熔化作用只涉及一種物料或物質。而溶解（第 30 頁）則涉及兩種或多種物質。（形容詞爲 molten）

熔化的（形） 描述一種材料或物質處於液態。此種材料或物質在室溫時爲固體。

固化（動） 使液體（↑）冷卻變爲固體（↑）。固化爲熔化的相反作用。此詞用於室溫時通常只涉及一種材料和物質。例如，熔融的鐵在約 1500°C 時固化。

凝結：凝結（他動） 指液體中的懸浮（第 86 頁）固體隨液體蒸發而形成一種固體。

凍結（動） 使液體（↓）在室溫以下冷卻變爲固體（↑）。該詞用於通常在室溫時爲液體的物質。例如，水凍結成冰。凍結是熔化的相反作用。（動詞爲 freezing，frozen）

液體（名） 物質狀態之一。液體有一定的（第 226 頁）質量和一定的體積，但無確定的形狀。例如，水和煤油在室溫時都是液體。液體取其容器（第 25 頁）的形狀。（動詞爲 liquefy，名詞 liquefaction，形容詞爲 liquid）

素沸（動） 加熱使液體（↑）變爲氣體（↓）。液體中產生氣泡（第 40 頁）同時釋放出（第 41 頁）氣體。沸騰過程中，液體的溫度保持不變（第 106 頁）。（形容詞爲 boiling，boiled，動詞爲 boil）

已煮沸的（形） 描述已煮沸一段時間的水。經煮沸的水已不含溶解的（第 30 頁）空氣。

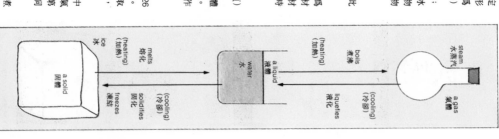

steam 水蒸汽 / a gas 氣體 / boils 煮沸 / (heating)(加熱) / (cooling)(冷卻) / liquefies 液化 / a liquid 液體 / water 水 / melts 熔化 / ice 冰 / (heating)(加熱) / (cooling)(冷卻) / solidifies 固化 / freezes 凍結 / a solid 固體

**liquefy** (v) to change a solid (↑) or a gas (↓) to a liquid (↑). A gas is liquefied by cooling. **liquefaction** (n).

**gas** (n) one of the states of matter (p.9). A gas has a definite (p.226) mass but no definite volume and no definite shape, e.g. air is a gas at room temperature and pressure. A gas expands to fill the volume of its containing vessel (p.25). **gaseous** (adj).

**gaseous** (adj) describes a substance in the state of a gas (↑), or a chemical reaction (p.62) between gases.

**vapour** (n) a substance in the gaseous (↑) state. A vapour can be changed to a liquid (↑) by increasing the pressure (p.102). A gas is called a vapour below its *critical temperature* (p.102). To compare *gas* with *vapour*: both are in the gaseous state, but a gaseous substance above its critical temperature is a *gas* and cannot be liquefied (↑) however great the pressure, while a gaseous substance below its critical temperature is a *vapour* and can be liquefied by a sufficient increase in pressure. **vaporize** (v), **vaporization** (n).

**vaporize** (v) to change a liquid (↑) to a vapour (↑) at a temperature lower than that at which it boils. For example, naphthalene vaporizes at room temperature. **vaporization** (n).

**evaporate¹** (v) to change a liquid (↑) to a vapour (↑) and so to cause the volume of the liquid slowly to become less. The important fact is the volume of liquid becoming less. *See evaporate²* (p.32). **evaporation** (n), **evaporated** (adj).

**condense** (v) to change a vapour (↑) to a liquid (↑) by cooling, or by increasing the pressure, or by both; change of a vapour to a liquid because it cools or because the pressure is increased. The word is used for materials and substances which are liquid at room temperature and the usual method of condensation is by cooling. **condensed** (adj), **condensation** (n).

**condensation¹** (n) the formation of a liquid from its vapour, e.g. the condensation of steam to water. A **fluid** (n) any liquid or gas is a fluid. A fluid is a substance that flows. **fluid** (adj), **fluidity** (n).

液化 (動) 使固體 (↑) 或氣體 (↓) 變爲液體 (↑)。氣體是通過冷卻液化。(名詞爲 liquefaction)

氣體 (名) 物態 (第 9 頁) 之一。氣體有一定的 (第 226 頁) 質量,但無確定的體積和形狀。例如,空氣在室溫和大氣壓力下是氣體。氣體可膨脹至充滿容器 (第 25 頁) 之體積。(形容詞爲 gaseous)

氣態的 (形) 描述處於氣體 (↑) 狀態的一種物質,或描述氣體之間的化學反應 (第 62 頁)。

蒸汽 (名) 指氣體 (↑) 狀態的一種物質。壓力 (第 102 頁) 升高可使蒸汽變爲液體 (↑)。氣體在溫度低於其臨界溫度 (第 102 頁) 時稱之爲蒸汽。氣體與蒸汽之比較:兩者均處於氣體狀態,但氣態物質在溫度高於其臨界溫度時是一種氣體,升高壓力不能使之液化;氣態物質在低於其臨界溫度時則是一種蒸汽,壓力升高到足夠高時可使之液化。(動詞爲 vaporize,名詞爲 vaporization)。

汽化 (動) 使液體 (↑) 在沸點以下溫度變爲蒸汽 (↑)。例如,使萘於室溫下汽化。(名詞爲 vaporization)

蒸發 (動) 使一種液體 (↑) 變爲蒸汽 (↑),液體的體積相應地慢慢地減少。重要的事實是液體的體積變少。參見 " 蒸發² " (第 32 頁)。(名詞爲 evaporation,形容詞爲 evaporated)

冷凝 (動) 藉冷卻或升高壓力,或兩者兼施,使蒸汽 (↑) 變爲液體 (↑);蒸汽受冷或受壓力升高而變爲液體。這個動詞用於室溫時爲液體的那些材料和物質,而常用的冷凝方法是冷卻。(形容詞爲 condensed,名詞爲 condensation)

冷凝作用 (名) 自液體的蒸汽形成液體。例如,使水蒸汽變成水的冷凝作用。流體 (名) 任何液體或氣體都是一種流體。流體係一種可流動的物質。(形容詞爲 fluid,名詞爲 fluidity)。

vaporization 汽化作用
vapour 蒸汽
liquid 液體
temperature below boiling point 溫度低於沸點

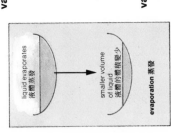

liquid evaporates 液體蒸發
smaller volume of liquid 液體的體積變少
evaporation 蒸發

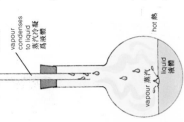

cold 冷
vapour condenses to liquid 蒸汽凝結爲液體
condensation 冷凝
hot 熱
vapour 蒸汽
liquid 液體

**boiling point** the temperature at which a liquid (p.10) boils (p.10). At the boiling point the *vapour pressure* (p.103) of the liquid is equal to the *atmospheric pressure* (p.102). The lower the atmospheric pressure, the lower is the boiling point of the liquid. The boiling point of water is 100°C at *standard atmospheric pressure* (p.102).

**melting point** the temperature at which a solid substance melts; at the melting point both solid and liquid substance exist (p.213) together. The melting point of a solid varies slightly with the ambient (p.103) pressure (p.102). The term melting point is used for substances which are solid at room temperature. *See freezing point* (↓).

**freezing point** the temperature at which a solid substance melts. The term freezing point is used for substances which are liquid at room temperature, e.g. the freezing point of water is 0°C but the melting point of naphthalene is 80°C.

**mass** (*n*) the property of a material or substance that causes it to be attracted by the earth. The *force* attracting an object, or any substance, to the earth is its *weight*. Mass is measured in *kilograms*, weight is measured in *newtons*.

**weight** (*n*) see mass (↑). **weigh** (*v*).

**volume** (*n*) the amount of space taken up by an object in three dimensions.

**density** (*n*) the mass (↑) of 1 m³ of a material or substance (p.8). Density = mass ÷ volume for any specimen (p.43) of a material or substance. Density is an intensive property (p.9) used in identification (p.225) of material and substances. Density is measured in kg/m³. **dense** (*adj*).

**relative density** the density of a material or substance divided by the density of water. Relative density has no units, it is a pure number.

**relative vapour density** the density of a gas or vapour divided by the density of hydrogen measured at the same temperature and pressure (p.102). Relative vapour density has no units, it is a pure number, and is independent of temperature and pressure. The relative vapour density of a substance is numerically equal to half its *molar mass* (p.114).

**vapour density** = relative vapour density (↑).

沸點　液體(第 10 頁)沸騰(第 10 頁)時的溫度。液體在沸點的蒸汽壓(第 103 頁)等於大氣壓(第 102 頁)。大氣壓越低，液體的沸點也越低。水在標準大氣壓(第 102 頁)下的沸點為 100℃。

熔點　固體物質熔化的溫度：在熔點時，固體物質與液體物質一齊存在(第 213 頁)。固體的熔點隨周圍壓力(第 102 頁)而略為不同。這個用詞用於室溫下呈固體的物質。參見"凝固點"(↓)。

凝固點　冰點。固體物質熔化時的溫度。這個用詞用於在室溫時呈液體的物質，例如，水的冰點為 0℃。萘的熔點為 80℃。

質量　(名)　材料或物質的屬性，此屬性引致材料或物質為地心所吸引。將物體或任何物質吸引地心的力即為該物質的重量。質量的量度單位為千克，而重量的量度單位為牛頓。

重量　(名)　見質量(↑)。(動詞局 weigh)

體積　(名)　一個物體三維空間中所佔據的空間總量。

密度　(名)　1 m³ 的材料或物質的質量(↑)。對一種材料或物質的任何試樣(第 43 頁)而言，密度 = 質量 ÷ 體積。密度是鑑別(第 225 頁)材料和物質所用的強度性質(第 9 頁)。密度的量度單位為 kg/m³。(形容詞局 dense)

相對密度　材料或物質的密度除以水的密度。相對密度為一純數，並無單位。

相對蒸汽密度　氣體或蒸汽的密度除以相同溫度和壓力(第 102 頁)下測得的氫氣的密度。相對蒸汽密度為一純數，並無單位；且與溫度及壓力無關。某一物質的相對蒸汽密度數值上等於其克分子質量(第 114 頁)的一半。

蒸汽密度　同相對蒸汽密度(↑)。

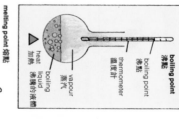
melting point 熔點
boiling point 沸點
thermometer 溫度計
melting point 熔點
melting solid 熔融的固體
boiling 沸騰
vapour 蒸汽
liquid 液體
heat 加熱

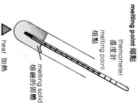
different densities 不同的密度
heat 加熱

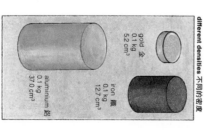
gold 金
0.1 kg
5.2 cm³
iron 鐵
0.1 kg
12.7 cm³
aluminium 鋁
0.1 kg
37.0 cm³

**physical change** a change in which no new materials or substances are formed. In a physical change, a material or substance may change its state, or some of its physical properties (p.9) may change, e.g. the change from water to steam is a physical change.

物理變化 指不生成新材料或新物質的一種變化。在物理變化中，材料或物質可以改變其狀態，或改變其某些物理性質（第 9 頁）。例如，由水變成水蒸汽是一種物理變化。

**state of division** a measure of the size of small particles (↓) into which a large piece of a solid has been divided, e.g. marble can be in lumps (↓), chips (↓), or powder (↓), three different states of division.

細分態 由大塊固體細分成小粒子（↓）大小的一個尺度。例如雲石可以細分爲塊（↓）、碎片（↓）或粉末（↓）三種不同的細分態。

**particle¹** (n) a very small piece of solid material or substance (p.8).

粒子 (名) 指微粒的固體物質（第 8 頁）。

**lump** (n) a large piece of a solid material or substance with an irregular (p.93) shape. **lumpy** (adj).

塊 (名) 指形狀不規則（第 93 頁）的大塊固體材料或物質。(形容詞爲 lumpy)

**chip** (n) a small piece of a solid material or substance, broken off from a large piece. A chip is smaller than a lump, but bigger than a granule.

碎塊 (名) 從大塊固體破碎成的小塊固體材料或物質。碎塊比塊小但比顆粒大。

**flake** (n) a small, flat lump (↑) of solid material or substance. A flake is similar in size to a chip (↑).

薄片 (名) 指小而扁平的固體材料或物質塊（↑）。薄片的大小與碎片（↑）相似。

**granule** (n) a small piece of a solid material or substance made up of several grains (↓). **granular** (adj).

顆粒 (名) 指由若干細粒（↓）構成的一小塊固體材料或物質。(形容詞爲 granular)

**grain** (n) a very small piece of solid material or substance; a particle (↑) that can be seen by the naked eye. Sand and salt consist of grains.

細粒 (名) 指一種極小顆的固體材料或物質；肉眼可觀察到的粒子（↑）。砂子和鹽就是由許多細粒組成的。

**powder** (n) a solid material or substance consisting of particles (↑) so small that they cannot be seen by the naked eye. **powdered** (adj), **powdery** (adj).

粉末 (名) 指由細小得肉眼觀察不到的粒子（↑）組成的固體材料或物質。(形容詞爲 powdered，powdery)

**filings** (n.pl.) small particles formed by rubbing a metal; they are similar in size to grains or granules (↑), but long and thin.

銼屑 (名，複) 鐼磨金屬所產生的細小粒子，其大小與細粒或顆粒（↑）相類似，但又長又薄。

**turnings** (n.pl.) particles formed by cutting a metal; they are much larger than filings (↑).

切屑 (名，複) 切削金屬所產生的粒子。其大小比銼屑（↑）大得多。

**fine** (adj) describes powders (↑) and filings (↑) in which the state of division (↑) is very small. **fineness** (n).

細微的 (形) 細小的（形）形容粉末（↑）和銼屑（↑），其細分態是（↑）極小的。(名詞爲 fineness)

**coarse** (adj) describes powders and filings which are bigger than those described as fine (↑). **coarseness** (n).

粗糙的 (形) 形容粉末和銼屑，但其大小比用"微細的"（↑）所形容者爲大。(名詞爲 coarseness)

**finely divided** (adj) describes a solid in powder form with very small particles in the powder, i.e. a fine powder.

細分的 (形) 形容一種粉末狀固體具有極微小的粉末粒子，亦即細粉末。

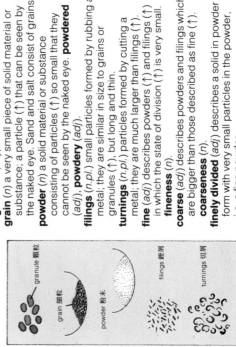

granule 顆粒

grain 細粒

powder 粉末

filings 銼屑

turnings 切屑

**texture** (n) the nature (p.19) of the surface of a solid, i.e. whether it is rough or smooth, is the texture of the surface. The texture of a powder, granules or grains, depends upon the fineness (p.13) or coarseness (p.13) of the particles, e.g. a surface can have a smooth texture or a powder a coarse texture.

**massive** (adj) describes a solid, particularly a metal, which consists of large pieces, including lumps, e.g. massive zinc consists of large pieces of zinc. Massive is the exact opposite of finely divided (p.13).

**elastic** (adj) describes a solid which can have its shape changed by a force and which returns to its original (p.220) shape when the force is removed, e.g. a piece of rubber is elastic. The property of such a solid is its **elasticity** (n).

**plastic¹** (adj) describes a solid material or substance which can have its shape changed by a force but which does not return to its original (p.220) shape when the force is removed, e.g. clay is plastic. The property of such a solid is its **plasticity** (n).

**brittle** (adj) describes a solid material or substance which breaks into small pieces under a force, e.g. glass is brittle, it breaks into small pieces when hit. The property of such a solid is its **brittleness** (n).

**ductile** (adj) describes a solid material or substance which can be drawn out to form a thin wire. Metals and alloys (p.55) are ductile. The property of such a solid is its **ductility** (n).

**malleable** (adj) describes a solid material or substance which can have its shape changed to a thin sheet by beating with a hammer, e.g. iron is malleable. Metals and alloys (p.55) are malleable. The property of such a solid is its **malleability** (n).

**abrasive** (adj) describes a material which wears away the surface of another material. **abrasion** (n).

**refractory** (adj) describes a solid material or substance which can be heated to a high temperature without changing its properties, e.g. some kinds of bricks are refractory. **refractoriness** (n).

（表面）質地：結構（名）指固體表面的本質（第 19 頁），即無論該固體表面是粗糙還是平滑的，都是指其表面的質地。（顯粒或或細地）取決於粒子的細度（第 13 頁）或或粗度（第 13 頁）。例如，一個表面可以具有平滑的質地而粉末的質地可以是粗糙的。

大塊的（形）形容一塊固體（尤指金屬）係由大塊包括角所組成的，例如，大塊的鋅是由大片的鋅幹組好（第 79 頁）和鋅幹碎（第 13 頁）相反。

彈性的（形）描述一種固體材料受力作用時可改變其形狀，但作用力消除之後又可回復其原始（第 220 頁）形狀的。彈性的材料的，例如，橡膠片是彈性的。**彈性**就是橡膠的屬性。

塑性的（形）描述固體材料或物質受力作用時可改變其形狀，但作用力消除之後不可回復其原始（第 220 頁）形狀。例如，新土是塑性的。**塑性**就是黏土的屬性。

脆的（形）描述固體材料或物質受力作用時便會破裂成小塊。例如，玻璃是脆的，當它受碰擊時便破裂成小塊。**脆性**就是玻璃的屬性。

延性的（形）描述一種固體材料或金屬和合金（第 55 頁）都具有延性。**延性**就是金屬和合金的屬性。

展性的（形）描述固體材料或物質可用鎚子捶打使其形狀變成一塊薄片。例如，鐵是具有展性的材料。金屬和合金（第 55 頁）都是有展性的材料。**展性**就是金屬和合金的屬性。

磨損性的（形）描述一種材料會損另一材料的表面。（名詞為 abrasion）

耐火的（形）描述一種固體材料或物質可被加熱至高溫而不改變其性質。例如，某種耐火磚。（名詞為 refractoriness）

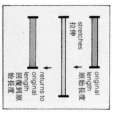

malleable solid 展性固體

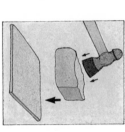

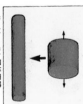

ductile solid 延性固體

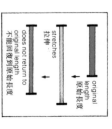

plastic solid 塑性固體

stretches
拉伸

does not return to
original length
不能回復到原始長度

elastic solid 彈性固體

stretches
拉伸

original
length
原始長度

returns to
original
length
回復到原始長度

**porous** (adj) describes a solid material full of very small holes, through which fluids can pass, e.g. a brick is porous. **porosity** (n).

**crystalline** (adj) describes a solid material or substance (p.8) which has molecules (p.77), atoms (p.110) or ions (p.123) arranged in a regular structure (p.82). A crystalline substance forms crystals (p.91); a metal is crystalline but it does not form regular crystals. *See metal crystals (p.95).*

**amorphous** (adj) describes a solid material or substance which has no crystalline (↑) structure. Glass, rubber and many plastics (p.210) are amorphous.

**coloured** (adj) describes a material or substance (p.8) which has colour, but without naming the colour, e.g. a coloured solution can be brown, blue, green, black, etc. A material or substance can be described as white or coloured, e.g. milk is a white liquid and lead (II) sulphide (p.30) is formed as a black precipitate (p.30) which is considered to be a coloured precipitate.

**colourless** (adj) describes a material or substance (p.8) which has no colour, e.g. water is colourless; air is colourless. Colourless is the opposite of coloured. To contrast *white* and *colourless*: the paper of this book is *white*; glass in a window is *colourless*.

**odour** (n) that property of a material or substance (p.8) which is recognized by smell, e.g. onions have a well-known odour.

**odourless** (adj) describes a material or substance which has no odour (↑).

**quality** (n) a property of a material or substance (p.8) which cannot be measured, but is an observation made only by the senses, e.g colour, odour and texture are some of the qualities of materials and substances.

**impart** (v) to give a quality (↑) or a quantity (p.81) to an object or to share a quality or a quantity with an object, e.g. sugar imparts sweetness to a drink; potassium salts impart a lilac colour to a flame; an explosion imparts energy to a rocket.

多孔的 (形) 描述一種固體材料滿佈着很多極細小的孔洞，流體可以流過這些細小孔洞，例如磚是多孔的。(名詞為 porosity)

晶體的 (形) 描述一種固體材料或物質 (第 8 頁) 所具有的分子 (第 77 頁)、原子 (第 110 頁) 或離子 (第 123 頁) 排列在一個規則的結構 (第 82 頁) 中。晶狀物質可形成晶體 (第 91 頁)；金屬雖爲晶狀的，但卻不形成規則晶體。參見 " 金屬晶體 " (第 95 頁)。

無定形的 (形) 描述一種固體材料或物質不具有晶體的 (↑) 結構。玻璃、橡膠和許多種塑膠 (第 210 頁) 是無定形的。

有色的 (形) 描述一種材料或物質 (第 8 頁) 具有顏色但沒有明確說出是何種顏色，例如一種有色的溶液可以指棕、藍、綠、黑等顏色。可將一種材料或物質形容成白色的或無色的，例如牛奶形容爲白色液體；二硫化鉛(II) 成爲一種黑色的沉澱物 (第 30 頁)，可視之爲有色的沉澱物。

無色的 (形) 描述一種材料或物質 (第 8 頁) 沒有顏色。例如，水是無色的，空氣也是無色的。無色爲有色的反面。白色和無色的比較：這本書的紙是白色的；窗玻璃是無色的。

氣味 (名) 可藉嗅覺辨認的材料或物質 (第 8 頁) 的屬性。例如，洋蔥有衆所週知的氣味。

無氣味的 (形) 描述一種材料或物質沒有氣味 (↑)。

品質 (名) 材料或物質 (第 8 頁) 的一種屬性，此種屬性不可測量，只可藉感覺作出觀察結果。例如，顏色、氣味和質地即爲材料或物質的某些品質。

給予 (動) 給某些物體以品質 (↑) 或數量 (第 81 頁) 或與某一物品共享一種品質或數量，例如，糖賦予飲品甜度；鉀鹽給予火焰淡紫色；爆炸賦予火焰能量。

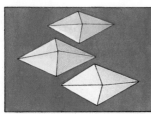

crystals 晶體

**surface** (*n*) the outside part of a solid object; it has length and breadth, but no thickness or depth. A liquid has a surface where it meets the air. Examples: a brick has six surfaces; the surface of water in a cup.

**granular** (*adj*) (1) describes a surface which appears to have many grains or granules. (2) describes a coarse powder consisting of granules.

**dull** (*adj*) describes a surface which does not shine when light falls on it. Dull is the opposite of bright. For example, the surface of wax is dull. **dullness** (*n*).

**lustre** (*n*) that property of a surface which causes it to shine when light falls on it. Lustre is a quality. For example, the surface of silver has a lustre. **lustrous** (*adj*).

**transparent** (*adj*) describes a solid object, material or substance which allows light to pass through it, so that a person can see through it, e.g. glass is transparent. **transparency** (*n*).

**translucent** (*adj*) describes a solid object, material or substance which allows light to pass through it, but a person cannot see clearly through it. For example, waxed paper is translucent, but not transparent; milk is a translucent liquid.

**opaque** (*adj*) describes an object, material or substance which does not allow light to pass through it, e.g. leather and paper are opaque; mercury is an opaque liquid. **opacity** (*n*).

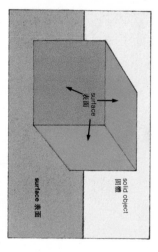

solid object
固體

surface
表面

surface
表面

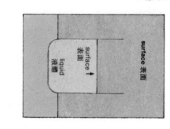

surface
表面

liquid
液體

**表面** (名) 指固體的外部，具有長度和寬度，但無厚度或深度。液體有一個表面與空氣接觸。例如，磚有六個表面；杯中的水只有一個表面。

**粒狀的** (形) (1) 描述一個表面看似有許多細粒和顆粒。(2) 描述一種由無數顆粒構成(第55頁)的粉末。

**暗淡的** (形) 描述一個表面當光線照射於其上時不閃耀。暗淡是光亮的反面。例如，蠟的表面無光澤。(名詞為 dullness)

**光澤** (名) 光線照射在表面上時引致該表面閃耀的性質。光澤是一種品質。例如，銀的表面有光澤。(形容詞為 lustrous)

**透明的** (形) 描述一種固體、材料或物質可讓光線透過它，因此人們可透過它視物。例如，玻璃是透明的。(名詞為 transparency)

**半透明的** (形) 描述一種固體、材料或物質可讓光線透過，但人們不能透過它清楚看見物體。例如，蠟紙是半透明的，而非透明的；牛奶是半透明的液體。

**不透明的** (形) 描述一種物體、材料或物質不能讓光線透過。例如，皮革和紙都是不透明的；水銀也是一種不透明的液體。(名詞為 opacity)

translucent 半透明的

translucent lamp shade
半透明的燈罩

清澈的 (形) 形容一種液體是透明的。例如，水是清澈的液體。清澈的液體可以有色或無色。例如，紊是褐色的清澈液體而煤油是無色的清澈液體。(名詞爲 clarity)

可溶的 (形) 描述一種固態或氣態的 (第 11 頁) 物質 (第 8 頁) 可溶解 (第 30 頁) 在一種液體中。液體一般是指水。物質可用極易溶解、輕微 (↓) 溶解、難 (↓) 溶解、不溶解 (↓) 或可溶解來描述它。例如，糖在水中可溶解 (糖可被溶於水)；石灰微溶於水。(名詞爲 solubility)

不溶解的 (形) 描述一種固態或氣態的物質 (第 8 頁) 不溶解在某一種液體中。不溶解之反面爲可溶解。完全不溶解的物質極少。

輕微的 (形) 描述某種物質可溶解 (↑) 於一種液體中的量很少。例如，石灰輕微溶於水，亦即只有少量石灰溶在水中。

很難的 (形) 描述只有極少量物質可溶解 (↑) 在液體中。甚至比微溶的 (↑) 物質的量更少。例如，空氣難溶於水。

絮凝的 (形) 描述一種沉澱物 (第 30 頁) 漂浮在液體中，狀如羊毛。例如，當二氧化碳通入石灰水中時，將二氧化碳溶解氫氧化鋁的沉澱物是絮凝的。

乳白色的 (形) 描述一種液體 (第 10 頁) 具有白色的沉澱物 (第 30 頁)，使液體狀如牛奶。沉澱物很輕。例如，將二氧化碳通入石灰水中時，輕質的碳酸鈣沉澱物使石灰水變成乳白色的液體。

乳脂狀的 (形) 描述比乳白色 (↑) 液體所形成沉澱物較重的一種白色沉澱物，此種沉澱物仍然漂浮於液體中。例如，氯化銀形成一種乳脂狀的沉澱物。

重的 (形) 形容一種沉澱物成一種重的沉澱物落在液體底部。硫酸鋇形成一種重的沉澱物。

**clear** (*adj*) describes a liquid which is transparent, e.g. water is a clear liquid. A clear liquid can be coloured or colourless, e.g. tea is a clear, brown liquid; kerosene is a clear, colourless liquid. **clarity** (*n*).

**soluble** (*adj*) describes a solid, or gaseous (p.11), substance (p.8) which can be dissolved (p.30) in a liquid; the liquid is usually water. A substance can be described as very soluble, slightly (↓) soluble, sparingly (↓) soluble, insoluble (↓) or soluble. For example, sugar is soluble in water (sugar can be dissolved in water); lime is slightly soluble in water. **solubility** (*n*).

**insoluble** (*adj*) describes a solid, or gaseous (p.11), substance (p.8) that does not dissolve in a liquid. It is the opposite of soluble. Very few substances are completely insoluble.

**slightly** (*adj*) describes **soluble** (↑) if only a small amount of substance dissolves in a liquid, e.g. lime is slightly soluble in water, that is, only a small amount of lime will dissolve in water.

**sparingly** (*adj*) describes **soluble** (↑) if an amount which dissolves is very small, even less than dissolves for a slightly soluble (↑) substance, e.g. air is sparingly soluble in water.

**flocculent** (*adj*) describes a precipitate (p.30) which has the appearance of wool floating in the liquid, e.g. a precipitate of aluminium hydroxide is flocculent.

**milky** (*adj*) describes a liquid (p.10) with a white precipitate (p.30) which gives the liquid the appearance of milk. The precipitate is very light, e.g. when carbon dioxide is passed into lime water, a light precipitate of calcium carbonate turns the lime water into a milky liquid.

**creamy** (*adj*) describes a white precipitate which is heavier than the precipitate forming a milky (↑) liquid; the precipitate still floats in the liquid, e.g. silver chloride forms a creamy precipitate.

**heavy** (*adj*) describes a precipitate (p.30) which sinks to the bottom of the liquid, e.g. barium sulphate forms a heavy precipitate.

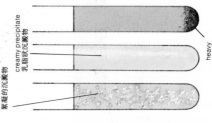

flocculent precipitate
絮凝的沉澱物

creamy precipitate
乳脂狀的沉澱物

heavy precipitate
重的沉澱物

**miscible** (*adj*) describes liquids that can mix in all proportions (p.76); the result looks like a single liquid, e.g. water and alcohol can mix completely and look like one liquid.

**immiscible** (*adj*) describes liquids that do not mix at all, e.g. oil and water are immiscible of liquid as oil and water form two layers (↓).

**layer** (*n*) a spread of material over the surface or thin, e.g. (p.16) of another material. A layer can be thick or thin, e.g. a layer of skin covers an orange; a sandwich has 3 layers – bread, meat, bread.

**film** (*n*) a thin layer (↑) of a substance. It can be a thin layer of a liquid, or a vapour, or a solid; a thin layer of one liquid on another liquid; a thin layer of a solid on another solid. For example, a thin film of oil on water; a thin film of oxide (p.48) on a metal.

**interface** (*n*) the meeting place between two layers of liquid, e.g. oil floats on water, where the oil and water meet is an interface.

**mobile** (*adj*) (1) describes a liquid that flows easily, e.g. kerosene is a mobile liquid. (2) describes an object that can be moved easily, e.g. a bicycle is mobile. For liquids, mobile is the opposite of viscous (↓). **mobility** (*n*).

**viscous** (*adj*) describes a liquid that does not flow quickly, e.g. engine oil is viscous. Viscous is the opposite of mobile (↑).

**volatile** (*adj*) describes a liquid that vaporizes (p.11) easily, e.g. petrol is a highly volatile liquid. **volatility** (*n*).

**viscosity** (*n*) the property of a liquid which prevents it from flowing readily, e.g. olive oil has a high viscosity; water has a very low viscosity.

less dense liquid
較不濃稠的液體

denser liquid
較濃稠的液體

interface
界面

immiscible
不溶混的

engine oil
機油

a viscous liquid
黏滯的液體

viscous 黏滯的

可溶混的 (形) 描述不同液體之間可按任何比例 (第 76 頁) 混合，所混合成的液體看來就像一種單一的液體，例如，水和酒精可完全混合，同時看來就像一種液體。

不溶混的 (形) 描述那些完全不溶混的液體，例如，油和水形成兩層 (↓) 液體，因爲這些液體是不溶混的。

層 (名) 指一種材料在另一種材料的表面 (第 16 頁) 上鋪開。層可厚可薄，例如，橙有一層皮包着；三明治分成蔴包，肉、蔴包共三層。

薄膜 (名) 指物質的一種薄層 (↑)。薄層可爲液體，或蒸汽，或固體的薄層；或指一種液體在另一種液體面上的薄層；例如，水面上的油另一種固體面上的氧化物 (第 48 頁) 薄層。

界面 (名) 指兩種液體之間的遇合面。例如油浮在水面上，油和水的遇合面就是一個界面。

流動的 (形) (1) 描述一種易流動。例如，煤油是一種流動的液體；(2) 描述一種容易活動的物體，例如，自行車是可移動的。對液體而言，易流動的爲黏滯的 (↓) 反面。(名詞爲 mobility)

黏滯的 (形) 描述一種不易流動快速。例如液體，例如，煤油是一種流動的液體。黏滯的爲易流動的 (↑) 的反面。(名詞爲 viscosity)

揮發性的 (形) 描述一種容易汽化 (第 11 頁) 的液體，例如油是一種易揮發性液體。(名詞爲 volatility)

黏度：黏性 (名) 係液體之性質，此性質阻止液體容易流動，例如，橄欖油的黏度很高，水的黏度則很低。

**chemical change** (n) a change in which new materials or new substances (p.8) are formed, e.g. when chalk is heated, lime and carbon dioxide are formed, this is a chemical change. When sulphur dioxide is passed into a solution of sodium hydroxide, sodium sulphite is formed, this is a chemical change.

**chemical property** a property which depends on the effect of other substances, e.g. a chemical property of sulphuric acid is that alkalis (p.45) neutralize the acid; a chemical property of water is that it is decomposed (p.65) by sodium.

**nature** (n) the essential (p.226) properties of a material or substance (p.8) form its nature, e.g. the nature of alkalis: they neutralize acids, and have a soapy feel when dissolved in water.

**action** (n) the effect of a substance on another material or substance, or the effect of heat, or an electric current, on a substance (p.8). For example, the action of sulphuric acid on calcium carbonate forms carbon dioxide and calcium sulphate; the action of heat decomposes (p.65) calcium carbonate.

**act on** to produce a chemical action (↑). e.g. hydrochloric acid acts on calcium carbonate, an action takes place (p.63), carbon dioxide is given off (p.41) and calcium chloride formed.

**active** (adj) (1) describes a substance (p.8) with many chemical properties, or with chemical properties that reacts vigorously, e.g. sulphuric acid is a very active substance, as many actions (↑) occur with other substances, and they are strong actions. (2) describes the part of a mixture (p.54) or molecule (p.77) which has the particular properties exhibited by the whole mixture or molecule, e.g. the hydroxide group (p.185) of an alcohol (p.175) is the active part of the molecule.

**inactive** (adj) describes a substance (p.8) with few chemical properties, e.g. the alkanes (p.172) are inactive substances. Compare stable (p.74).

**inert** (adj) describes: (1) a substance (p.8) which has no chemical properties, e.g. krypton is an inert gas; (2) an atmosphere preventing oxidation. Compare stable (p.74).

化學變化 (名) 生成新材料或新物質 (第 8 頁) 的一種變化。例如，將白堊加熱產生石灰和二氧化碳是一種化學變化。將二氧化硫通過氫氧化鈉溶液形成亞硫酸鈉亦爲化學變化。

化學性質 和其他一些物質的效應有關的性質。例如硫酸的化學性質是可使鹼 (第 45 頁) 中和；水的一種化學性質是可爲鈉所分解 (第 65 頁)。

本性：本質 (名) 材料或物質 (第 8 頁) 的本性就是它的基本 (第 226 頁) 性質。例如，鹼的本性是可中和酸，而溶於水時則有膩滑感。

作用 (名) 指一種物質在另一材料或物質 (第 8 頁) 上的效應，又指熱或電流在一種物質 (第 8 頁) 上的效應。例如，硫酸和碳酸鈣產生二氧化碳和硫酸鈣的作用；熱使碳酸鈣分解的作用。

作用於 使產生一種化學作用 (↑)。例如，鹽酸作用於碳酸鈣，發生一種化學作用 (第 63 頁) 作用放出 (第 41 頁) 二氧化碳同時生成氯化鈣。

活潑的 (形) (1) 描述一種物質 (第 8 頁) 具有多種化學性質或者具有反應劇烈的化學性質。例如硫酸是一種極活潑的物質，因硫酸可與其他一些物質發生多種作用 (↑) 而且作用強烈；(2) 形容混合物 (第 54 頁) 的一部分或分子 (第 77 頁) 的一部分具有全部或混合物或分子所顯現的各種特有性質。例如醇 (第 175 頁) 的氫氧基 (第 185 頁) 爲醇分子中的活性部分。

不活潑的 (形) 描述具有很少化學性質的一種物質 (第 8 頁)。例如，烷烴 (第 172 頁) 爲不活潑的物質。與 "穩定的" (第 74 頁) 作比較。

惰性的 (形) (1) 描述不具有化學性質的一種物質 (第 8 頁)。例如，氪是一種惰性氣體；(2) 形容阻止氧化作用之環境。與 "穩定的" (第 74 頁) 作比較。

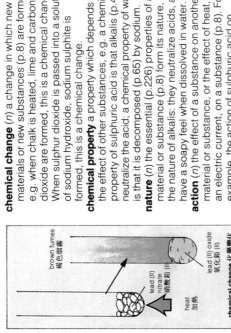

brown fumes 褐色煙霧
lead (II) oxide (II) 氧化鉛(II)
lead (II) nitrate 硝酸鉛(II)
heat 加熱
chemical change 化學變化

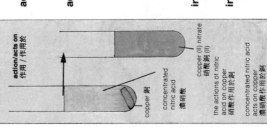

action/acts on 作用/作用於
copper (II) nitrate (II) 硝酸銅(II)
copper 銅
concentrated nitric acid 濃硝酸
the actions of nitric acid on copper 硝酸作用於銅
concentrated nitric acid on copper 濃硝酸作用於銅

**pure** (adj) describes a substance (p.8) which has no other substances mixed with it, e.g. pure silver has no other substances mixed with it. **purify** (v), **purification** (n).

**impure** (adj) describes a substance (p.8) which is not pure (↑) as it has other substances mixed with it; these other substances are impurities. The amount of the impurities is usually small, e.g. iron usually contains a small amount of carbon, and is therefore impure iron. **impurity** (n).

**trace** (n) a very small amount of any substance (p.8) present (p.217) in a mixture or in the earth. Also, a very small amount of an impurity, e.g. a trace of arsenic in sulphur is a very small amount of impurity; a trace element in the earth is an element (p.8) present in very small amounts.

**contaminate** (v) to make a substance impure (↑) with a small amount of an unwanted impurity (↑), e.g. drinking water contaminated with small amounts of dissolved lead; aluminium contaminated with silicon. **contamination** (n), **contaminated** (adj).

**chemical**[1] (adj) describes a property or action concerned with the making of new materials or substances (p.8).

**chemical**[2] (n) an element (p.8) or compound (p.8) which takes part in a chemical change, e.g. sodium hydroxide is a chemical; sulphuric acid is a chemical.

**fine chemical** a chemical which is pure (↑). Fine chemicals are used in analysis (p.82) and for other special purposes. Only small quantities are made of some fine chemicals.

**pharmaceutical chemical** a pure substance used for medicinal purposes.

**technical chemical** a reasonably pure substance, not as pure as a fine chemical, but purer than a coarse chemical. Technical chemicals are used for most industrial (p.171) purposes. See heavy chemical (p. 171).

**coarse chemical** an impure substance used in industry and agriculture. Suitable for the purpose used and cheap to manufacture.

純的(形) 描述沒有其他物質與之混合的一種物質，例如純銀中沒有其他物質與之混合。(動詞爲 purify，名詞爲 purification)

不純的(形) 描述其他物質有與之混合(↑)的一種物質(第8頁)，因而不是純的。雜質的量通常都含少。例如鐵中通常都含少量碳，這些的。(名詞爲 impurity)

微量：微量(名) 指在一種混合物或土壤中存在(第217頁)量極少的任何一種物質(第8頁)，也指含極少量的雜質。例如，硫中有微量的砷即爲微量的雜質；土壤中存在的微量元素是存在量極少的元素(第8頁)。

玷染(動) 由於有少量有害雜質或作用與製造新材料質不純(↑)。例如，飲用水受少量溶解鉛玷染；鋁被矽所玷染。(名詞爲 contamination，形容詞爲 contaminated)

化學的(形) 描述一種性質或作用與製造新材料或物質(第8頁)有關。

化學品(名) 參加化學變化的一種元素(第8頁)或化合物(第8頁)。例如，氫氧化鈉是一種化學品；硫酸也是一種化學品。

藥物化學品 指用於醫藥用途的一種純物質。

精細化學品 指一種純的(↑)化學品。精細化學品既用於分析(第82頁)，也用於其他特殊用途。精細化學品的製造量很少。

工業化學品 指一種相當純的物質，雖不如精細化學品那麼純但卻比粗製化學品純。工業化學品用於大部分的工業(第157頁)用途。參見"重化學品"(第171頁)。

粗製化學品 指一種供工業和農業使用的不純的物質。粗製化學品適於其所用的目的，且製造成本低廉。

trace 微量

no other substances 無其他物質
pure 純的
small amount of carbon 少量的碳
iron 鐵
trace of arsenic 痕量的砷
impure 不純的
sulphur 硫

pharmaceutical chemical 藥物化學品
aspirin tablets 阿司匹靈藥片

**corrosive** (*adj*) describes any chemical (↑) which attacks the surface (p.16) of solids and of living things, and destroys these surfaces. Examples of corrosive chemicals are strong mineral acids (p.55), such as concentrated sulphuric acid, which destroy skin. *Compare corrosion (p.61).*

**caustic** (*adj*) describes a chemical (↑) which attacks and destroys the surface of living things. Strong alkalis (p.45) are described as caustic, e.g. sodium hydroxide is called caustic soda and it burns skin when it comes into contact with it.

**mild** (*adj*) describes an alkali which is not caustic (↑), but is stronger than a weak alkali. *Mild* describes a condition which is between strong and weak, e.g. mild steel is less strong than hard steel, but harder than iron.

**bland** (*adj*) describes any chemical (↑) which does not irritate (p.22) nor cause discomfort to people. Bland usually describes food and pharmaceutical chemicals (↑), e.g. olive oil is bland.

**passive** (*adj*) describes the surface of a substance (p.8) which has been made inactive (p.19). A passive substance may also be inactive because its surface is covered with a thin film (p.18) of oxide (p.48), as on some metals. Usually the surface is made inactive by the attack of a corrosive (↑) chemical. e.g. iron is made passive by the action of concentrated nitric acid. Aluminium is passive because its surface is covered with a thin film of oxide.
**passivity** (*n*).

**activated** (*adj*) describes a substance (p.8) which has been made active (p.19), e.g. activated charcoal has had its power to absorb (p.35) made greater than that of ordinary charcoal.

**flammable** (*adj*) describes a substance (p.8) which readily bursts into flames.

**inflammable** (*adj*) another word for flammable (↑).

**non-flammable** (*adj*) describes a substance which does not burst into flames, and does not burn. The opposite of flammable (↑).

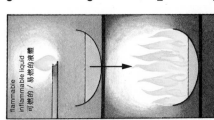

flammable
inflammable liquid
可燃的 / 易燃的液體

**flammable/inflammable 可燃的/不易燃的**

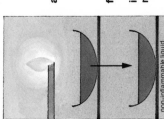

**non-inflammable 不易燃的**

non-inflammable liquid
不易燃的液體

腐蝕性 (形) 描述任何一種化學品 (↑) 會侵蝕固體物質和生物體之表面 (第 16 頁) 並破壞這些表面。濃硫酸之類強無機酸 (第 55 頁) 就是腐蝕性化學品的例子，濃硫酸會損傷皮膚。參見 "腐蝕" (第 61 頁)。

苛性的 (形) 描述一種化學品 (↑) 會侵蝕和破壞生物體表面。強鹼 (第 45 頁) 被認爲是苛性的。例如，氫氧化鈉的被稱爲苛性鈉，皮膚與之接觸會受灼傷。

溫和的 (形) 描述一種鹼不是苛性的 (↑) 但強於弱鹼，也形容介乎平強和弱之間的一種情況。例如軟鋼係一種較軟於硬而硬於鐵的鋼材。

刺激性少的：溫和的 (形) 描述任何一種化學品 (↑) 亦不致令人感到不舒服 (第 22 頁) 既不刺激。通常是金屬表面所見，因而也可以是不活潑的。通常是使用腐蝕性 (↑) 化學品使表面受腐蝕而變爲不活潑。例如，利用濃硝酸的腐蝕作用使鐵的表面變爲鈍態。鋁的表面包覆有一層氧化物膜，因而是鈍態的。(名詞爲 passivity)

鈍態的 (形) 描述一種物質 (第 8 頁) 的表面已被變爲不活潑 (第 19 頁)。鈍態物質的表面已覆蓋著一層氧化物 (第 48 頁) 薄膜 (第 18 頁)，如在某些金屬表面所見，因而也可以是不活潑的。

活化的 (形) 描述一種物質 (第 8 頁) 已使之成爲活性 (第 19 頁)。例如，活性炭所具之吸收 (第 35 頁) 能力比普通炭更强。

可燃的 (形) 描述一種物質 (第 8 頁) 容易燒着發出火焰。

易燃的 (形) 係可燃的 (↑) 的另一詞語。

不易燃的 (形) 描述一種物質不易燒着及不燃燒發不出火焰。其反意詞爲可燃的 (↑)。

**odoriferous** (*adj*) describes a material or substance which produces an odour (p.15), e.g. onions are odoriferous.

**irritate** (*v*) to make parts of animals painful; to be unpleasant to the senses, particularly touch. For example, wood smoke irritates the eyes and the nose of a person; the eyes become painful, and the nose has an unpleasant feeling. **irritation** (*n*), **irritating** (*adj*).

**pungent** (*adj*) describes an odour (p.15) which has a strong effect on the nose or the tongue, e.g. vinegar has a pungent odour and taste. The odour is neither pleasant nor unpleasant.

**acrid** (*adj*) describes an odour (p.15) which is like the smell of wood smoke. It is an irritating (↑) odour.

**choking** (*adj*) describes an odour which is more irritating (↑) than an acrid (↑) odour; it is felt at the back of the throat and is very unpleasant, e.g. the smell of hydrogen chloride gas is a choking odour.

acrid smell
刺鼻嗆喉的氣味

wood smoke
燻木材的煙

**acrid** 刺鼻嗆喉的

wood fire
燻木材

eyes, nose
irritates
刺激鼻、眼

**malodorous** 惡臭的

bad food
腐敗的食物

**fragrant** 芬芳的

fragrant
芬芳的

**malodorous** (*adj*) describes a material or substance (p.8) which has a very unpleasant odour, e.g. bad food is malodorous.

**fragrant** (*adj*) describes an odour which is pleasant to the senses, e.g. the odours of fruit and flowers are often fragrant.

**dense** (*adj*) (1) describes fumes (p.33) which are opaque (p.16), e.g. a fire can give off dense smoke and it is not possible to see through the smoke. (2) describes any material or substance (p.8) with a high density (p.12).

有氣味的 (形)：產生某種氣味（第15頁）的材料或物質。例如洋蔥含有氣味。

刺激 (動)：使動物身體部分疼痛；使感覺不適，尤其是觸覺不舒服。例如燻木材發出的煙刺激人的眼和鼻。使眼覺刺痛、鼻有不舒服的感覺。（名詞為 irritant，形容詞為 irritating）

刺鼻和舌的 (形)：描述一種氣味（第15頁）對鼻子和舌頭有強烈影響。例如，醋有刺鼻和刺激舌頭的氣味。此種氣味既非令人舒服亦非令人不舒服。

刺鼻嗆喉的 (形)：描述一種似燻木材的氣味（第15頁）。這是一種刺激性（↑）氣味。

窒息的 (形)：描述一種氣味比刺鼻嗆喉（↑）氣味更具刺激性（↑）；這是在咽喉後部所引起的感覺。例如，氯化氫氣體的難聞氣味是一種窒息性氣味。

惡臭的 (形)：描述一種材料或物質（第8頁）發出極難聞氣味。例如，腐敗的食物是惡臭的。

芬芳的 (形)：一種令人感覺舒適的氣味。例如，水果和花朵的氣味總是芬芳的。

濃的 (形)：（1）描述那些不透明（第16頁）的煙霧（第33頁）。例如，火焰可以發出濃煙而且不能透過這種煙見物；（2）描述任何材料或物質（第8頁）具有高的密度（第12頁）。

**device** (*n*) an object made for a special purpose, e.g. a telescope is a device for seeing distant objects; a thermostat is a device for keeping a constant (p.106) temperature.

**tool** (*n*) a device (↑) which is held in the hand and used to help a person who is working with his hands, e.g. a saw, a spanner, a screwdriver.

裝置（名） 指爲專門用途而製造的一種物件。例如，望遠鏡是用於觀察遠距離物體的一種裝置；恆溫器是用於保持溫度恆定（第 106 頁）的一種裝置。

工具（名） 一種手持式的裝置（↑），用於幫助用手工作的人。例如鋸子、扳手、螺絲刀。

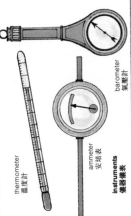

thermometer
溫度計

barometer
氣壓計

ammeter
安培表

**instruments**
儀器儀表

**instrument** (*n*) a device (↑) used for measuring, recording (p.39) or detecting (p.225), e.g. a thermometer, a barometer, an ammeter, a spectroscope. The work done with an instrument is more accurate (p.227) than the work done with a tool (↑) and needs more knowledge.

儀器儀表（名） 一種用於測量、記錄（第 39 頁）或檢測（第 225 頁）的裝置（↑）。例如，溫度計、氣壓計、安培計、分光計。用儀器工作比用工具（↑）更準確（第 227 頁），而需要更多知識。

**apparatus** (*n*) (*apparatus n.pl.*) all the objects, devices (↑), tools (↑) and instruments (↑) used for work in chemistry, or science, e.g. thermometers, beakers, ammeters, thermostats, supports.

器具（名） 包括在化學或科學方面使用的所有物件、裝置（↑）、工具（↑）和儀器儀表（↑）。例如，溫度計、燒杯、安培計、恆溫器、支架等。

**laboratory** (*n*) a room in which scientific experiments (p.42) are carried out.

實驗室（名） 指人們從事科學實驗（第 42 頁）的工作室。

**tank** (*n*) a large container for liquids.

槽（名） 盛放液體之大型容器。

**tap** (*n*) a device (↑) which controls the flow of a liquid or gas from a pipe.

龍頭（名） 控制管內液體或氣體體流量的一種裝置（↑）。

**pinchcock** (*n*) a device (↑) which closes a tube (p.29) or a pipe.

彈簧夾（名） 封閉管或喉管（第 29 頁）或喉管用之一種裝置（↑）。

**stopcock** (*n*) a kind of tap (↑) used with glass tubes (p.29).

活栓（名） 供與玻璃管（第 29 頁）一起使用的一類龍嘴（↑）。

**practical** (*adj*) describes work with chemicals on experiments, which is opposite to work with writing and ideas.

實驗的（形） 描述用各種化學品在實驗中工作，其反意爲以書寫或思惟進行工作。

**theoretical** (*adj*) describes work with ideas, usually recorded (p.39) in writing. Theoretical work is the opposite of practical (↑) work.

理論的（形） 描述以書進行工作，一般是以書面作記錄（第 39 頁）。理論工作爲實驗（↑）工作之反面。

**tools 工具**

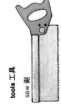

saw 鋸

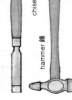

chisel 鑿

hammer 鎚

**tank 槽**

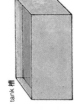

tap
龍頭、旋塞

**pinchcocks**
彈簧夾

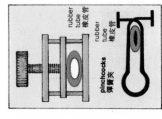

rubber
tube
橡皮管

rubber
tube
橡皮管

**stopper** (n) an object which closes a hole in a pipe, a bottle, or a flask (↓), e.g. a glass stopper in a bottle.

**cork** (n) a stopper (↑) made from the bark of a cork tree.

**bung** (n) a stopper (↑) made of rubber, or a large stopper made of wood.

**delivery tube** a glass tube (p.29) connecting different pieces of apparatus (p.23); it conducts fluids (p.11) from one piece of apparatus to another.

**connect** (v) to join together different pieces of apparatus, so that fluids can flow from one part of the apparatus to another.

**disconnect** (v) to separate different pieces of apparatus so that they are not connected (↑). Disconnect is the opposite of connect.

**trough** (n) (1) a flat container for liquids. (2) the hollow between the crests of two waves.

**pneumatic trough** a flat container used in the collection of gases.

**gas-jar** (n) a tall vessel (↓) for the collection of gases.

**beehive** (n) the stand for a gas-jar which is put in a pneumatic trough (↑).

**aspirator** (n) a large vessel used to supply air or water to apparatus.

**eudiometer** (n) a glass tube (p.29), with a scale, used to measure the volume of gases.

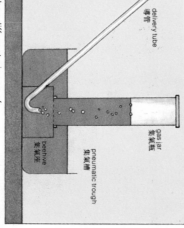

delivery tube 導管
gas jar 集氣瓶
beehive 蜂巢座
pneumatic trough 集氣槽

塞（名） 封閉管、瓶、燒瓶（↓）用的一種物件。例如插有塞的玻璃瓶的玻璃塞子。

軟木塞（名） 木栓質（用木栓質樹皮所製的）塞（↑）。

塞（↑）。 用橡膠製成的一種塞（↑），或木製之大塞。

腰導管（名） 用橡膠製成的一種塞（↑），或木製之大塞。

導管（名）（第 29 頁） 用於連接各件儀器（第 11 頁）的玻璃管。導管使液體可從一部儀器傳導到另一部儀器。

連接（動） 使各部儀器連接在一起，令液體可從儀器的某一部分流到另一部分。

拆開（動） 使各部儀器分開不連接（↑）。拆開是連接的反意詞。

槽（名）：波谷 （1）盛放液體用的平底容器；（2）兩個波峰之間的凹陷處。

集氣槽（名） 收集氣體用的平底容器。

集氣瓶（名） 收集氣體用的高身容器（↓）。

蜂巢座（名） 供集氣瓶（↑）內的集氣瓶放入的架子。

吸氣器（名） 向儀器設備供氣或供水的一種大型容器。

量氣管（名） 管上有刻度，用於測量氣體積的一種玻璃管（第 29 頁）。

water in 進水
air out 出氣
two glass tubes connected by a rubber tube 用橡膠管連接兩根玻璃管
rubber tube 橡皮管
glass tube 玻璃管
connect 連接
glass stopper 玻璃塞
rubber bung 橡膠塞
cork bung 軟木塞

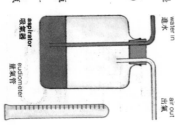

aspirator 吸氣器
eudiometer 量氣管

**receptacle** (*n*) a container in which solids, liquids, or gases are put. Any cup, beaker (↓), crucible (p.27), flask (↓) or receiver (p.28) is a receptacle.

**vessel** (*n*) a receptacle (↑) for containing liquids.

**rim** (*n*) the outside edge of the opening to a receptacle (↑).

**spout** (*n*) a pipe leading from a vessel (↑), through which a liquid passes when leaving the vessel.

**leak** (*v*) to lose fluids from a receptacle or pipe through a small hole in the receptacle or pipe, e.g. water leaks from a hole in a water pipe.

**beaker** (*n*) an upright vessel for containing liquids and also in which liquids can be heated.

**flask** (*n*) a round-shaped vessel or bottle with a long narrow neck; it is used to contain liquids for experiments (p.42). Flasks can have round bottoms or flat bottoms.

rim 邊緣

spout 嘴脊

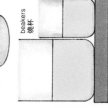

beakers 燒杯

**conical flask** a vessel with a narrow neck and a flat bottom; it is used in volumetric analysis (p.82).

**Erlenmeyer flask** a large conical flask (↑) containing more than 500 cm³ liquid.

**U-tube** (*n*) a glass tube (p.29) in the shape of the letter U.

**Woulfe bottle** a glass bottle with two necks; it is used to pass gases through a liquid.

U-tube U 型管

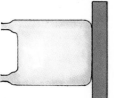

Woulfe bottle 渥耳夫瓶

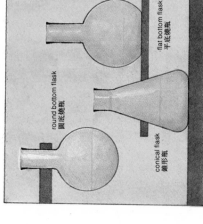

round bottom flask 圓底燒瓶

conical flask 錐形瓶

flat bottom flask 平底燒瓶

貯藏器：容器（名）盛放固體、液體或氣體用的容器。任何一種杯子、燒杯（↓）、坩堝（第 27 頁）、燒瓶（↓）或接受器（第 28 頁）都是一種貯藏器。

容器；器皿（名）盛裝液體用的一種貯藏器（↑）。

邊緣（名）貯藏器（↑）開口的外側邊緣。

嘴管（名）從容器（↑）引出的一根管，液體經由此管離開容器。

洩漏（動）液體經由貯藏器或管子的小孔從貯藏器或管流失。例如，水經由水管的小孔洩漏。

燒杯（名）盛裝液體用的一種直立式容器，或用以加熱液體的容器。

燒瓶（名）瓶頸細長的一種圓形容器或瓶子。燒瓶供盛裝做實驗（第 42 頁）用的液體。分圓底燒瓶和平底燒瓶。

錐形瓶：三角瓶 指有狹頸和平底的一種容器，供容量分析（第 82 頁）用。

愛倫美氏燒瓶 盛裝液體容量超過 500 cm³ 的大型錐形燒瓶（↑）。

U 形管（名）一種 U 字形玻璃管（第 29 頁）。

渥耳夫瓶 一種有兩個瓶頸可讓液體流過的玻璃瓶。

**graduation** (*n*) a mark on an instrument, or measuring vessel, which is equally spaced from other marks to show a scale (↓) for measurement. **graduated** (*adj*).

**scale** (*n*) a set of marks, with numbers beside them, rising from a low value to a high value, e.g. the scale on a thermometer rising from 0°C to 100°C. Each mark is a graduation (↑) on the scale.

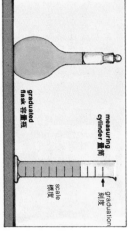

graduated flask 容量瓶

measuring cylinder 量筒

graduation 刻度

scale 標度

**measuring cylinder** a tall, narrow vessel, with a scale (↑) for volume; it is used to measure the volume of liquids. Measuring cylinders can measure volumes of 100 cm³, 1 dm³, etc.

**burette** (*n*) a long, narrow vessel with a spout and stopcock; used for measuring the volumes of a liquid which is allowed to run out of the burette. Burettes are used in volumetric analysis (p.82).

**pipette** (*n*) a vessel of a particular shape which delivers a fixed amount of liquid. Common sizes of pipettes are: 10 cm³, 20 cm³, 25 cm³, and 50 cm³.

**graduated flask** a flask with a very long and very thin neck. It has a graduation (↑) on the neck and measures the volume of a liquid very accurately. Graduated flasks commonly have sizes of 100 cm³, 250 cm³, 500 cm³ and 1 dm³. They are used in volumetric analysis (p.82).

**calibrate** (*v*) to make a scale (↑) on a device, instrument, or other measuring apparatus, so that it measures accurately, e.g. to pass a known electric current through an ammeter (p.123) and to mark the value of the current on a scale, and then to complete the scale by marking other graduations. **calibration** (*n*).

刻度（名）　儀器或量器上的細線。刻度與其旁標示一個由低數值至等，作為量度用之標度（↓）。（形容詞為 graduated）

標度（名）　一組刻線，其旁標示一個由低數值至高數值的數字，例如，溫度計上的數值由 0°C 升至 100°C。每一條刻線都是標度上的一個分度（↑）。

量筒　一種有體積標度（↑）的細長容器，用於量度液體的體積。量筒可測量的體積有：100 cm³、1 dm³ 等等。

滴定管（名）　一種有嘴管和活栓的細長容器，用於測量流出滴定管的液體之體積。滴定管用於容量分析（第 82 頁）。

移液管：吸管（名）　一種具有特別形狀，用於移出特定量液體的容器。常用的移液管規格有：10 cm³、20 cm³、25 cm³ 和 50 cm³。

容量瓶　一種瓶頸細長的瓶。瓶頸上刻有分度（↑），供極準確量測液體積釀用。量瓶的常用規格有：100 cm³、250 cm³、500 cm³ 以及 1 dm³。量瓶用於容量分析（第 82 頁）。

作校準（動）　於一種裝置、儀器或其他量測儀器上作一個標度（↑），使之能準確測量，例如，令已知強度之電流流過一具安培計（第 123 頁）並於度上標記下電流值，隨後標示其他刻線完成標度。（名詞為 calibration）

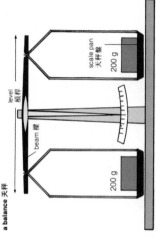

a balance 天平

top loading balance
托盤天平

**funnel** (*n*) a piece of apparatus with a wide mouth and a thin pipe; it is used to put liquids into bottles, flasks and other vessels.

**separating funnel** a kind of funnel (↑) used to separate immiscible (p.18) liquids.

**porous pot** (*n*) a vessel (p.25) with porous (p.15) sides; it is used with gases and liquids, and these fluids (p.11) can pass through the sides.

**crucible** (*n*) a receptacle (p.25) used for heating solids to a high temperature.

**generator** (*n*) any piece of apparatus which gives a supply of a required gas, e.g. a generator of hydrogen.

**Kipp's apparatus** a generator used to give a supply of a gas, especially hydrogen sulphide.

**thermostat** (*n*) a device used to keep a liquid at a constant (p.106) temperature.

**balance¹** (*n*) any piece of apparatus used to weigh solids, and liquids contained in vessels. The *mass* (p.12) of a solid is measured by a beam balance or a lever balance. The *weight* of a solid is measured by a spring balance.

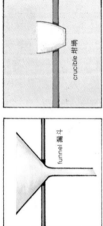

funnel 漏斗

crucible 坩堝

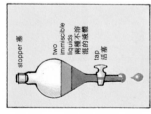

separating funnel 分液漏斗

stopper 塞
two immiscible liquids 兩種不溶的液體
tap 活塞

漏斗 (名) 一種有寬口和細管的儀器，用於將液體灌入瓶子、燒瓶或其他容器。

分液漏斗 供分離不溶混 (第 18 頁) 液體用的一種漏斗 (↑)。

素燒瓶；多孔瓶 (名) 一種側面多孔 (第 15 頁) 的容器 (第 25 頁)，素燒瓶供和氣體及液體一起使用，這些流體 (第 11 頁) 可流過素燒瓶的側面。

坩堝 (名) 將固體物質加熱至高溫用的一種容器 (第 25 頁)。

發生器 (名) 指提供所需氣體發生器用的任何設備。例如氫氣發生器。

吉普氏氣體發生器 指提供氣體，尤其是供硫化氫氣體的一種發生器。

恆溫器 (名) 保持液體溫度不變 (第 106 頁) 用的一種裝置。

天平 (名) 供秤量固體物質和盛於容器內液體的質量。固體的質量。固體質量係用槓桿式天平或槓桿式天平稱量。固體的重量則用彈簧天平稱量。

level 槓桿
beam 樑
scale pan 天秤盤
200 g
200 g

pan 天秤盤
scale 標度

**distillation flask** a flask used when distilling (p.33) liquids; it has a side-arm on its neck.

**condenser** (*n*) a long glass tube (↓) which has a larger glass tube around it. Vapour passes down the inner tube and cold water passes through the outer tube. The cold water condenses (p. 11) the hot vapour.

**receiver** (*n*) a vessel (p.25) put at the end of a condenser (↑) to receive the condensed (p.11) liquid.

**support** (*n*) a wooden or iron rod which takes the weight of a piece of apparatus and prevents the apparatus from falling. **support** (*v*).

**thermometer** (*n*) an instrument for measuring temperature (p.102). Most thermometers use mercury to measure the temperature on the Celsius scale. On this scale, water freezes at 0°C and boils at 100°C.

**lag** (*v*) to put a material (p.8) round a pipe, or piece of apparatus, to prevent heat escaping. **lagging** (*n*).

**retort** (*n*) a vessel (p.25) with a long neck, used in distillation (p.33).

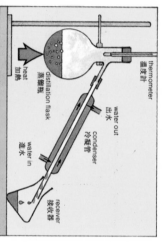

thermometer 溫度計

distillation flask 蒸餾瓶

heat 加熱

condenser 冷凝管

water out 出水

water in 進水

receiver 接收器

retort 曲頸瓶

蒸餾瓶 供蒸餾（第 33 頁）液體用的一種燒瓶，瓶頸部位有一根側臂管。

冷凝管 (名) 一根玻璃長管套着一個口徑較大的玻璃管。蒸汽自上面下流過內管，冷水則流過外管。冷水使熱的蒸汽冷凝（第 11 頁）。

接受器 (名) 置於冷凝管 (↑) 末端接收冷凝（第 11 頁）液體的容器（第 25 頁）。

支架 (名) 供支承儀器設備重量並防止它跌落的木或鐵枝製造的樣架。（動詞爲 support）

溫度計 (名) 量測溫度（第 102 頁）用之儀表。大多數溫度計是使用水銀，按照攝氏標度測量溫度。在此度上，水於 0℃ 結冰，於 100℃ 沸騰。

加套於…(動) 將一種材料（第 8 頁）包在管道或一件器具的周圍，以防熱量散逸。（名詞爲 lagging）

曲頸瓶 (名) 具長頸，供蒸餾（第 33 頁）用的一種容器（第 25 頁）。

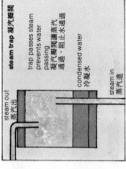

steam trap 凝汽瓣閥
trap passes steam prevents water passing 凝汽瓣閥讓蒸汽通過，阻止水通過
steam out 蒸汽出
condensed water 冷凝水
steam in 蒸汽進

pipe 管　jet 噴嘴
spatula 刮鏟

**trap** (n) a device (p.23) which allows one fluid (p.11) to pass through, but prevents another kind of fluid from passing through, e.g. a steam trap allows steam to pass through, but prevents water passing through.

閥閥；汽水閥（名）指可讓一種流體（第 11 頁）流過，而又阻止另一種流體流過的設備（第 23 頁）。例如凝汽瓣閥能讓蒸汽流過而又阻止水流過。

**jet** (n) (1) a very small hole at the end of a pipe. (2) a stream of fluid which comes out of a jet, e.g. a jet of water.

噴嘴；射流（名）（1）在管子端的一個極細的孔洞；（2）自噴嘴噴出的一股流體。例如水射流。

**blowpipe** (n) a piece of apparatus like a pipe, ending in a jet (↑). It can be used to blow a small, very hot flame onto an object.

吹管（名）一根形似管子、終端為一個噴嘴（↑）的儀器，用於吹細小而極熱的火焰到一個物體上。

**spatula** (n) a tool shaped like a spoon, or which has a flat surface, used for adding and removing small quantities of solids.

刮鏟（名）一種狀似匙羹的工具；或者一種表面平坦，用作加入或去除少量固體的工具。

**tongs** (n.pl.) an article used to hold hot objects.

鉗（名，複）支持熱物體用的一種物件。

**tube** (n) a long, hollow article, through which fluids can flow. Chemical apparatus uses glass and rubber tubes.

管（名）一種長形的中空製品，流體可在其中流動。化學儀器上使用玻璃管或橡膠管。

**tubing** (n) different kinds of tubes, either glass or rubber, may be of different bores (↓).

管類（名）不同種類的各種管，可以是玻璃管或橡膠管，有各種口徑（↓）。

**stout-walled** (adj) describes tubes (↑) with thick walls relative to the diameter of the bore (↓).

厚壁的（形）描述管子（↑）相對於其孔的直徑（↓）而言具有厚的管壁。

**thin-walled** (adj) describes tubes (↑) with thin walls relative to the diameter of the bore (↓).

薄壁的（形）描述管子（↑）相對於孔口徑（↓）而言具有薄的管壁。

**bore** (n) the size of the hole in a tube (↑), e.g. a bore of 5 mm.

孔徑（名）指管子（↑）孔口的大小，例如孔徑為 5 毫米。

**ground glass** (n) glass which has been rubbed to give a very smooth surface.

磨砂玻璃（名）車磨至表面極光滑的玻璃。

**quick-fit** (adj) describes apparatus which uses ground-glass (↑) stopper (p.24) and joints.

磨口接頭的（形）描述使用磨砂玻璃（↑）塞（第 24 頁）和接頭的儀器。

**diagram** (n) a drawing, using lines, to show how apparatus (p.23) is connected. A diagram is used because it is simpler to draw a diagram than to draw a picture.

簡圖（名）用線條表示儀器（第 23 頁）如何連接的一張圖。簡圖因比圖畫易繪製而被使用。

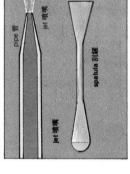

tongs 鉗
tubing 管類
tube 管
stout walled 厚壁的
bore 口徑
thin walled 薄壁
drawing of a beaker 燒杯圖

diagram 簡圖
diagram of a beaker 燒杯簡圖

**dissolve** (v) (1) to mix a solid (p.10) or a gas (p.11) with a liquid resulting in a solution (p.86), e.g. sugar is dissolved in water, air is dissolved in water. (2) to mix with a liquid, e.g. solids and gases dissolve in water; sugar dissolves in water. The solid or gas is soluble (p.17) if it dissolves.

**precipitate** (n) a solid (p.10) which appears in a solution when two solutions (p.86) are mixed. It is the result of a chemical reaction (p.62), e.g. when a solution of sulphuric acid is added to a solution of barium chloride, a precipitate appears. Precipitates are described as flocculent, creamy or heavy (see p.17).

**precipitate** (v), **precipitated** (adj).

**filter** (v) to separate an insoluble (p.17) solid from a liquid by pouring through a filter in a funnel. The filter can be filter paper or glass wool. **filter** (n), **filtered** (adj), **filtrate** (n), **filtration** (n).

**filtrate** (n) the liquid that passes through a filter (↑).

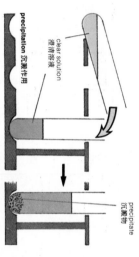

precipitation 沉澱作用

clear solution 澄清溶液

precipitate 沉澱物

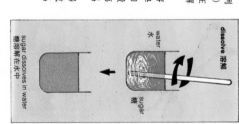

dissolve 溶解

water 水

sugar 糖

sugar dissolves in water 糖溶解在水中

**溶解**（動） (1) 使固體（第 10 頁）或氣體（第 11 頁）與液體混合形成一種溶液（第 86 頁）。例如，糖溶解在水中，空氣溶解在水中；(2) 與一種液體混合。例如，固體和氣體溶解在水中；糖溶解在水中。如果固體或氣體溶解了，它就是可溶的（第 17 頁）。

**沉澱物**（名） 指兩種溶液（第 86 頁）混合時，溶液中出現的一種固體（第 10 頁）。這是化學反應（第 62 頁）的結果。例如，硫酸溶液加入氯化鋇溶液中就會出現一種沉澱物。沉澱物可形容為絮凝的、乳脂狀的或重的（見第 17 頁）。（動詞為 precipitate，形容詞為 precipitated）

**過濾**（動） 將一種液體倒入漏斗，讓它流過濾器分離出不溶的（第 17 頁）固體。可用濾紙或玻璃棉作為濾器。（名詞為 filter，形容詞為 filtered，名詞為 filtrate，filtration）

**濾液**（名） 從濾器（↑）流出的液體。

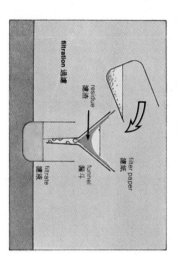

filtration 過濾

residue 濾渣

filter paper 濾紙

funnel 漏斗

filtrate 濾液

**residue** (*n*) (1) the solid that does not pass through a filter (↑). A precipitate (↑) is collected as a residue during filtration. (2) the material (p.8) or substances (p.8) left behind after any process, e.g. the residue left in a flask after distillation (p.33); the residue left in an evaporating basin after evaporation (p.11).

**suspended** (*adj*) describes light particles (p.13) of an insoluble (p.17) solid which float at all levels in a liquid or gas (p.11).

**settle** (*v*) to fall slowly through a fluid; particles (p.13) of a solid fall slowly to the bottom of a fluid when they settle. For example, insoluble particles fall to the bottom of a liquid; dust settles from the air onto a surface (p.16).

**sediment** (*n*) solid particles (p.13) that settle (↑) to the bottom of a vessel (p.25) form sediment. **sedimentation** (*n*).

**decant** (*v*) to pour off a clear (p.17) liquid, leaving any sediment (↑) at the bottom of the vessel (p.25). Before decanting liquid, any suspended (↑) material is allowed to settle as a sediment. *See supernatant (p.90).*

濾渣：殘渣（名）（1）指不能流過濾過器（↑）的固體。過濾過程中，沉澱物（↑）作爲一種濾渣而被收集；（2）經任何過程之後餘留下的材料（第 8 頁）或物質（第 8 頁）。例如經蒸餾作用（第 33 頁）之後，燒瓶中留下餘渣；蒸發作用（第 11 頁）之後，蒸發皿中留下餘渣。

懸浮的（形）描述不溶解的（第 17 頁）固體輕粒子（第 13 頁）漂浮在液體或氣體（第 11 頁）的所有層面上。

沉降（動）（慢慢經過）一種流體而降落；慢慢沉降時，固體粒子（第 13 頁）沉降到流體的底部。例如，不溶解的粒子降落到液體的底部；塵埃從空氣中降落到一個表面（第 16 頁）上。

沉積物（名）沉降（↑）到容器（第 25 頁）底部的固體粒子（第 13 頁）形成沉降物。（名詞爲 sedimentation）

傾出（動）傾出澄清的（第 17 頁）液體，使容器（第 25 頁）底部留下任何沉積物（↑）。液體傾析之前先讓懸浮的（↑）材料沉降成爲沉降物。參見"浮面的"（第 90 頁）。

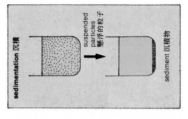

**sedimentation** 沉積
suspended particles 懸浮的粒子
sediment 沉積物

decanting 傾出
clear liquid 澄清液體
sediment left in flask 沉積物留在燒瓶中

**evaporate²** (v) to heat a liquid so that it boils and vapour is given off. This makes the volume of the liquid become less. A liquid can be evaporated to dryness, with all the liquid changed to vapour and dissolved (p.30) solids left as a residue (p.31). **evaporated** (adj).

**evaporating basin** a vessel (p.25) used for evaporating (↑) liquids.

**concentrate** (v) to boil away liquid from a solution (p.86) so that the same amount of solid is dissolved (p.30) in less liquid. The concentration (p.81) of the solution is then increased. **concentration** (n).

**digest** (v) to make a solid dissolve (p.30) in a liquid by adding the solid to the hot liquid. The hot liquid is usually stirred (↓). To contrast *digest* and *dissolve*: when a solid is *digested* a chemical action (p.19) takes place between the solid and the liquid; when a solid is *dissolved* a physical change (p.13) takes place, and there are no new substances formed. For example, copper (II) oxide is digested in hot sulphuric acid to make copper (II) sulphate. **digestion** (n).

**stir** (v) to move a glass rod, or other article, in a circular path in a liquid or powder (p.54) in order to mix the constituents (p.54) together. **stirrer** (n).

**fuse** (v) (1) to melt a powdered solid so that it forms one solid mass. **fusible** (adj), **fusion** (n), **fusibility** (n).

**calcine** (v) (1) to heat a solid to a high temperature to drive off volatile (p.18) substances. (2) to heat a metal to a high temperature to form the oxide (p.48) of the metal. **calcined** (adj).

**ignite** (v) to set a material or substance on fire so that it burns (p.59). The ignition temperature of a substance is the lowest temperature at which it will catch fire. **ignition** (n), **ignited** (adj).

**deflagrate** (v) to set a material or substance alight so that it burns with flame. **deflagration** (n) the bursting into flame of a substance caused by chemical action, e.g. the deflagration of phosphorus in chlorine. **deflagrate** (v).

蒸發 (動) 將液體加熱至沸騰而放出蒸汽，結果使液體的容量變少。液體可以蒸發至乾，所溶解的（第31頁）固體則留下來成爲殘渣（第31頁）。（形容詞爲 evaporation，evaporated）

蒸發皿 (動) 用以蒸發 (1) 液體用的一種容器（第25頁）。

濃縮 (動) 使液體從溶液（第86頁）中蒸發掉水而使固體溶解（第30頁）於較少量的液體中，從而提高溶液的濃度（第81頁）。（名詞爲 concentration）

熱解：蒸解 (動) 將固體加入熱的液體使之溶解（第30頁），或熱液體通常要攪拌與溶解之比較：固體熱解時，與液體溶解時只發生化學作用（第19頁）；而固體溶解時只發生物理變化（第13頁），沒有形成新的物質。例如，氧化銅 (II) 在熱硫酸中熱解生成硫酸銅 (II)。（名詞爲 diges-tion）

攪拌 (動) 在液體或粉末（第13頁）中以環形路徑移動一根玻璃棒或其他物件以使各成分（第54頁）混合在一起。（名詞爲 stirrer）

熔合 (動) (1) 使固體加熱至高溫而使之溶化（第18頁）。(2) 將金屬加熱至高溫使之生成其氧化物（第48頁）。（形容詞爲 calcined）

點燃：着火 (動) 將材料或物質置於火中使之燃燒（第59頁）。着火溫度是物質開始燃燒的最低溫度。（名詞爲 ignition，形容詞爲 ignited）

爆燃 (動) 將固體物質加熱化成一塊固體（第18頁）中以使還形成一固體塊狀物質 fusible。（名詞爲 fusion, fusibility）（形容詞爲 fusible，名詞爲 fusion, fusibility）

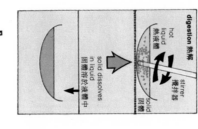

evaporation 蒸發
evaporating basin 蒸發皿
solution concentrated 溶液濃縮
volume less 體積減少
heat 加熱
liquid 液體

digestion 熱解
hot liquid 熱液體
stirrer 攪拌器
solid dissolves in liquid 固體溶於液體中
solid 固體

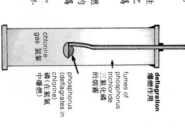

deflagration 爆燃作用
funes of phosphorus trichloride 三氯化磷的煙霧
phosphorus (deflagrates in chlorine) 磷（在氯氣中爆燃）
chlorine gas 氯氣

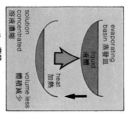

**pyrolysis** (*n*) the decomposition (p.65) of a chemical compound (p.8) by heat. Pyrolysis is usually used for organic (p.55) compounds; the organic compound is decomposed to simpler compounds.

**distil** (*v*) to change a liquid to a vapour (p.11) by heating and then to condense (p.11) the vapour back to a liquid. **distillation** (*n*).

**distillation** (*n*) the process (p.157) of distilling a liquid. Distillation is used to separate two or more liquids which have different boiling points. It is also used to purify (p.43) liquids.

**distilled** (*adj*) describes a liquid which has been purified (p.43) by distillation, e.g. distilled water is very pure water.

**bumping** (*n*) the forming of very large bubbles (p.40) when a liquid boils. As a bubble rises, the vessel containing the liquid jumps up and down; this event is called boiling by bumping. To prevent bumping, small pieces of porous pot (p.27) are put in the liquid.

**sublime** (*v*) to change a solid to a vapour (p.11) by heating, and then to cool the vapour so that it changes directly back to a solid, e.g. to sublime sulphur. **sublimation** (*n*), **sublimed** (*adj*).

**sublimation** (*n*) the process (p.157) of subliming a solid. Sublimation is used to purify solids. Not many solids sublime.

**sublimate** (*n*) a solid substance that has been formed (p.41) by sublimation.

**fumes** (*n.pl.*) (1) small particles (p.13) in the air, looking like smoke, are fumes. (2) vapour given off by an acid and combining with water vapour in the air to give the appearance of smoke. (3) any visible (p.42) vapour, especially a vapour which irritates (p.22) the nose and eyes. **fumes** (*v*).

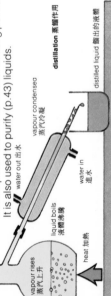

distillation 蒸餾作用

vapour condensed 蒸汽冷凝

distilled liquid 蒸餾出的液體

water out 出水

water in 進水

liquid boils 液體沸騰

vapour rises 蒸汽上升

heat 加熱

sublimation 昇華作用

cool surface 冷表面

vapour rises 蒸汽上升

sublimate 昇華物

solid 固體

heat 加熱

熱裂解：裂解作用 (名) 加熱使一種化合物 (第 8 頁) 分解 (第 65 頁) 的作用。熱裂解通常應用於有機 (第 55 頁) 化合物，使有機化合物分解成較簡單的化合物。

蒸餾 (動) 藉加熱使液體化成蒸汽 (第 11 頁)，然後再將水蒸汽冷凝 (第 11 頁) 成爲液體。(名詞爲 distillation)

蒸餾作用 (名) 將液體蒸餾的過程 (第 157 頁)。蒸餾作用可用於分離兩種或多種不同沸點的液體，也用於提純 (第 43 頁) 液體。

蒸餾過的 (形) 描述已藉蒸餾作用而提純的 (第 43 頁) 一種液體。例如，蒸餾水是很純淨的水。

突沸：爆沸 (名) 當液體沸騰時形成很大的氣泡 (第 40 頁)，盛液體的容器隨氣泡上升而上、下跳動，此情形稱爲突沸沸騰。爲防止突沸，可在液體中放置一個素燒瓶 (第 27 頁)。

昇華 (動) 加熱使固體化爲蒸汽 (第 11 頁) 之後使蒸汽冷卻直接變回固體。例如，使硫黃昇華。(名詞爲 sublimation，形容詞爲 sublimed)

昇華作用 (名) 使一種固體昇華的過程 (第 157 頁)。昇華作用可用於提純固體。可昇華的固體並不多。

昇華物 (名) 昇華作用所生成 (第 41 頁) 的固體。

烟霧 (名，複) (1) 指在空氣中像烟烟似的小粒子 (第 13 頁)；(2) 指由酸放出並與空氣中的水蒸汽結合在一起而看似的蒸汽；(3) 指任何可見的 (第 42 頁) 蒸汽，尤指會刺激 (第 22 頁) 鼻和眼的蒸汽。(動詞爲 fumes)

**separation** (*n*) the way in which substances are separated from each other, especially liquids. Immiscible (p.18) liquids are separated by using a separating funnel (p.27). Miscible (p.18) liquids are separated by distillation (p.33). Different methods are used for the separation of solid mixtures (p.54). **separate** (*v*). **separable** (*adj*).

**extraction** (*n*) (1) the process (p.157) of taking one substance from a mixture (p.54) of substances, e.g. the extraction of a solid by using a solvent (p.86) such as the extraction of iodine from its solution in water by using tetrachloromethane as a solvent. (2) the process of obtaining an element (p.8) from the earth; in a few cases the element can be extracted directly; in most cases an ore (p.154) is taken from the earth and the element is extracted from the ore. **extract** (*v*).

**dialysis** (*n*) a process (p.157) for the separation (↑) of a colloid (p.98) from a crystalloid (p.91). A mixture of a colloid and a crystalloid is put in a receptacle (p.25) made from a membrane (p.99), usually parchment or cellophane. Water is passed round the membrane and the crystalloid passes through the membrane and is taken away by the water. The colloid remains in the membrane. **dialyze** (*v*).

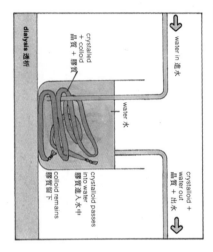

dialysis 透析

crystalled + colloid
晶質 + 膠質

water in 進水

water 水

water out
水出去

crystalloid + water out
晶質 + 出水

crystalloid passes into water
膠質進入水中

colloid remains
膠質留下

分離 (名) 使混合物中的各種物質 (第 8 頁) 尤其是液體彼此分離開的方法。不溶混的 (第 18 頁) 液體可用分液漏斗 (第 27 頁) 分離。可溶混的 (第 18 頁) 液體則利用蒸餾 (第 33 頁) 的方法分離。固體混合物 (第 54 頁) 可採用不同的方法分離。(動詞爲 separate (第 54 頁)，形容詞爲 separable)

萃取法：抽提法 (名) (1) 自物質的混合物 (第 54 頁) 中取出一種物質的方法 (第 157 頁)。例如，利用溶劑 (第 86 頁) 萃取一種固體物質，像利用四氯甲烷爲溶劑自碘的水溶液中萃取碘；(2) 指自土壤中獲取某種元素 (第 8 頁) 的方法。在少數情況下是先從土壤中取得一種礦石 (第 154 頁)，再由礦石中提取元素。(動詞爲 extract)

滲析 (名) 自晶質 (第 91 頁) 中分離 (↑) 出膠體 (第 98 頁) 的一種方法 (第 157 頁)。將膠體和晶體的混合物置於一個由膜 (第 99 頁) 構成的容器 (第 25 頁) 之中，通常是用半皮紙或玻璃紙作爲膜。水在膜內通過，晶體或晶質穿過膜並由水帶走。膠體則留於膜內。(動詞爲 dialyze)

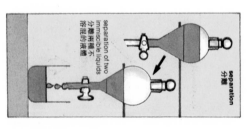

separation 分離

separation of two immiscible liquids
分離兩種不溶混的液體

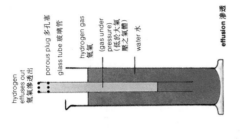

hydrogen effuses out 氫氣滲出
porous plug 多孔塞
glass tube 玻璃管
hydrogen gas 氫氣
(gas under pressure) (低於大氣壓之氣體)
water 水

**effusion** 滲透

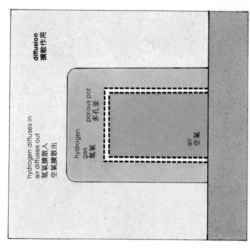

**diffusion** 擴散作用

hydrogen diffuses in air diffuses out 氫氣擴散入 空氣擴散出

hydrogen gas 氫氣
porous pot 多孔釜
air 空氣

**diffusion** (*n*) (1) the process (p.157) of one gas spreading through another gas, e.g. petrol vapour spreads through the air in a room by diffusion. (2) when two miscible (p.18) liquids are put together, they spread through each other by diffusion. (3) when a soluble solid is put in a liquid, the solid dissolves and spreads through the liquid by diffusion. (4) gases pass through a porous (p.15) membrane (p.99) by diffusion if the pressure on each side of the membrane is the same. **diffuse** (*v*).

**effusion** (*n*) the process (p.157) of a gas passing through a porous (p.15) membrane (p.99) or through a small hole when the gas goes from a higher pressure to a lower pressure. Compare diffusion (↑) where the pressure is the same on both sides of the membrane or hole. **effuse** (*v*).

**absorption** (*n*) (1) the process (p.157) of a solid taking in a liquid or a gas, e.g. the absorption of a gas by charcoal, the absorption of water by a gel (p.100). (2) the process of a liquid taking in a gas, e.g. sodium hydroxide solution absorbs carbon dioxide from the air.

**擴散作用 (名)** (1) 一種氣體散佈到另一種氣體中的過程 (第 157 頁)。例如：汽油蒸汽藉擴散散佈到室內空氣中；(2) 兩種可溶混的 (第 18 頁) 液體放在一起時，它們藉擴散而相互散佈；(3) 可溶的固體置於一種液體中，該固體溶解並擴散散佈到液體可藉擴散而透過多孔膜兩孔 (第 15 頁) 膜 (第 99 頁)。(助詞為 diffuse)

**滲透 (名)** 當氣體從較高壓力進入較低壓力狀態時，氣體穿過多孔 (第 99 頁) 或穿過小孔的過程 (第 157 頁)。參比 "擴散" (↑)，擴散時孔膜兩側的壓力相同。(助詞為 effuse)

**吸收作用 (名)** (1) 指固體吸入一種液體或氣體的過程 (第 157 頁)。例如，炭吸收氣體，凝膠體 (第 100 頁) 吸入水分；(2) 指液體吸入一種氣體的過程。例如，氫氧化鈉溶液吸收空氣中的二氧化碳。

**chromatography** (*n*) a method of separating a mixture of solutes, by using a solvent and a separating medium which can be paper or an acid material such as silica or alumina, or an inert (p. 19) support coated with a solvent. In gas chromatography, volatile constituents (p. 54) are made to flow through a column packed with an inert support coated with a suitable solvent by an inert gas.

色層分離法 (名)　用溶劑和分離介質分離可混合物的一種方法。材料如矽石或礬土或塗覆以溶劑的一種惰性材料（第 19 頁）載體。在氣相色層分析法中，讓揮發性組分（第 54 頁）流過一種以惰性載體塗覆的柱。載體則塗覆以一種合適的溶劑。

cork 軟木塞
glass support 玻璃支架
paper strip 條紙
glass vessel 玻璃容器
mixture in solution 混合物溶液
second solvent 第二溶劑

**paper chromatography** 濾紙層析法

**paper chromatography** a solid mixture is dissolved in a suitable solvent and drops of this solution are placed on marks on a strip of filter paper. The paper strip is put in a corked container with one end in a second solution of a reagent which forms a coloured derivative (p. 200) of a constituent. This action, to produce different coloured derivatives, is called *development*. **develop** (*v*).

濾紙層析法　將固體混合物溶解於一種合適的溶劑內，取此溶液數滴滴於一條濾紙條的各標記位置。取此無色者，則條紙須嗆放入一隻有軟木塞的容器內。其一端浸在第二溶劑中（見圖）。第二溶劑沿條紙向上移動並攜帶混合物的各個成分（第 54 頁）上移。不同成分所上升的高度不同。

**development** (*n*) if the constituents of a mixture used in paper chromatography (↑) are colourless, then the paper is sprayed with a dilute solution of a reagent which forms a coloured derivative (p. 200) of a constituent. This action, to produce different coloured derivatives, is called *development*. **develop** (*v*).

顯色 (名)　濾紙層析法中所用之混合物的各成分若為無色者，則條紙須嗆以稀的試劑溶液，此種試劑能和一個成分形成有色的衍生物（第 200 頁）。產生不同顏色的衍生物的作用稱為"顯色"。（動詞為 develop）

**solvent front** the greatest height reached by a solvent in paper chromatography (↑). The constituents of a mixture never reach this height.

溶劑前沿　溶劑沿紙條上升達到之最高高度。各個成分絕不會達到此高度。

**chromatogram** (*n*) a paper strip, or a column, in chromatography (↑) with the individual constituents marked by coloured spots, using development (↑) if necessary, after separation by a suitable solvent.

色層譜：層析譜 (名)　色層分析中，用一種合適的溶劑分離並以色點（必要時可用顯色（↑）法）標示各個成分的圖譜。

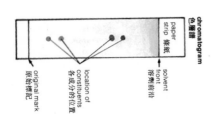

**chromatogram** 色層譜
paper strip 條紙
solvent front 溶劑前沿
location of constituents 各成分的位置
original mark 原始標記

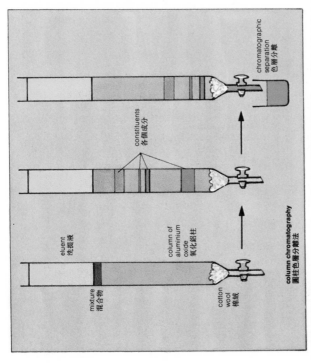

eluent
洗提液

mixture
混合物

column of
aluminium
oxide
氧化鋁柱

cotton
wool
棉絨

constituents
各個成分

chromatographic
separation
色層分離

**column chromatography**
圓柱色層分離法

**column chromatography** a glass tube is packed with an inert (p.19) solid, usually aluminium oxide, forming a column. A mixture is placed on the top of the column. A suitable solvent is added above the mixture, *see diagram*. The solvent flows slowly down the column taking the constituents (p.54) of the mixture with it. Different constituents travel different distances. By continually adding solvent, all the constituents will eventually pass out of the column and can be separately collected.

**eluent** (*n*) the solvent used to separate a mixture in column chromatography (↑). This method of separation is called *elution*.

**location** (*n*) the process of finding the positions of constituents (p.54) in a chromatogram (↑). For colourless constituents, development (↑) or ultra-violet light is used to produce colours.

**locate** (*v*).

**圓柱色層分離法** 一根玻璃管內填充入一種惰性 （第 19 頁）固體（通常是氧化鋁）形成一根分 離柱。柱之頂部放入一種混合物，再於混合 物之上加入適當之溶劑（見圖）。溶劑徐徐流 向柱底，並將混合物中的各成分（第 54 頁） 帶下。不同的成分所移動的距離不同，隨溶 劑不斷地加入，最終都使全部成分流出柱並 可分別收集之。

**洗提液**（名） 在圓柱色層分離法（↑）中，用於分 離混合物的一種溶劑。此種分離方法稱爲 "洗提"。

**定位**（名） 找出各個成分（第 54 頁）在色層譜（↑） 中之位置的過程。對於無色成分，需使用顯 色（↑）或紫外光以產生顏色。（動詞爲 locate）

**grind** (v) to change large lumps of a solid into a powder by rubbing in a pestle and mortar (↓). **ground** (adj).

**pestle** (n) an article with a handle and a rounded glass or stone end, used to grind (↑) substances in a mortar (↓).

**mortar** (n) a receptacle (p.25) for holding substances to be ground by a pestle.

**triturate** (v) to mix by grinding (↑). Two or more solids, or a solid and a liquid can be triturated. **trituration** (n).

**treat** (v) to add a reagent (p.63) or any chemical to a material or substance to cause a chemical change; e.g. to treat iron with concentrated nitric acid to make the iron passive; to treat cotton with sodium hydroxide solution to give the cotton a shiny surface (p.16). **treatment** (n).

**acidify** (v) to add acid (p.45) to a solution (p.86) so that there is excess (p.230) acid present. **acidified** (adj).

**indicator** (n) a chemical substance (p.8) that shows whether a solution is acidic (p.45), alkaline (p.45) or neutral (p.45). **indicate** (v).

**indicate** (v) to show clearly, or to make a sign. **indicate** (v).

**pH** (abbr) a way of describing how acid or alkaline a solution is.

**pH value** (n) a number on a scale of 1 to 14 showing the strength or concentration of an acid or alkali. A value of pH = 1 indicates (↑) a strong acid. A value of pH = 14 indicates a strong alkali. A value of pH = 7 indicates a neutral solution.

litmus paper 石蕊試紙
indicator 指示劑
in acids 在酸中
in alkalis 在鹼中

pH scale pH標度
strong 強　0　2　weak 弱　4　6　neutral 中性　7　8　weak 弱　10　12　strong 強　14
acid 酸　　　　　　　　　　　　　　　alkali 鹼

pestle 研杵
mortar 研缽

研磨（動）　將大塊固體置於研缽（↓）中用研杵和磨擦成粉末。（形容詞爲 ground）

研杵（名）　一種具手柄和圓球形玻璃端或石頭端的物件，用於研磨（↓）中研磨（↑）物質。

研缽（名）　一種用於盛載用研杵所研磨物質的容器（第 25 頁）。

研碎（動）　藉研磨（↑）以混合。兩種或多種固體，或一種固體和一種液體均可研碎。（名詞爲 trituration）

處理（動）　在一種材料或物質中加入一種試劑（第 63 頁）或任何一種化學藥劑而引致發生化學變化。例如，用濃硫酸處理鐵使鐵變鈍；用氫氧化鈉溶液處理棉使棉具有光澤的表面（第 16 頁）。（名詞爲 treatment）

酸化（動）　在一種溶液（第 86 頁）中加入酸（第 45 頁），使該溶液含酸過量（第 230 頁）（形容詞爲 acidified）

指示劑（名）　指示溶液是酸性（第 45 頁）、鹼性（第 45 頁）或中性（第 45 頁）的一種化學物質（第 8 頁）。（動詞爲 indicate）

pH（略）　表示溶液酸度或鹼度的一種方法。

pH 值（名）　在 0-14 範圍的一個標度上表示酸或鹼的強度或濃度的數值。pH = 1 的值指示（↑）一種強酸。pH = 14 指示一種強鹼，pH = 7 的值指示一種中性的溶液。

**titration** (n) the process (p.157) of letting a solution (p.86) flow from a burette (p.26) into another solution held in a conical flask (p.25) until a chemical reaction (p.62) is complete. A common titration is between an acid (p.45) and an alkali (p.45) using an indicator (p.38). **titrate** (v).

**end point** the sign that a titration (↑) is complete as the chemical reaction has been completed. In an acid and alkali titration the end point is shown by a change in the colour of an indicator.

**titre** (n) the volume of a solution from a burette needed to reach the end point in a titration (↑).

**reading** (n) the value on a scale of an instrument (p.23) or a piece of apparatus (p.23) which is taken as a measurement, e.g. the reading of the level of the solution in a burette from the scale of volume on the burette; mercury in a thermometer shows the temperature on the thermometer scale, this is the reading on the thermometer.

**record** (v) to write down a reading (↑) or an observed (p.42) effect. **record** (n).

**result** (n) a change which is measured or observed (p.42). To contrast *result* and *effect*: when iron is exposed to air and water it rusts (p.61) forming iron oxide; rusting is the effect, and iron oxide is the result. **result in** (v).

**tabulate** (v) to write down results (↑) in a table. **tabulation** (n).

滴定法 (名) 使溶液 (第86頁) 從滴定管 (第26頁) 流入錐形瓶 (第25頁) 內的另一種溶液中直至完成化學反應 (第62頁) 的過程 (第157頁)。普通滴定法是在酸 (第45頁) 和鹼 (第45頁) 之間作用一種指示劑 (第38頁) 進行的。(動詞爲 titrate (v)。

終點 表示化學反應完成而滴定 (↑) 已結束的標誌。在酸鹼滴定中,藉指示劑的顏色變化指示終點。

滴定度 (名) 達到一次滴定 (↑) 終點所需的滴定管溶液量 (體積)。

讀數 (名) 儀表 (第23頁) 或取作量度的一台器具 (第23頁) 的標度上的數值。例如,自滴定管體積標度上的滴定管內溶液液面的讀數;溫度計內的水銀在溫度計標度上指示溫度,這是在溫度計上的讀數。

記錄 (動) 寫下一個讀數 (↑) 或一個觀察到的 (第42頁) 現象。(名詞爲 record (n)。

結果 (名) 指測量或觀察到 (第42頁) 的一種變化。result (結果) 與 effect (效果) 之比較:鐵暴露於空氣和水時,鐵生銹 (第61頁) 生成氧化鐵;生銹是產生的效果而氧化鐵是生成的結果。(動詞爲 result in (v)。

列表 (動) 在表格中寫下結果 (↑)。(名詞爲 tabulation (n)。

**table of result**
結果表格

| | cm³ | cm³ | cm³ | cm³ |
|---|---|---|---|---|
| 2nd reading 第二次讀數 | 24.1 | 46.7 | | 26.3 |
| 1st reading 第一次讀數 | 1.4 | 24.1 | | 3.6 |
| titre 滴定度 | 22.7 | 22.6 | | 22.7 |

圖線 (名) 表示兩個變量之間的關係的線,例如氣體壓力與體積的關係、物質的溶解度 (第87頁) 與溫度的關係的圖線。(形容詞爲 graphical)。

**graph** (n) a line drawn to show the relation between two changing quantities, e.g. pressure and volume of a gas, solubility (p.87) of a substance and temperature. **graphical** (adj).

**plot** (v) to make a graph (↑) by putting marks for results (↑) and connecting the marks by a line.

繪圖 (動) 將結果 (↑) 作記號並用一條線將各個記號連接繪成一條圖線。

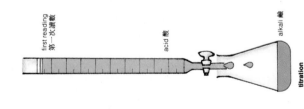

titration 滴定

**evolve** (v) to form bubbles (↓) of gas and give off (p.41) the gas. To contrast *form*, *give off*, and *evolve* gases: *form* and *give off* can be used with both physical and chemical changes, *form* describes a weaker effect than *give off*, *evolve* is used only with chemical changes and describes a stronger effect than *give off* and the gas or vapour evolved can be collected. Evolution of gases are described as *steady*, *brisk* and *rapid* as the quantity of gas evolved is increased. **evolution** (n).

**effervesce** (v) to evolve (↑) a gas rapidly with the formation of many bubbles (↓) at a liquid surface (p.16), e.g. when a dilute acid is added to a carbonate, the mixture effervesces. **effervescence** (n).

**bubble¹** (n) a small quantity of a gas or vapour with liquid around it. The liquid can be a thin film (p.18) round the bubble or the liquid can be in a vessel with bubbles in the liquid.

**bubble²** (v) to cause bubbles (↑) of gas to go through a liquid.

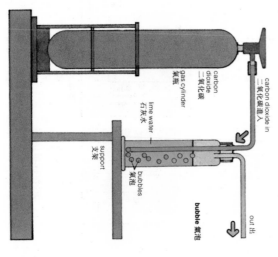

carbon dioxide 二氧化碳
gas cylinder 氣瓶
carbon dioxide in 二氧化碳進入
lime water 石灰水
support 支架
bubbles 氣泡
out 出
**bubble** 氣泡

放出(動)　形成氣體之氣泡(↓)並散發(第 41 頁)該氣體。形成(form)氣體、散發(give off)氣體、放出(evolve)氣體之間的區別：form 和 give off 既可用於物理變化亦可用於化學變化，form 所描述之效應比 give off 弱；而 evolve 這一詞只用於化學變化，所描述之效果比 give off 強，所放出的氣體或蒸汽可以收集。當放出的氣體量增加時，放出氣體可描述為穩定、冒泡和迅速。(名詞爲 evolution)

氣泡(動)　隨着液體表面(第 16 頁)所包圍的氣體一層薄膜的形成而迅速放出(↑)氣體。例如，在稀酸中加入一種稀酸，混合物即泡騰。(名詞爲 effervescence)

泡騰(動)　少量爲液體所包圍着氣泡的少量氣體或蒸汽。液體可爲包圍着氣泡的一層薄膜(第 18 頁)，或者是在一個容器中含有氣泡的液體。

起泡(動)　使氣體之氣泡(↑)通過一種液體。

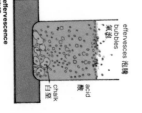

**effervescence** 泡騰
effervesces bubbles 氣泡
acid 酸
chalk 白堊

**form** (v) to cause to come into being, e.g. when two solutions (p.86) are mixed, a precipitate (p.30) is formed; kerosene forms a vapour; hydrogen is formed when calcium metal reacts with water. **formation** (n).

**give off** of an object, living thing, or chemical action, to cause a gas, vapour (p.11), or odour (p.15) to come from itself. For example, when iron is added to dilute sulphuric acid, hydrogen (a gas) is given off by the reaction; when water is boiled, steam is given off by the water; a flower gives off an odour. See liberate (p.69).

**generate** (v) to produce (p.62) large quantities of a gas for a particular purpose. See generator (p.27). **generator** (n).

**collect** (v) (1) to obtain a gas in a vessel (p.25), usually a gas-jar (p.24). (2) to obtain a distilled liquid in a receiver (p.28). (3) to obtain a specimen (p.43) of crystals (p.91).

**pass over** to cause a gas to flow over a solid, e.g. hydrogen is passed over heated copper (II) oxide.

形成：產生（動）使出現。例如，兩種溶液（第86頁）混合時，產生沉澱物（第30頁）；煤油形成蒸汽；鈣金屬與水反應產生氫氣。（名詞爲 formation）

發出：散發 指某種物體、生物體、或化學反應本身產生一種氣體、蒸氣（第11頁）或氣味（第15頁）。例如，將鐵加入稀硫酸中時，發生化學反應發出氫（一種氣體）；由水發出蒸汽；花朵發出一種氣味。參見"釋放出"（第69頁）。

發生（動）爲某一特定目的而產生（第62頁）大量氣體。參見"發生器"（第27頁）。（名詞爲 generator）

收集（動）（1）在一個容器（第25頁），通常是一個集氣瓶（第24頁）中收取一種氣體；（2）在一個接受器（第28頁）中取得餾出的液體；（3）獲得晶體（第91頁）的一個樣本（第43頁）。

流過：使氣體從固體上面流過。例如……上流過，使氫氣從熱的氧化銅(II)上面流過。

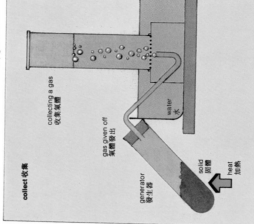

collect 收集
collecting a gas 收集氣體
gas given off 氣體發出
generator 發生器
solid 固體
heat 加熱
water 水

displaced air out 排代的氣體出
dense gas in 濃的氣體入
collect 收集
collecting a dense gas 收集濃的氣體

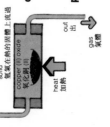

pass over 流過
hydrogen gas in 氫氣入
passing hydrogen gas over a heated solid 氫氣在熱的固體上流過
copper (II) oxide 氧化銅(II)
heat 加熱
gas out 氣體出

**investigate** (v) to study carefully by means of experiments (↓), tests (↓) and recorded (↓) facts, e.g. to investigate the properties (p.9) of sulphuric acid is to find and record the properties by making a careful study of them. **investigation** (n).

**experiment** (n) to work with instruments and apparatus in a practical (p.23) investigation on the behaviour and nature of substances. See perceptible (↓).

**visible** (adj) describes anything that can be seen. See perceptible (↓).

**observation** (n) the use of the senses for a particular purpose, e.g. when a chemical reaction (p.62) takes place, a person makes observations on any changes of state, colour, or odour, and on any new substances formed. Observations are recorded as results (p.39). **observe** (v), **observable** (adj).

**perceptible** (adj) describes any changes, usually physical changes, which can be observed (↑) by the senses, although such changes are small. Observable changes can be noticed more easily than perceptible changes. Changes can be readily, barely or hardly perceptible, showing increasing difficulty of perception. **perception** (n), **perceive** (v).

**test** (n) the use of a chemical reagent (p.225) to identify (p.125) a substance, a metal, a cation or anion (p.125) or any group of substances. The test gives a result and from the result an inference (↓) is made. Some tests give a definite (p.226) result so the identification is made without doubt. **test** (v).

**confirmatory** (adj) describes a test which makes an identification without doubt. **confirm** (v), **confirmation** (n).

**test paper** (n) a piece of paper with a reagent (p.63) on it; the paper changes colour when testing particular substances and thus identifies (p.225) these substances, e.g. litmus paper.

**demonstrate** (v) to show clearly by practical examples, e.g. to demonstrate the properties of chlorine gas by showing in an experiment (↑) the action of various substances on the gas. **demonstration** (n).

調查（動） 藉助實驗(↓)、試驗(↓)和記錄的(第39頁)事實作精細的研究。例如，為了研究硫酸的性質(第9頁)必須對硫酸作深入的精細研究以發現並記錄下其性質。（名詞為investigation）

實驗（名） 在實際的(第23頁)調查研究中，使用儀器和器具研究各種物質的形狀和本質。參見 可覺察的（形）。

可見的（形） 描述肉眼可看見的任何事物。參見 可覺察的（形）。

觀察（名） 使感覺應用於某種特殊目的，例如，在有關生化學反應(第62頁)時，人們可對有關狀態、顏色或氣味等作記錄，同時觀察生成的一些新物質。將觀察到的事實作為結果(第39頁)記錄下來。（動詞為observe，形容詞為observable）

可覺察的（形） 形容任何一種變化，通常是物理變化，儘管這些變化很細微，但眼睛仍可觀察(↑)到。可觀察到的變化比可覺察到的變化更易引起人們注意。所謂變化可以分成"易覺察到的"、"僅可覺察到的"或"很難覺察到的"，表明可覺察的難度逐漸加大了。（名詞為perception，動詞為perceive）

試驗（名） 用化學試劑(第63頁)鑑別(第225頁)一種物質、金屬、陽離子或陰離子(第125頁)或任何一組物質。試驗可提出結果並由該結果作出推論(↓)。某些試驗可提供確定的(第226頁)結果，因而可作出明確的鑑別。（動詞為test）

證實的（形） 形容一種試驗能的無疑地作出鑑定。（動詞為confirm，名詞為confirmation）

試紙（名） 指各有一種試劑(第63頁)的一張紙；當用於試驗特定物質時，試紙會變色因而可鑑別(第225頁)出這些物質，石蕊試紙即為試紙之一例。

驗證（動） 用實例清楚地顯示。例如，在一項實驗(↑)中通過顯示各種不同物質對氯氣的作用以驗證氯氣的性質。（名詞為demonstration）

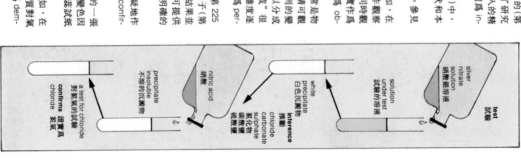

silver nitrate solution 硝酸銀溶液
solution under test 試驗的溶液
test 試驗
white precipitate 白色沉澱物
inference 推斷
nitric acid 硝酸
precipitate insoluble 不溶的沉澱物
chloride carbonate sulphate 氯化物 碳酸鹽 硫酸鹽
confirms 證實為
chloride 氯化物
a test for chloride 對氯氣的試驗

**inference** (n) the use of results (p.225) to decide on the identification (p.225) of a substance. An inference is only an opinion, a confirmatory (↑) test has to be made to be sure. **infer** (v).

**conclusion** (n) (1) the use of results (p.39) to decide on the identification (p.225) of a substance without doubt, e.g. inferences (↑) test from tests together with a confirmatory (↑) test allow a conclusion to be made. (2) the use of results to decide on the relation (p.232) between observations (↑), e.g. a set of results on pressure and volume allows a conclusion to be drawn on the relation between the pressure and the volume of a gas. **conclude** (v).

**technique** (n) an accepted way of carrying out a process (p.157) which needs practical (p.23) skill and a knowledge of chemistry, e.g. if a person knows the technique of distillation (p.33), he can connect suitable apparatus, heat to the correct temperature and collect a pure distillate.

**isolate** (v) to obtain a pure substance from a mixture (p.54) of substances. The pure substance can be a compound (p.8) or an element (p.8), e.g. to isolate bromine from sea water.

**preparation** (n) (1) a substance made in a laboratory (p.23) for a particular purpose, e.g. a student makes crystals of copper (II) sulphate, the substance is a preparation. (2) to make ready the apparatus and any solutions (p.86) needed in an experiment (↑). **prepare** (v).

**specimen** (n) (1) a small quantity of a material or substance (p.8) which is isolated (↑) from a mixture. (2) a quantity of a substance which is used as an example of that substance, i.e. the properties of the specimen are the properties of all quantities of the substance. For example, tests carried out (p.157) on a specimen of copper demonstrate (↑) the properties of the specimen and thus the properties of copper.

**purification** (n) a process (p.157) or processes carried out (p.157) to remove impurities (p.20) from a substance, e.g. the purification of silver by which impurities such as lead are removed (p.215). **purify** (v).

推斷：推論（名）　用實驗結果（第 39 頁）去決定一種物質的鑑別（第 225 頁）。推斷只是一種意見，必須作證實（↑）試驗方可確定。（動詞為 infer）

結論（名）（1）利用實驗結果（第 225 頁）去決定對物質所作的鑑別（第 225 頁）的無可疑。例如，利用試驗作出的推斷（↑）再加確證（↑）的試驗便可作出一種結論；（2）用結果去決定各種觀察（↑）之間的關係（第 232 頁）。例如，由壓力和體積的一組試驗結果可對壓力與氣體體積之間的關係作出結論。（動詞為 conclude）

技術（名）　實現（第 157 頁）一個過程（第 157 頁）的認可方法，為此必須具備有實際（第 23 頁）技能和化學知識。例如，掌握了蒸餾（第 33 頁）的技術，就可以將合適的器器相連接，然後加熱到合適的溫度，收集純淨的餾出物。

分離：離析（動）　自物質的混合物（第 54 頁）獲取一種純物質。純此化合物可以是一種化合物（第 8 頁）或一種元素（第 8 頁）。例如，自海水中分離出溴。

製品：配製品：預製品（名）（1）在實驗室（第 23 頁）中為特定用途而裂造的一種物質。例如，一位學生製造硫酸銅（II）的晶體，此種物質就是一種製劑；（2）準備好實驗（↑）所需的一套用具、儀器和任何溶液（第 86 頁）。（動詞為 prepare）

樣本：標本（名）（1）自一種混合物中分離出的（↑）少量材料或物質（第 8 頁）；（2）一定量的物質用作該種物質所具有的例子，亦即樣本的性質代表全量物質所具有的性質。例如，對銅的樣本進行（第 157 頁）的試驗，驗證（↑）了銅樣本的性質，因而是銅的性質。

提純（名）　去除一種物質中的雜質（第 20 頁）所進行（第 157 頁）的一個過程（第 157 頁）或多個過程。例如，銀之提純使其純質如鉛得以除去（第 215 頁）。（動詞為 purify）

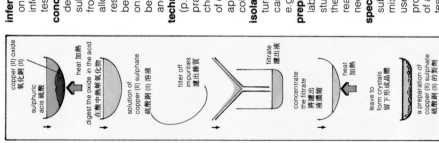

preparation 製劑

copper (II) oxide 氧化銅(II) — heat 加熱 — sulphuric acid 硫酸 — digest the oxide in the acid 在酸中解氧化物 — solution of copper (II) sulphate 硫酸銅(II)溶液 — filter off impurities 濾出雜質 — filtrate 濾出液 — concentrate the filtrate 濃縮濾出液 — heat 加熱 — leave to form crystals 留下形成晶體 — a preparation of copper (II) sulphate 硫酸銅(II)的製劑

**chemical nomenclative**
化學命名法

| | | |
|---|---|---|
| trivial name 俗名 | quick lime 生石灰 | blue vitriol 藍礬 |
| | calcium oxide 氧化鈣 | cupric sulphate 硫酸銅 |
| | calcium oxide 氧化鈣 | copper (II) sulphate 硫酸銅 (II) |
| traditional name 慣用名 | green vitriol 綠礬 | oil of vitriol 礬油 |
| | ferrous sulphate 硫酸亞鐵 | sulphuric acid 硫酸 |
| systematic name 學名 | iron (II) sulphate 硫酸鐵 (II) | sulphuric acid 硫酸 |
| | common salt 食鹽 | potassium permanganate 高錳酸鉀 |
| | sodium chloride 氯化鈉 | potassium manganate (VII) 高錳酸鉀 (VII) |
| | sodium chloride 氯化鈉 | |

**nomenclature** (*n*) a way of naming chemical substances (p.8).

**trivial name** a nomenclature (↑) for chemical substances which was used before chemistry was studied properly. Examples of trivial names are alum, blue vitriol, lime, and chalk.

**traditional name** a nomenclature (↑) for chemical substances which shows their chemical composition (p.82); this method was used before system-atic names (↓) were used. Some traditional names are still used, and some are the same as systematic names. Examples of traditional names are cupric sulphate and lead nitrate.

**systematic name** the modern nomenclature (↑) for chemical substances. For inorganic (p.55) compounds, the oxidation number (p.78) of the metal is given, and the acid radical (↓) or anion (p.125) is described with the oxidation number for its important element (p.8), e.g. copper (II) sulphate, iron (III) sulphate, potassium manganate (VII). For organic (p.55) compounds, the systematic name is taken from a suitable alkane (p.172), e.g. *ethanoic* acid is taken from *ethane*; ethanoic acid is the system-atic name of acetic acid (traditional name).

**binary compound** a compound (p.8) formed from the chemical combination (p.55) of two elements (p.8). The systematic name (↑) for such compounds ends in *-ide*, e.g. lead (II) oxide, calcium carbide, phosphorus trichloride.

**命名法：名稱 (名)** 給各種化學物質 (第 8 頁) 取名的一種方法。

**俗名** 在真正研究化學之前，各種化學物質曾取用的名稱 (↑)。例如明礬、膽礬、石灰、白堊均是俗名。

**慣用名** 用以表示化學物質組成 (第 82 頁) 之名稱 (↑)：先是用慣用名，後來才使用學名 (↓)。某些慣用名現仍沿用，而有些則與學名相同。例如，硫酸銅、硝酸鉛既是學名也是慣用名。

**學名** 化學物質的現代化學命名法 (↑)。無機 (第 55 頁) 化合物的學名，示出其金屬的氧化值 (第 78 頁)，酸根 (↓) 或陰離子 (第 125 頁) 則描述其主要元素 (第 8 頁) 的氧化值。例如，硫酸銅 (II)、硫酸鐵 (III)、高錳酸鉀 (VII) 酸鉀。有機 (第 55 頁) 化合物的學名係取自一合適的烷屬烴 (第 172 頁)。例如，乙酸取名於乙烷；乙酸是醋酸 (慣用名) 的學名。

**二元化合物** 由兩種元素 (第 8 頁) 化學結合 (第 64 頁) 所形成的一種化合物 (第 8 頁)。此種化合物的學名 (↑) 其詞尾皆為-*ide*。例如，氧化鉛 (II)、碳化鈣、三氧化磷之英名學名詞尾均為-*ide*。

**radical** (n) a group of atoms, which are part of a molecule (p.77) or form an ion (p.123); the radical often remains unchanged throughout a chemical reaction (p.62) or a series of reactions. The atoms of a radical which forms an ion are held together by covalent bonds (p 136). Some radicals are also functional groups (p.185). Examples of radicals are the sulphate ion, the nitrate ion, the manganate (VII) ion and the ammonium ion.

**acid radical** a radical (↑) combined (p.64) with hydrogen in an acid (↓).

**acid** (n) a substance (p.8) which contains (p.55) hydrogen which can be replaced (p.68) by a metal, or by a base (p.46). An acid is a covalent (p.136) substance, which when dissolved in water produces hydrogen ions (p.123) in the solution. The strength of an acid is measured by its pH value (p.38). **acidify** (v), **acidic** (adj).

**alkali** (n) a soluble base (p.46). The solution of an alkali contains hydroxide (p.132) ions; it reacts with an acid (↑) to produce a salt (p.46) and water only. **alkaline** (adj), **alkalinity** (n).

**acidic** (adj) (1) describes a substance having the nature (p.19) of an acid (↑). (2) describes a solution (p.86) containing an acid. (3) describes a compound (p.8) which forms an acid when dissolved in water, e.g. sulphur dioxide is an acidic oxide.

**alkaline** (adj) describes a solution (p.86) with the properties of an alkali, e.g. sodium hydroxide solution is alkaline.

**neutral** (adj) describes a substance or a solution which is neither acidic (↑) nor basic (p.46). A neutral solution has a pH value (p.38) of 7. **neutralize** (v).

根 (名) 指構成分子 (第77頁) 的一部分或形成一個離子 (第123頁) 的一組原子;根在一個化學反應 (第62頁) 或一連串的反應過程中始終保持不變。構成離子的一個根的各原子靠共價鍵 (第136頁) 結合在一起。某些根也是官能團 (第185頁)。例如,硫酸根離子、硝酸根離子、高錳 (VII) 酸根離子和銨根離子都是根的例子。

酸根 指酸 (1) 分子中與氫相結合 (第64頁) 的根。

酸 (名) 含有氫而可為金屬或鹽 (第46頁) 置換的一種物質 (第8頁)。酸是一種共價 (第136頁) 物質。酸溶解於水中時,溶液中產生氫離子 (第123頁)。酸的強度由其 pH 值 (第38頁) 量度。(動詞為 acidify,形容詞為 acidic)

強鹼 (名):鹼是一種可溶性鹼 (第46頁)。強鹼的溶液含氫氧根離子 (第132頁);強鹼與酸 (1) 反應只生成鹽 (第46頁) 和水。(形容詞為 alkaline,名詞為 alkalinity)

酸性的 (形) 描述一種具有酸 (1) 的本質 (第19頁) 的物質;(2) 描述一種含有酸的溶液;(3) 描述一種溶解於水可形成酸的化合物 (第8頁)。例如,二氧化硫是一種酸性的氧化物。

鹼性的 (形) 描述一種溶液 (第86頁) 具有鹼的性質。例如,氫氧化鈉的溶液是鹼性的。

中性的 (形) 描述一種物質或溶液既非酸性 (1) 亦非鹼性 (第46頁)。中性溶液的 pH 值 (第38頁) 為7。(動詞為 neutralize)

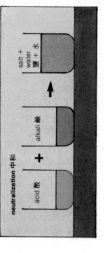

neutralization 中和

acid 酸 + alkali 鹼 → salt + water 鹽 + 水

**base** (n) a substance which reacts (p.62) with an acid (p.45) to produce a salt (↓) and water only. Bases are usually the oxides (p.48) or hydroxides (p.48) of metals. Many bases are insoluble. **basic** (adj.). **basicity** (n).

**basic** (adj.) describes a substance having the nature of a base (↑), e.g. copper (II) oxide is a basic oxide.

**amphoteric** (adj.) describes a substance which has both acidic (p.45) and basic (↑) properties, e.g. aluminium hydroxide reacts with **acids** to form a salt and water and it also reacts with **alkalis** to form a salt and water.

**salt** (n) a compound made by replacing some or all of the hydrogen of an acid (p.45) by a metal. A base or an alkali or a metal reacts with an acid to replace the hydrogen by a metal or an ion (p.123) such as the ammonium ion. Salts in solution usually form ions (p.123). Examples of salts are copper (II) sulphate (in which copper is combined (p.64) with the sulphate radical (p.45) as it has replaced the hydrogen of sulphuric acid H₂SO₄), iron (III) chloride and lead (II) nitrate.

**basicity** (n) the number of hydrogen atoms which can be replaced (p.68) in one molecule of an acid; also the number of hydrogen ions (p.123) formed from one molecule of an acid. For example, hydrochloric acid HCl has one atom of hydrogen that can be replaced in one molecule of the acid, so it has a basicity of one; sulphuric acid $H_2SO_4$ has two replaceable hydrogen atoms, therefore it has a basicity of two.

**monobasic** (adj.) describes an acid with a basicity (↑) of one.

**dibasic** (adj.) describes an acid with a basicity (↑) of two.

**tribasic** (adj.) describes an acid with a basicity (↑) of three.

**normal salt** a salt formed when all the hydrogen atoms in an acid (p.45) molecule have been replaced by a metal or an cation (p.125) of a metal, e.g. sodium chloride, copper (II) sulphate, sodium carbonate.

鹼：鹽基（名） 與酸（第 45 頁）反應，生成鹽（↓）和水的一種物質。鹼通常都是金屬的氧化物（第 48 頁）或金屬的氫氧化物（第 48 頁），許多鹼都是不溶的。（形容詞屬 basic，名詞屬 basicity）

鹼性的（形） 描述一種物質具有鹼的本質。例如，氧化銅(II)是一種鹼性的氧化物。

（酸鹼）兩性的（形） 形容一種物質既具有酸性（第 45 頁）中的一部分又具有鹼性(↑)的性質。當鹼或金屬或酸根離子(第 123 頁)置換時，金屬在溶液中通常形成離子(第 123 頁)。例如，硫酸銅(II)（其中銅置換硫酸中的氫而與硫酸根（第 45 頁）結合（第 64 頁）），氯化鐵(III)、硝酸鉛(II)都是鹽。

鹼度（名） 指一個酸（第 68 頁）氫原子之數目：又指自一個酸分子形成的氫離子(第 123 頁)之數目。例如，鹽酸(HCl)的一個氫原子中有一個氫原子可被置換，故其鹼度為 1；硫酸($H_2SO_4$)有兩個可被置換的氫原子，故其鹼度為 2。

一元的（形） 描述一種具有鹼度(↑)為 1 的酸。

二元的（形） 描述一種具有鹼度(↑)為 2 的酸。

三元的（形） 描述一種具有鹼度(↑)為 3 的酸。

正鹽 一個酸（第 45 頁）分子中的全部氫原子已為金屬或金屬的陽離子（第 125 頁）所置換而形成的一種鹽。例如：氯化鈉、硫酸銅(II)、碳酸鈉。

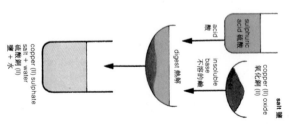

sulphuric acid 硫酸 / acid 酸 / insoluble base 不溶的鹼 / copper (II) oxide 氧化銅(II) / digest 熱解 / copper (II) sulphate 硫酸銅(II) / salt 鹽 / salt + water 鹽＋水

**acid salt** a salt formed when not all the hydrogen atoms in an acid (p.45) molecule have been replaced by a metal or a cation (p.125), e.g. potassium hydrogensulphate, calcium hydrogencarbonate. Only dibasic (↑) and tribasic (↑) acids can form acid salts.

**basic salt** a salt formed when not all the base (p.45) reacts with the acid (p.45) in a reaction. The base combined (p.64) with the normal salt (↑) to form an insoluble basic salt, e.g. basic lead carbonate, formed from lead (II) carbonate and lead (II) oxide.

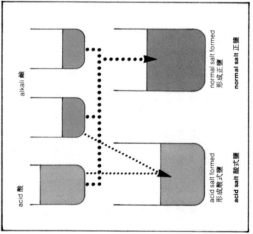

aikali 鹼　　acid 酸

normal salt formed 形成正鹽　　**normal salt 正鹽**

acid salt formed 形成酸式鹽　　**acid salt 酸式鹽**

**double salt** a compound (p.8) of two normal salts (↑) which forms crystals (p.91), e.g. aluminium potassium sulphate, a compound of aluminium sulphate and potassium sulphate, known as alum.

**complex salt** a compound (p.8) containing a complex ion (p.132); either the metal or the ion (p.123) of a normal salt, or both, can form a complex ion. An example is tetraammine copper (II) sulphate (cuprammonium sulphate); see also hexacyanoferrate (p.53).

**酸式鹽** 酸（第 45 頁）分子中的氫原子並非全部為金屬或陽離子（第 125 頁）置換所形成的鹽，例如硫酸氫鉀、硫酸氫鈣。只有二元（↑）酸和三元（↑）酸才能形成酸式鹽。

**鹼式鹽** 在一反應中並非全部鹼（↑）都與酸（第 45 頁）反應所生成的鹽。鹼與正鹽（↑）結合（第 64 頁）生成不溶的鹼式鹽。例如，鹼式碳式酸鉛是由碳酸鉛（II）與氧化鉛反應生成的。

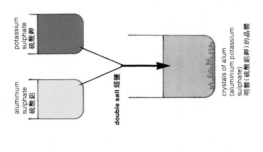

aluminium sulphate 硫酸鋁　　potassium sulphate 硫酸鉀

crystals of alum (aluminium potassium sulphate) 明礬（硫酸鋁鉀）的晶體

**double salt 複鹽**

**複鹽** 形成晶體（第 91 頁）的兩種正鹽（↑）的化合物（第 8 頁）。例如，硫酸鋁鉀是硫酸鋁和硫酸鉀的化合物，通稱為礬。

**錯鹽** 絡鹽 含有錯離子（或稱絡離子）（第 132 頁）的一種化合物（第 8 頁）；金屬或正鹽的離子（第 123 頁）或兩者都可形成一個錯合離子的一例是四氨合銅（II）硫酸四氨合銅（即硫酸銅銨）為錯合鹽。參見"六氰合鐵鹽"（第 53 頁）。

**hydrated** (adj) describes a salt (p.46) with water chemically combined (p.64) with the compound in its crystals (p.91). The combined water is called water of crystallization, e.g. copper (II) sulphate .5H₂O, in which five molecules of water are combined with one of the salt in its crystals.

**water of crystallization** see hydrated (↑).

**anhydrous** (adj) describes a salt with no water of crystallization (↑), e.g. sodium chloride crystals have no water of crystallization.

**anhydride** (n) (1) a compound (p.8) formed by removing (p.215) the elements (p.8) of water from a substance, but no water of crystallization (↑), e.g. ethanoic acid (acetic acid) can have the elements of water removed to form ethanoic anhydride. (2) a compound which, when added to water, forms a new chemical compound; the anhydride is named after this new chemical compound, e.g. ethanoic anhydride, when added to water, forms ethanoic acid.

**oxide** (n) a binary compound (p.8) formed by the combination of an element (p.8) with oxygen. The element can be a metal or a non-metal (p.116), e.g. lead (II) oxide, carbon dioxide.

**hydroxide** (n) a compound (p.8) formed by the reaction of a basic (p.46) oxide with water and having the radical (p.45) —OH. A soluble hydroxide forms hydroxyl (p.132) ions in water, e.g. sodium hydroxide NaOH, calcium hydroxide Ca(OH)₂ (formed by the action of water on calcium oxide, a basic oxide).

**peroxide** (n) an oxide which reacts (p.62) with cold dilute sulphuric acid to produce hydrogen peroxide. Peroxides in water form the ion (p.123) (O—O)²⁻. Example: sodium peroxide.

**higher oxide** (n) an oxide with an element showing a higher oxidation number (p.78) than usual, e.g. manganese (VII) oxide is a higher oxide with manganese (IV) oxide the usual oxide.

水合的(形) 描述一種會與化合物(第64頁)晶體中(第91頁)含有的水稱為"結晶水"。結合的水稱為"結晶水",例如五水硫酸銅(II)的晶體中,一個分子銅鹽與結合五個水分子。

結晶水 參見"水合的"(↑)。

無水的(形) 描述一種鹽與無結晶水(↑)。例如,氯化鈉晶體不含結晶水。

酐(名) 去除(第215頁)一種物質中水的元素(第8頁),而不是去除結晶水(↑)所形成的化合物(第8頁)。例如,乙酸(醋酸)含有水的元素,去除其水後形成乙酸酐;(2)指加入水後形成新化合物的一種化合物,酐是根據此新化合物命名。例如,乙酸酐,當加入水時形成乙酸。

氧化物(名) 指某一元素(第8頁)與氧化合生成的一種二元化合物(第8頁)。此元素可為一種金屬或非金屬(第116頁)。例如,氧化鉛(II)、二氧化碳。

氫氧化物(名) 由鹼性(第46頁)氧化物與水反應產生含有氫氧根(第45頁)—OH的一種化合物(第8頁)。可溶的氫氧化物在水中形成氫氧(第132頁)離子。例如,氫氧化鈉(NaOH)、氫氧化鈣(Ca(OH)₂)(係由一種鹼性氧化物氧化鈣和水作用形成)。

過氧化物(名) 可與冷的稀硫酸反應(第62頁)生成過氧化氫的一種氧化物。過氧化物在水中形成(O—O)²⁻離子(第123頁)。例如,過氧化鈉。

高級氧化物(名) 指某一種帶有含元素比正常顯示較高氧化值(第78頁)的氧化物。例如,七氧化二錳(VII)是高級氧化物,而二氧化錳(IV)是普通氧化物。

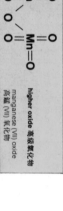

higher oxide 高級氧化物
manganese (VII) oxide 錳酸(VII) 氧化物

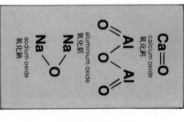

oxides 氧化物
calcium oxide 氧化鈣
aluminium oxide 氧化鋁
sodium oxide 氧化鈉

hydroxide 氫氧化物
sodium hydroxide 氫氧化鈉

peroxide 過氧化物
sodium peroxide 過氧化鈉

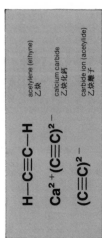

H−C≡C−H  acetylene (ethyne) 乙炔

Ca²⁺(C≡C)²⁻  calcium carbide 乙炔化鈣

(C≡C)²⁻  carbide ion (acetylide) 乙炔離子

**carbide** (n) a binary compound (p.44) containing a metal combined (p 64) with carbon. There are several different kinds of carbides, including the acetylides (↓).

**acetylide** (n) a carbide (↑) which reacts with water to form acetylene (p.174), $C_2H_2$. The metals of Group I and Group II of the periodic system (p.119) form acetylides. The compounds are ionic (p.123); an acetylide ion with an electrovalency of 2 is formed. Examples of acetylides are $Na_2C_2$ (sodium acetylide) and $CaC_2$ (calcium carbide or calcium acetylide).

**carbonate** (n) a compound (p.8) of a metal with a carbonate ion (p.123). Carbonates are salts (p.46) of carbonic acid, a very weak acid formed by dissolving carbon dioxide gas in water. The ion, see diagram, is trigonal planar (p.83) in structure (p.82), with three coordinate bonds (p.136).

**hydrogen carbonate** (n) an acid salt (p.47) of carbonic acid; only one hydrogen atom is replaced (p.68) by a metal, e.g. sodium hydrogencarbonate, $NaHCO_3$.

**bicarbonate** (n) the traditional name (p.44) for a hydrogencarbonate.

碳化物 (名) 含有金屬與碳結合 (第 64 頁) 的二元化合物 (第 44 頁)。碳化物有多種，包括乙炔化物 (↓)。

乙炔化物 (名) 和水反應能生成乙炔 (第 174 頁)($C_2H_2$) 的一種碳化物 (↑)。週期系 (第 119 頁) I 族和 II 族中的金屬可形成乙炔化物。這些化合物屬離子型 (第 123 頁)，可形成電價 (價為 2 的乙炔離子。$Na_2C_2$ (乙炔鈉) 或 $CaC_2$ (碳化鈣或乙炔化鈣) 為乙炔化物的例子。

碳酸鹽 (名) 金屬與碳酸根離子 (第 123 頁) 結合而成的化合物 (第 8 頁)。碳酸鹽是碳酸的鹽 (第 46 頁)。碳酸係二氧化碳溶解在水中而成的一種很弱的酸 (第 82 頁)。其離子為平面三角形 (第 83 頁) (見圖)，含三個配位鍵 (第 136 頁)。

碳酸氫鹽：酸式碳酸鹽 (名) 碳酸的一種酸式鹽 (第 47 頁)；只有一個氫原子可為金屬置換 (第 68 頁)。例如碳酸氫鈉 ($NaHCO_3$)。

重碳酸氫鹽 (名) 為碳酸氫鹽的慣用名 (第 44 頁)。

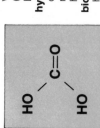

sodium hydrogen carbonate 碳酸氫鈉

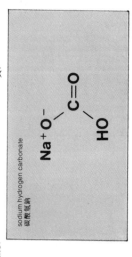

carbonate ion 碳酸根離子

carbonic acid 碳酸

**chloride** (*n*) a compound (p.8) of an element (p.8) and chlorine. Metals form ionic (p.123) chlorides; non-metals form covalent (p.136) chlorides, e.g. Na⁺Cl⁻ (ionic chloride); CCl₄ (covalent chloride). *See chloro group (p.187).* Chlorides are salts (p.46) of hydrochloric acid.

**bromide** (*n*) a compound (p.8) of an element (p.8) and bromine. Metals usually form ionic (p.123) bromides; non-metals form covalent (p.136) bromides. *See bromo group (p.187).* Bromides are salts (p.46) of hydrobromic acid.

**iodide** (*n*) a compound (p.8) of an element (p.8) with iodine. Metals usually form ionic (p.123) iodides; non-metals form covalent (p.136) iodides. *See iodo group (p.187).* Iodides are salts (p.46) of hydroiodic acid.

**halide** (*n*) a compound (p.8) which is a chloride, a bromide, an iodide, or a fluoride. It is formed from an element (p.8) and a halogen (p.117).

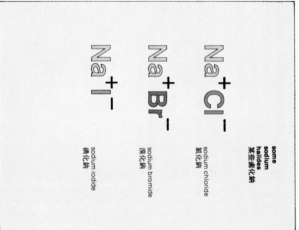

Na⁺Cl⁻
sodium chloride
氯化鈉

Na⁺Br⁻
sodium bromide
溴化鈉

Na⁺I⁻
sodium iodide
碘化鈉

some
sodium
halides
某些鹵化鈉

氯化物（名）由一種元素（第 8 頁）和氯生成而成的一種化合物（第 8 頁）。金屬則生成離子（第 123 頁）型的氯化物，例如，Na⁺Cl⁻（離子型氯化物）；CCl₄（共價型氯化物）。多見"氯根"（第 187 頁）。氯化物係鹽酸的鹽類（第 46 頁）。

溴化物（名）由一種元素（第 8 頁）和溴生成而成的一種化合物（第 8 頁）。金屬和溴通常是形成離子型（第 123 頁）溴化物。非金屬則形成共價（第 136 頁）溴化物。多見"溴根"（第 187 頁）。溴化物係氫溴酸的鹽類（第 46 頁）。

碘化物（名）由一種元素（第 8 頁）與碘化合而成的一種化合物（第 8 頁）。金屬和碘通常形成離子型（第 123 頁）碘化物。非金屬則形成共價（第 136 頁）碘化物。多見"碘根"（第 187 頁）。碘化物係氫碘酸的鹽類（第 46 頁）。

鹵化物（名）指包括氯化物、溴化物、碘化物的一類化合物。係由一種元素（第 8 頁）和鹵素（第 117 頁）化合而成。

**sulphide** (n) a binary compound (p.8) of an element (p.8) with sulphur. Metals form ionic (p.123) sulphides, most of which are insoluble. Non-metals form covalent (p.136) sulphides. Sulphides are salts (p.46) of the weak acid, hydrogen sulphide, $H_2S$.

**sulphate** (n) a compound (p.8) of a metal with a sulphate ion (p.123). Sulphates are salts (p.46) of sulphuric acid, a strong acid which is dibasic, forming normal and acid salts. The sulphate ion, see diagram, it tetrahedral (p.83) in structure (p.82) with four coordinate bonds (p.136).

**hydrogen sulphate** (n) an acid salt of sulphuric acid, i.e. an acid sulphate (↑). For example, sodium hydrogensulphate $NaHSO_4$ which forms the ion $HSO_4^-$.

**bisulphate** (n) traditional name (p.44) for hydrogen sulphate.

**sulphite** (n) a compound (p.8) of a metal with a sulphite ion (p.123). Sulphites are salts (p.46) of sulphurous acid, a dibasic acid forming normal and acid salts. The sulphite ion, see diagram, is trigonal pyramidal (p.84) in structure (p.82) with three coordinate bonds (p.136) and a lone pair (p.133) of electrons.

**hydrogen sulphite** (n) an acid salt of sulphurous acid, i.e. an acid sulphite, e.g. sodium hydrogensulphite $NaHSO_3$ with the ion $HSO_3^-$.

**bisulphite** (n) traditional name for hydrogen-sulphite (↑).

**thiosulphate** (n) a compound (p.8) of a metal with a thiosulphate ion (p.123). Thiosulphates are the salts (p.46) of thiosulphuric acid $H_2S_2O_3$. The thiosulphate ion, is tetrahedral (p.83) in structure (p.82) with four coordinate bonds (p.136).

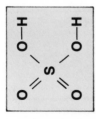

sulphuric acid 硫酸
(covalent)（共價）

sulphurous acid 亞硫酸
(covalent)（共價）

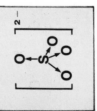

sulphate ion 硫酸根離子
(tetrahedral)（四面體）

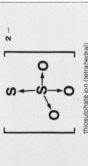

lone pair 不共用電子對

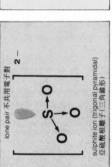

sulphite ion (trigonal pyramidal)
亞硫酸根離子（三角錐形）

thiosulphate ion (tetrahedral)
硫代硫酸根離子（四面體）

硫化物（名）由一種元素（第8頁）與硫化合而成的二元化合物（第44頁）。金屬與硫化合形成離子（第123頁）硫化物，大多數是難溶解的硫化物。非金屬與硫化合則形成共價（第136頁）硫化物。各種硫化物都是弱酸——硫化氫（$H_2S$）的鹽類（第46頁）。

硫酸鹽（名）金屬與硫酸根離子（第123頁）所形成的化合物（第8頁）。各種硫酸鹽都是硫酸的鹽類（第46頁）。硫酸是一種二元強酸，可生成正鹽和酸式鹽。硫酸根離子（見圖）為四面體（第83頁）結構，有四個配價鍵（第136頁）。

硫酸氫鹽（名）硫酸的一種酸式鹽，即酸式硫酸鹽（↑）。例如：可形成$HSO_4^-$離子的硫酸氫鈉（$NaHSO_4$）。

酸式硫酸鹽（名）硫酸氫鹽的慣用名（第44頁）。

亞硫酸鹽（名）金屬與亞硫酸根離子（第123頁）所形成的化合物（第8頁）。各種亞硫酸鹽都是亞硫酸的鹽類（第46頁）。亞硫酸是一種二元酸，可生成正鹽和酸式鹽。亞硫酸根離子（見圖）為三角錐形（第84頁）結構，有三個配價鍵（第136頁）和一個不共用電子對（第133頁）。

亞硫酸氫鹽（名）亞硫酸的一種酸式鹽，即酸式亞硫酸鹽，例如：含有$HSO_3^-$離子的亞硫酸氫鈉（$NaHSO_3$）。

酸式亞硫酸鹽（名）亞硫酸氫鹽（↑）的慣用名。

硫代硫酸鹽（名）金屬與硫代硫酸根離子（第8頁）的鹽類（第46頁）。各種硫代硫酸鹽是硫代硫酸（$H_2S_2O_3$）的鹽類（第46頁）。硫代硫酸根離子為四面體（第83頁）結構，有四個配價鍵（第136頁）。

**nitride** (n) a binary compound (p.44) of a metal with nitrogen. The common nitrides are formed with the metals of groups I, II and III of the periodic system (p.119). These nitrides react (p.62) with water to form ammonia and the hydroxide of the metal:

**nitrate** (n) a compound (p.8) of a metal with a nitrate ion (p.123). Nitrates are salts (p.46) of nitric acid, a strong acid which is monobasic (p.46). The nitrate ion, *see diagram*, is trigonal planar (p.83) in structure (p.83) with three coordinate bonds (p.136).

**nitrite** (n) a compound (p.8) of a metal with a nitrite ion (p.123). Nitrites are salts (p.46) of nitrous acid, a monobasic acid (p.45). The nitrite ion, *see diagram*, is non-linear (p.83) with two coordinate bonds (p.136).

**chromate (VI)** (n) a compound (p.8) of a metal with the chromate (VI) ion (p.123). Chromic acid, the acid of the salts (p.46), cannot be isolated (p.43). Although the acid is dibasic, only the normal salts (p.46) can be made. The chromate (VI) ion, *see diagram*, is tetrahedral (p.83) in structure (p.82) with four coordinate bonds (p.136).

**chromate** (n) traditional name (p.44) for chromate (VI).

**dichromate (VI)** (n) a compound of a metal with the dichromate (VI) ion (p.123). Dichromic acid, the acid of the salts (p.46), cannot be isolated (p.43). Although the acid is dibasic, only the normal salts (p.46) can be made. The dichromate (VI) ion, *see diagram*, has the structure of two tetrahedra with six coordinate bonds (p.136) and two covalent bonds (p.136).

**dichromate** (n) traditional name (p.44) for dichromate (VI).

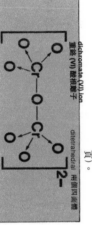

dichromate (VI) ion
重鉻 (VI) 酸根離子

氮化物（名）金屬與氮化物而成的二元化合物（第44頁）。常見的氮化物都是和週期系（第119頁）I、II、III 族中的金屬化合而成的。這些氮化物和水反應（第62頁）生成氨和該金屬的氫氧化物。

硝酸鹽（名）一種化合物（第8頁）所形成的鹽類（第46頁）。硝酸根離子（見圖）為平面三角形（第83頁）結構（第82頁），有三個配價鍵（第136頁）。

亞硝酸鹽（名）金屬與亞硝酸根離子（第123頁）所形成的一種化合物（第8頁）。亞硝酸鹽都是亞硝酸的鹽類（第46頁）。亞硝酸根離子（見圖）是一元酸（第45頁）。各有兩個配價鍵非直線型的（第83頁）。

鉻（VI）酸鹽（名）金屬與鉻（VI）酸根離子（第123頁）所形成的一種化合物（第8頁）。鉻酸鹽（第46頁）的酸——鉻酸是二元酸，但只能製出來。雖然鉻酸是二元酸，但只能得正鹽（第46頁）。鉻（VI）酸根離子（見圖）具有四面體的結構，有4個配價鍵（第136頁）。

鉻酸鹽（名）鉻（VI）酸鹽的慣用名（第44頁）。

重鉻（VI）酸鹽（名）金屬與重鉻（VI）酸根離子（第123頁）所形成的一種化合物（第8頁）。重鉻酸鹽（第46頁）的酸——重鉻酸是二元酸，但只能得正鹽（第46頁）。重鉻（VI）酸根離子（見圖）有兩個（個個四面體的結構，有6個配價鍵（第136頁）和兩個共價鍵（第136頁）。

重鉻（VI）酸鹽（名）重鉻（VI）酸鹽的慣用名（第44頁）。

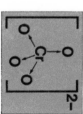

dichromate (VI) ion tetrahedral
鉻 (VI) 酸根離子四面體

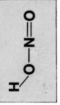

chromate (VI) ion tetrahedral
chromate (VI) 鉻 (VI) 酸根離子四面體

nitrous acid
亞硝酸
(non-linear) 非直線型

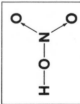

nitrite ion
亞硝酸根離子

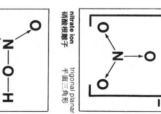

nitrate ion
硝酸根離子
trigonal planar 平面三角形

nitric acid
硝酸

**manganate (VI)** (n) a compound (p.8) of a metal with the manganate (VI) ion (p.123). The acid of these salts (p.46) cannot be isolated (p.43); no acid salts are formed, only normal salts, although the acid is dibasic (p.46). The manganate (VI) ion, see diagram, is tetrahedral (p.83) in structure (p.136) with four coordinate bonds (p.136). The manganese atom has an oxidation number (p.78) of 6. An example of a salt is potassium manganate (VI) $K_2MnO_4$.

**manganate** (n) traditional name (p.44) for manganate (VI).

**manganate (VII)** (n) a compound (p.8) of a metal with the manganate (VII) ion (p.123). The acid of these salts (p.46) cannot be isolated (p.43); it is monobasic (p.46). The manganate (VII) ion, see diagram, is tetrahedral (p.83) in structure (p.82) with four coordinate bonds (p.136). The manganese atom has an oxidation number of 7. An example of a salt is potassium manganate (VII) $KMnO_4$.

**permanganate** (n) traditional name (p.44) for manganate (VII).

**hexacyanoferrate (II)** (n) a compound (p.8) of a metal and the hexacyanoferrate (II) ion (p.123). There is no acid for these salts (p.46). The hexacyanoferrate (II) ion, see diagram, is octahedral (p.83) in structure (p.82) with six coordinate bonds (p.136). The iron atom has an oxidation number of 2. An example of a salt is potassium hexacyanoferrate (II) $K_4Fe(CN)_6$. The hexacyanoferrate ion is a complex ion (p.132).

**ferrocyanide** (n) the traditional name (p.44) for hexacyanoferrate (II).

**hexacyanoferrate (III)** (n) a compound (p.8) of a metal and the hexacyanoferrate (III) ion. There is no acid for these salts (p.46). The hexacyanoferrate (III) ion has the same structure as the hexacyanoferrate (II) ion (↑), but the iron atom has an oxidation number (p.78) of 3; it is a complex ion. An example of a salt is potassium hexacyanoferrate (III) $K_3Fe(CN)_6$.

**ferricyanide** (n) the traditional name (p.44) for hexacyanoferrate (III).

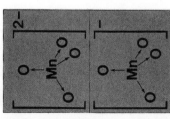

manganate (VI) ion
tetrahedral
錳(VI)酸根離子
四面體

manganate VII ion
tetrahedral
高錳(VII)酸根離子
四面體

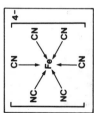

hexacyanoferrate (II) ion
六氰鐵(II)合離子

錳(VI)酸鹽(名)　由金屬與錳(VI)酸根離子(第123頁)所形成的一種化合物(第8頁)。這些鹽(第46頁)的酸不能分離(第43頁)出來;雖然錳酸是二元(第46頁)酸,卻不生成酸式鹽,只生成正鹽。錳(VI)酸根離子(見圖)為四面體(第83頁)結構(第136頁),有四個配價鍵(第136頁)。其中錳原子的氧化值(第78頁)為6。錳(VI)酸鹽的一個例子是錳酸鉀($K_2MnO_4$)。

錳酸鹽(名)　錳(VI)酸鹽的慣用名(第44頁)。

高錳(VII)酸鹽(名)　金屬與高錳(VII)酸根離子(第8頁)所形成的一種化合物(第8頁)。這些鹽(第46頁)的酸不能分離(第43頁)出來;高錳(VII)酸是一元(第46頁)酸。高錳酸根離子(見圖)為四面體(第83頁)結構(第82頁),有四個配價鍵(第136頁)。其中錳原子的氧化值(第78頁)為7。例如,高錳(VII)酸鉀(KMnO₄)是高錳(VII)酸鹽的慣用名(第44頁)。

高錳酸鹽(名)　錳(VII)酸鹽的慣用名(第44頁)。

六氰合鐵(II)酸鹽(名)　由金屬和六氰合鐵(II)離子(第123頁)所形成的一種化合物(第8頁)。這些鹽(第46頁)沒有相應的酸。六氰合鐵(II)離子(見圖)為八面體(第83頁)結構(第82頁),有6個配價鍵(第136頁)。其中鐵原子的氧化值(第78頁)為2。六氰合鐵(II)酸鉀[$K_4Fe(CN)_6$]是其鹽的一個例子。六氰鐵酸離子是一個複離子(第132頁)。

亞鐵氰化物(名)　六氰合鐵(II)酸鹽的慣用名(第44頁)。

六氰合鐵(III)酸鹽(名)　金屬和六氰合鐵(III)離子所形成的一種化合物(第8頁)。這些鹽(第46頁)沒有相應的酸。六氰合鐵(III)離子的結構與鐵六氰合鐵(II)離子(↑)相同,但鐵原子的氧化值(第78頁)為3,也是錯離子。六氰合鐵(III)酸鉀[$K_3Fe(CN)_6$]是此種鹽的慣用名(名)。

鐵氰化物(名)　六氰合鐵(III)酸鹽的慣用名(第44頁)。

**mixture** (n) different substances (p.8), put together, form a mixture. The substances can be elements (p.8), compounds (p.8) or materials (p.8), e.g. a mixture of charcoal (the element, carbon), sulphur (an element) and potassium nitrate (a compound) forms a mixture called gunpowder. A particular substance can be separated (p.34) from a mixture. Mixtures can be solid, liquid, or gaseous (p.11). **mix** (v).

**constituent** (n) (1) a member of a mixture (↑), e.g. a mixture of sulphur and copper has two constituents; charcoal, sulphur and potassium nitrate mixed together form gunpowder, sulphur is one of the constituents of gunpowder. Each constituent keeps its own properties in a mixture. A liquid mixture can be a true solution (p.86), a colloidal solution (p.98) or a suspension (p.86). (2) a part of a compound (p.8), e.g. a molecule of sulphur dioxide consists of (p.55) one atom of sulphur and two atoms of oxygen, it has three constituent atoms. Constituents of a compound are elements (p.8), radicals (p.45), functional groups (p.185) and ions (p.123).
**constitute** (v), **constitution** (n).

**ingredient** (n) a substance (p.8) needed for a mixture (↑) before it is put in the mixture, e.g. charcoal, sulphur and potassium nitrate are ingredients of gunpowder before they are mixed to form gunpowder; after mixing they are constituents (↑).

**homogenous** (adj) describes a material or substance (p.8) which is the same throughout in its properties (p.9) and composition (p.82). Homogenous also describes a chemical reaction (p.62) with all substances in the same state of matter (p.9), e.g. all in the gaseous state, or all in the liquid state. **homogenize** (v), **homogeneity** (n).

**heterogenous** (adj) describes a material, substance (p.8), or chemical reaction (p.62) which is not the same throughout in its properties (p.9), composition (p.82), or state of matter (p.9). It is the opposite of homogenous (↑). **heterogeneity** (n).

混合物（名） 不同的物質（第 8 頁）混合在一起就形成一種混合物。所謂混合物質可以是元素（第 8 頁）、化合物（第 8 頁）或材料（第 8 頁）。例如炭（元素碳）、硫（一種元素）和硝酸鉀（一種化合物）混合而成的混合物稱為火藥（一種個別的物質可從混合物中分離（第 34 頁）出來。混合物可以是固體或液態，也可以是氣態（第 11 頁）。（動詞爲 mix）

組分（名） （1）指混合物（↑）中的一個組成部分。例如硫和銅的混合物有兩個組分：炭、硫和硝酸鉀的混合物組成火藥，硫是火藥的組分之一。混合物的每一個組分都保持其本身的性質。液體混合可以是真溶液（第 86 頁）、膠體溶液（第 98 頁）或懸浮液（第 86 頁）。（2）指化合物（第 8 頁）的一部分。例如一個二氧化硫分子由一個硫原子和二個氧原子組成（第 55 頁），它有三個組分原子。化合物的成分包括元素（第 8 頁）、根（第 45 頁）、官能團（第 185 頁）和離子（第 123 頁）。（動詞爲 constitute，名詞爲 constitution 意混合物構造）

配合成分（名） 混合物（↑）中所需的物質（第 82 頁），例如炭、硫和硝酸鉀在混合物組成火藥之前就爲火藥的配合成分而混合之後就成爲混合物的組分（↑）。

均勻的（形） 描述性質（第 9 頁）和組成（第 82 頁）完全相同的材料或物質（第 8 頁）；也描述一個化學反應（第 62 頁）中所具的全部物質都處於相同物態（第 9 頁）。例如都處於氣態或全部處於液態。（名詞爲 homogenity，動詞爲 homogenize）

不均勻的（形） 描述性質（第 9 頁）並非完全相同的材料、物質（第 8 頁）或化學反應（第 62 頁）。其反意詞爲均勻的（↑）。（名詞爲 heterogenity）

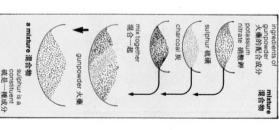

ingredients of gunpowder 火藥的配合成分

potassium nitrate 硝酸鉀
sulphur 硫磺
charcoal 炭

mix together 混合一起

gunpowder 火藥

a mixture 混合物

sulphur is a constituent 硫是一種成分

mixtures of iron and sulphur varied proportions 鐵和硫比例不同的混合物

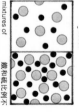

mixture 混合物

compound of iron and sulphur constant proportions 鐵和硫比例恆定的化合物

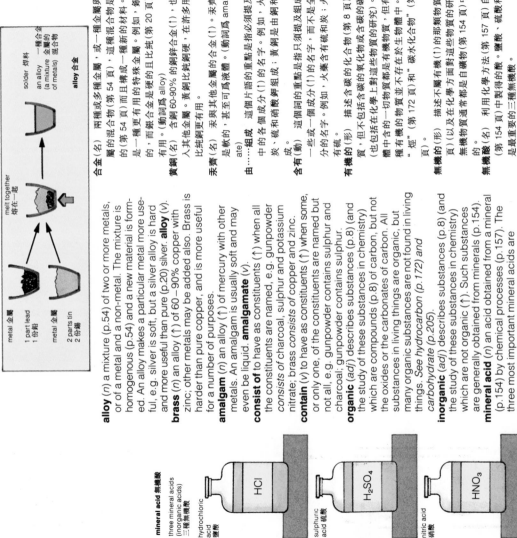

metal 金屬 1 part lead 1份鉛

metal 金屬 2 parts tin 2份錫

melt together 熔在一起

solder 焊料
an alloy 一種合金 (a mixture of metals) 混合物

alloy 合金

---

mineral acid 無機酸

three mineral acids (inorganic acids) 三種無機酸

hydrochloric acid 鹽酸 — HCl

sulphuric acid 硫酸 — H₂SO₄

nitric acid 硝酸 — HNO₃

---

**alloy** (n) a mixture (p.54) of two or more metals, or of a metal and a non-metal. The mixture is homogenous (p.54) and a new material is formed. An alloy makes a particular metal more useful, e.g. silver is soft, but a silver alloy is hard and more useful than pure (p.20) silver. **alloy** (v).

**brass** (n) an alloy (↑) of 60−90% copper with zinc; other metals may be added also. Brass is harder than pure copper, and is more useful for a number of purposes.

**amalgam** (n) an alloy (↑) of mercury with other metals. An amalgam is usually soft and may even be liquid. **amalgamate** (v).

**consist of** to have as constituents (↑) when all the constituents are named, e.g. gunpowder consists of charcoal, sulphur and potassium nitrate; brass consists of copper and zinc.

**contain** (v) to have as constituents (↑) when some, or only one, of the constituents are named but not all, e.g. gunpowder contains sulphur and charcoal; gunpowder contains sulphur.

**organic** (adj) describes substances (p.8) (and the study of these substances in chemistry) which are compounds (p.8) of carbon, but not the oxides or the carbonates of carbon. All substances in living things are organic, but many organic substances are not found in living things. See hydrocarbon (p.172) and carbohydrate (p.205).

**inorganic** (adj) describes substances (p.8) (and the study of these substances in chemistry) which are not organic (↑). Such substances are generally obtained from minerals (p.154).

**mineral acid** (n) an acid obtained from a mineral (p.154) by chemical processes (p.157). The three most important mineral acids are hydrochloric, sulphuric and nitric acids.

---

合金 (名) 兩種或多種金屬，或一種合金屬與非金屬的混合物 (第54頁)。這種混合物是均相的 (第54頁) 而且構成一種新的材料。合金是一種更有用的特殊金屬。例如，銀是軟的，而銀合金是硬的且比純 (第20頁) 銀更有用。合金 (動詞為 alloy)。

黃銅 (名) 含銅 60-90% 的銅鋅合金 (↑)，也可加入其他金屬。黃銅比純銅硬，在許多用途上比純銅更有用。

汞齊 (名) 汞與其他金屬的合金 (↑)。汞齊通常是軟的，甚至可為液體。(動詞為 amalgamate)

由……組成 這個片語的重點是指它的名字中的各個成分 (↑) 的名字。例如，火藥由炭、硫和硝酸鉀組成；黃銅由銅和鋅組成。

含有 (動) 這個詞的重點是指只須提及組成中的一些或一個成分 (↑) 的名字，而不是全部成分的名字。例如，火藥含有硫和炭；火藥含有硫。

有機的 (形) 描述含碳有機 (第8頁) 的物質，但不包括含氧化物或含碳的碳酸鹽 (也包括在化學上對這些物質的研究)。生物體中含的一切物質都是有機物質，但有許多種有機的物質並不存在於生物體中。參見 "烴" (第172頁) 和 "碳水化合物" (第205頁)。

無機的 (形) 描述不屬有機 (↑) 的那類物質 (第8頁) (以及在化學方面對這些物質的研究)。無機物質通常都是自礦物 (第154頁) 中取得的。

無機酸 (名) 利用化學方法 (第154頁) 自礦物 (第157頁) 中製得的酸。鹽酸、硫酸和硝酸是最重要的三種無機酸。

**air** (n) a mixture (p.54) of gases which forms the atmosphere and is the cause of atmospheric pressure (p.102). Air contains (p.55) about 20% oxygen, 79% nitrogen, 1% noble gases (↓) and 0.3% carbon dioxide; this is the composition (p.82) of dry air. In addition air always contains water vapour (p.11).

**diluent** (n) a substance added to a solution or to a mixture of solids or gases to reduce (p.219) the concentration of the solution or to reduce the proportion (p.76) of one of the constituents (p.54) of the mixture, e.g. water can be added to a concentrated (p.88) alkali as a diluent; nitrogen in the air is a diluent for oxygen.

**noble** (n) (1) *noble gases* of the air (↑) describes the gases helium, neon, argon, krypton and xenon. The gases are generally considered inert (p.19), (2) *noble* metals describes gold, platinum and other metals which do not react with the usual mineral acids.

**pollution** (n) undesirable substances (p.8) in the air, water or earth; the surroundings (p.103) become unhealthy or impure, e.g. the pollution of the air by smoke.

**water** (n) a liquid substance formed by the combination (p.64) of hydrogen and oxygen (formula: H₂O).

**water vapour pressure** all water on the Earth's surface evaporates (p.11) and so air contains water vapour. The vapour exerts (p.106) a pressure, which is part of atmospheric pressure (p.102).

**saturated water vapour pressure** The water vapour pressure (↑) when the air is saturated (p.87) with water vapour. It is the highest water vapour pressure at a particular temperature. Saturated water vapour pressure increases with a rise in temperature.

**water cycle** a cycle (p.64) in which water evaporates (p.11) from the Earth's surface and forms clouds, the clouds break to form rain, the rain passes through the Earth to rivers, lakes and finally the sea. Water evaporates from the rivers, lakes and sea. Respiration (p.61) also helps in the cycle.

空氣 (名) 形成大氣層的各種氣體的混合物（第54頁）的。空氣是大氣壓（第102頁）的成因。空氣含有（第55頁）約大氣壓20%、氮79%、惰性空氣的組成（第82頁），此外，還含有水蒸汽（第11頁）。

稀釋劑 (名) 加入溶液中或加入固體或氣體的混合物中某一成分（第54頁）的濃度或減少混合物中某一成分（第54頁）的比例（第76頁）的一種物質。例如水可作為一種稀釋劑加入濃（第88頁）鹼中；空氣中的氮是氧的稀釋劑。

稀有氣體 (名) 指在空氣、水或土壤中存在的不良物質（第8頁）；此種物質令周圍環境、第103頁）變成不衛生或不純淨。例如空氣受煙污染。

污染 (名) 由氫和氧化合（第64頁）而成的一種液體物質（水的分子式為H₂O）。

水蒸汽壓力 地球表面上的水分都在蒸發（第11頁）使空氣中含有水蒸汽。此蒸汽對大氣施加（第106頁）一種壓力，成為大氣壓（第102頁）的一部分。

飽和水蒸汽壓力 空氣為水蒸汽所飽和（第87頁）時的水蒸汽壓力（↑）。這是在特定溫度時的最高水蒸汽壓力。飽和水蒸汽壓力隨溫度升高而加大。

水循環 水自地球表面因蒸發（第11頁）形成雲下降成為雨，雨水經地面流入河流、湖泊，最後匯流入海洋，一個循環（第64頁）。水自河流、湖泊或海洋蒸發；呼吸作用（第61頁）也有助於循環。

atmosphere
大氣

air forms the atmosphere
空氣構成大氣

air 空氣

Earth 地球

air 空氣

water cycle
水循環

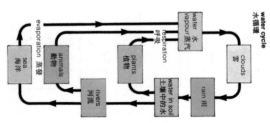

water vapour 水蒸汽

clouds 雲

respiration 呼吸

plants 植物

water in soil 土壤中的水

rivers 河流

rain 雨

animals 動物

sea 海洋

evaporation 蒸發

**hard water** water which does not form a lather (↓) easily with soap. It contains (p.55) salts of calcium and magnesium, or else such salts have been removed (p.215). **soften** (v).

硬水　指不易與肥皂形成泡沫 (↓) 的水。硬水含有 (第 55 頁) 鈣鹽和鎂鹽，這些鹽與肥皂形成不溶性的鹽。(名詞爲 hardness)

**soft water** water which forms a lather (↓) easily with soap. It does not contain salts of calcium and magnesium, or else such salts have been removed (p.215). **soften** (v).

軟水　易與肥皂形成泡沫 (↓) 的水。軟水不含鈣鹽和鎂鹽，或者已經去除 (第 215 頁) 這些鹽。(動詞爲 soften)

**lather** (n) a large quantity of very small bubbles (p.40) formed when soap is mixed with water. **lather** (v).

泡沫 (名)　肥皂與水混合時產生的大量很細小的氣泡 (第 40 頁)。(動詞爲 lather)

**temporary hardness** a hardness of water (↑) which can be removed (p.215) by boiling the water. It is caused by calcium hydrogen-carbonate being dissolved in the water.

暫時硬度　將水煮沸即可去除 (第 215 頁) 的硬度 (↑)。暫時硬度的成因是水中溶解有碳酸氫鈣。

**permanent hardness** a hardness of water (↑) which cannot be removed (p.215) by boiling. It is removed by (a) adding sodium carbonate, (b) by detergents (p.171), or (c) by zeolite (↓). It is caused by dissolved salts of calcium and magnesium, such as the sulphate.

永久硬度　經煮沸不能去除 (第 215 頁) 的水 (↑) 的硬度，可採用以下方法去除此種硬度：(a) 加入碳酸鈉；(b) 使用去垢劑 (第 171 頁)；或 (c) 使用沸石 (↓)。此種硬度是水中溶有硫酸鹽的鈣鹽或鎂鹽，如硫酸鈣或鎂引起的。

**water softening** any process by which hard water (↑), with either temporary or permanent hardness (↑), is changed to soft water (↑).

水的軟化　指使軟水 (↑) 的硬度或永久硬度 (↑) 變成軟水 (↑) 的任何過程。

**zeolite process** a process (p.157) using minerals (p.154), called zeolites, to soften (↑) water. A zeolite contains sodium ions (p.123) which can be replaced by other metal ions. In water softening, the zeolite removes (p.215) the calcium and magnesium ions from hard water and replaces (p.68) them with sodium ions.

沸石法　使用稱之爲沸石的礦物 (第 154 頁) 使水軟化 (↑) 的過程 (第 157 頁)。沸石中的鈉離子 (第 123 頁) 可爲其他金屬離子置換。在水軟化的過程中，沸石中的鈉離子置換 (第 68 頁) 並去除 (第 215 頁) 硬水中的鈣離子和鎂離子。

**contamination** (n) the presence in water, food, or any other substance or material (p.8) of causative agents of disease, e.g. the presence of viruses, bacteria, protozoa, etc. which cause disease.

沾染 (名)　指水、食物或任何其他物質或材料 (第 8 頁) 中有可致病的病原體。例如存在可致病的病毒、細菌、原生動物等。

soft water
軟水

soap
肥皂

water
水

lather
泡沫

hard water
硬水

soap
肥皂

water
水

scum,
no lather
浮渣，
無泡沫

water softening
水軟化

water softener
水軟化劑

soft water out
軟水出

zeolite
沸石

hard water in 硬水入

**combustion** (*n*) any chemical reaction (p.62) in which heat, and usually light, is produced. Combustion commonly is the burning of organic (p.55) substances during which oxygen from the air (p.56) is used to form carbon dioxide and water vapour. **combustible** (*adj*).

**rapid combustion** combustion (↑) in which heat is produced (p.62) at a high temperature, usually with flames, by the combination (p.64) of substances with oxygen.

**slow combustion** combustion (↑) in which heat is produced (p.62) at a low temperature, without flames, by the combination (p.64) of substances with oxygen.

**explosion** (*n*) a chemical reaction which takes place very quickly, forming large volumes of gases and releasing (p.69) energy (p.135). The energy causes a rise in temperature. An explosion produces (p.62) sound and often light. The increase in pressure caused by the large volume of gas causes destruction in the surroundings. **explode** (*v*) **explosive** (*adj*).

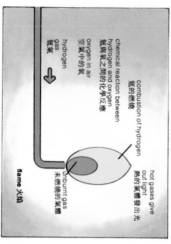

chemical reaction between
hydrogen and oxygen
氫與氧之間的化學反應

oxygen in air
空氣中的氧

hydrogen
gas
氫氣

combustion of hydrogen
氫的燃燒

hot gases
give
out light.
熱的氣體發出光。

Unburnt gas
未燃燒的氣體

**flame** 火焰

**flame** (*n*) combustion (↑) of gases produces (p.62) a flame. The hot gases give out light.

**spontaneous combustion** combustion (↑) which takes place without any apparent (p.223) cause, e.g. the spontaneous combustion of phosphorus in oxygen.

**incombustible** (*adj*) describes any substance which does not undergo (p.213) combustion (↑).

燃燒（名） 指產生熱而且通常發光的任何化學反應（第 62 頁）。燃燒通常是指有機（第 55 頁）物質的燃燒，在此過程中，空氣（第 56 頁）中的氧被用於生成二氧化碳和水蒸汽。（形容詞爲 combustible）

速燃 物質與氧化合（第 64 頁），在高溫下產生熱（第 62 頁），通常伴有火焰的燃燒（↑）。

緩燃 物質與氧化合（第 64 頁），在低溫下產生熱（第 62 頁），沒有伴有火焰的燃燒（↑）。

爆炸（名） 指極快速發生，產生大量氣體並釋放出（第 69 頁）的能量（第 135 頁）的一種化學反應。釋放出的能量使溫度上升。爆炸產生（第 62 頁）聲音，通常還發出光。由於產生大量氣體導致環境壓力升高而造成破壞。（動詞爲 explode，形容詞爲 explosive）

火焰（名） 氣體燃燒（↑）產生（第 62 頁）火焰。熱的氣體發出光。

自發燃燒 指無任何明顯的（第 223 頁）原因而發生的燃燒（↑）。例如磷在氧氣中的自發燃燒。

不能燃燒的（形） 描述那些不遭受（第 213 頁）燃燒作用（↑）的任何物質。

**burn** (v) to undergo (p.213) combustion (↑) producing (p.62) heat, flame and sometimes smoke, e.g. wood burns with a smoky flame; to burn coal to produce heat.

**glow** (v) to give out light because of a chemical reaction (p.62) or because a substance (p.8) is heated strongly, e.g. a piece of wood glows when there are no flames; a piece of iron glows when heated strongly, **glow** (n).

**smoulder** (v) to give out smoke because of a chemical reaction (p.62) taking place at a low temperature without a flame.

**warm** (v) to supply only enough heat to make substances react (p.62) or a process (p.157) to take place. To raise the temperature of a liquid or a solid so that it can still be touched without discomfort. For example, to warm a solution so that crystals dissolve. **warm** (adj).

**heat** (v) to supply enough heat to raise the temperature so that a substance is hot enough for a chemical reaction (p.62) to take place or a process (p.157) to take place. The temperature is too high for the substance to be touched with comfort. For example, to heat zinc in a stream of oxygen. **hot** (adj), **heat** (n), **heater** (n).

**char** (v) to change an organic (p.55) substance to carbon by heating it or by using a strong dehydrating (p.66) agent. For example, to char wood by heating it, forming charcoal; to char sugar by adding concentrated sulphuric acid.

**water-bath** (n) a vessel (p.25) containing water, which is used to heat (↑) apparatus (p.23). This way of heating does not allow the temperature to rise above 100°C.

燒 (動) 遭受 (第 213 頁) 燃燒 (↑) 時能產生 (第 62 頁) 熱、火焰，有時還發出煙。例如木材燃燒着時發出有煙的火焰；燒煤產生熱量。

發光：灼熱 (動) 因化學反應 (第 62 頁) 或因物質 (第 8 頁) 受強烈加熱而發出光。例如，木頭灼熱而沒有火焰時會發光；一塊鐵被強烈加熱時會發光。(名詞爲 glow)

發煙燜燒 (動) 由於在低溫下發生化學反應 (第 62 頁) 而火焰因而發出煙。

升溫 (動) 僅提供足夠的熱量使物質能進行反應 (第 62 頁) 或進行一個過程 (第 157 頁)。使液體或固體溫度升高但仍可觸摸而不會感到不適。例如，使液體升溫讓晶體溶解。(形容詞爲 warm)

加熱 (動) 提供足夠的熱量以升高溫度，使一種物質熱到足於發生化學反應 (第 62 頁) 或進行一個過程 (第 157 頁)。如果溫度過高，就不能觸摸該物質。否則引起不適。例如，在氧氣流中加熱。(形容詞爲 hot，名詞爲 heat、熱量，heater 加熱器)

焦化：炭化 (動) 藉加熱或使用強脫水 (第 66 頁) 劑使一種有機 (第 55 頁) 物質化爲炭。例如，將木柴加熱焦化使之形成木炭，或把糖中加入濃硫酸使之焦化。

水浴 (名) 一種盛有水，用於加熱 (↑) 儀器 (第 23 頁) 的容器 (第 25 頁)。這種加熱方法，溫度升高不會超過 100℃。

glow 發光

smoulder 發煙燜燒

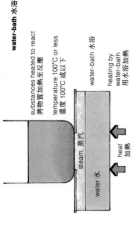

**water-bath** 水浴

substances heated to react
將物質加熱至反應

temperature 100°C or less
溫度 100℃ 或以下

water-bath 水浴

steam 蒸汽

water 水

heat 加熱

heating by water-bath
用水浴加熱

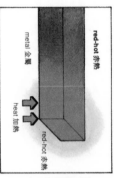

red-hot 赤熱

metal 金屬

heat 加熱

red-hot 赤熱

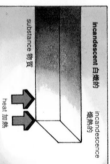

incandescent 白熾的

substance 物質

heat 加熱

incandescence 熾熱的

**red-hot** (*adj*) describes a substance, usually a metal, heated to such a high temperature that it glows (p.59) red.

**incandescent** (*adj*) describes a substance heated to such a high temperature that it gives out white light. The temperature of an incandescent substance is higher than the temperature of a red-hot (↑) substance. **incandescence** (*n*).

**light¹** (*n*) a form of energy whose effect on the eye causes the sense of seeing. White light is composed (p.82) of all colours. Different colours can cause different chemical effects. Some substances, e.g. silver chloride, are decomposed by light falling on them. See *photochemical* (p.65).

赤熱的（形）描述一種物質，通常是一種金屬，被加熱到高溫，發出（第 59 頁）紅光。

白熾的（形）描述一種物質被加熱到高溫，發出白光。白熾物質的溫度比赤熱（↑）物質的溫度高。（名詞屬 incandescence）

光（名）使眼睛產生視覺效應的一種能量形式。白光係由各種光所組成（第 82 頁）。不同的顏色可產生不同的化學效應。某些物質，例如氯化銀，受光照射時發生分解。多見"光化學的"（第 65 頁）。

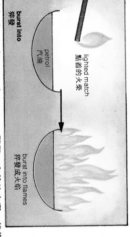

burst into 迸發

lighted match 點着的火柴

petrol 汽油

burst into flames 迸發成火焰

**burst into** to appear suddenly, as flames do, e.g. when a lighted match is put near petrol, the petrol bursts into flames. *Burst into* is used when a lot of flames appear; with a small flame, the substance *catches fire*.

**light²** (*v*) to put a flame to a supply of a gas, so that the gas catches fire, i.e. flames appear, e.g. to light a supply of hydrogen from a jet.

**extinguish** (*v*) to cause a flame to stop burning.

迸發 突然地出現，就像出現火焰一樣。例如擦着一根火柴靠近汽油，汽油即迸發成火焰。burst into（迸發）這組詞的詞義只用於出現大火焰時；只要有小的火焰物質就會"着火"。

點火（動）將火苗放到一個氣體源中，使氣體着火，亦即出現火焰。例如，將噴嘴的氫氣源點火。

使熄滅（動）使火焰停止燃着。

candle extinguished 蠟燭熄滅

candle alight 蠟燭點着

light 點火    extinguish 熄滅

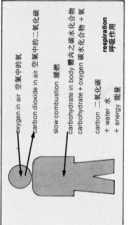

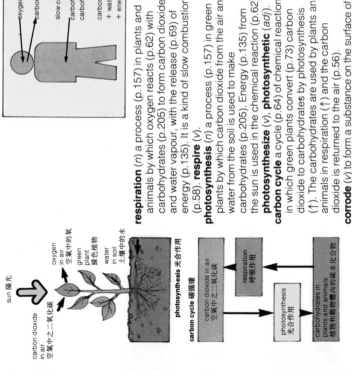

**respiration** (*n*) a process (p.157) in plants and animals by which oxygen reacts (p.62) with carbohydrates (p.205) to form carbon dioxide and water vapour, with the release (p.69) of energy (p.135). It is a kind of slow combustion (p.58). **respire** (*v*).

**photosynthesis** (*n*) a process (p.157) in green plants by which carbon dioxide from the air and water from the soil is used to make carbohydrates (p.205). Energy (p.135) from the sun is used in the chemical reaction (p.62). **photosynthesize** (*v*), **photosynthetic** (*adj*).

**carbon cycle** a cycle (p.64) of chemical reactions in which green plants convert (p.73) carbon dioxide to carbohydrates by photosynthesis (↑). The carbohydrates are used by plants and animals in respiration (↑) and the carbon dioxide is returned to the air (p.56).

**corrode** (*v*) to form a substance on the surface of a metal. The metal reacts (p.62) with particular gases in the air to form the substance; in many cases an oxide is formed. When a metal corrodes, small holes appear on the surface and the strength of the metal becomes less. Oxygen, carbon dioxide, sulphur dioxide or hydrogen sulphide in the air cause a metal to corrode. **corrosion** (*n*), **corrosive** (*adj*).

**rust** (*n*) corrosion (↑) of iron, forming a red dust which is the rust on the surface of the iron. Rust is an oxide of iron. **rust** (*v*), **rusty** (*adj*).

**tarnish** (*v*) to corrode if the metal has a shiny surface, e.g. silver tarnishes by the forming of a layer of silver sulphide on the surface: silver sulphide is a corrosion, or tarnish. **tarnish** (*n*).

**呼吸作用** (名) 植物和動物體內進行的一個過程 (第157頁)，呼吸過程中氧氣與碳水化合物 (第205頁) 起作用 (第62頁) 生成二氧化碳和水蒸汽，同時釋放出 (第69頁) 能量 (第135頁)。呼吸作用是一種緩慢燃燒 (第58頁) 作用。(動詞為 respire)

**光合作用** (名) 綠色植物內進行的一個過程 (第157頁)，此過程中的水分用於製造碳水化合物 (第205頁)。從太陽吸收的能量 (第135頁) 被用於這造化學反應 (第62頁)。(動詞為 photosynthesize，形容詞為 photosynthetic)

**碳循環** 化學反應中的一個循環 (第64頁)。在此循環中，綠色植物藉光合作用 (↑) 將二氧化碳轉化 (第73頁) 成碳水化合物。植物、動物在呼吸 (↑) 中又利用碳水化合物，而二氧化碳則返回空氣 (第56頁) 中。

**腐蝕** (動) 在金屬表面形成某種物質。金屬與空氣中的某一些氣體起反應 (第62頁) 生成此種物質；通常是生成氧化物。當一種金屬受腐蝕時會在表面出現很多小孔，令其強度下降。空氣中的氧、二氧化碳、二氧化硫或硫化氫都會引起金屬腐蝕。(名詞為 corrosion，形容詞為 corrosive)

**銹** (名) 鐵的氧化作用 (↑)，使鐵之表面上形成一層紅色粉末，即鐵表面上的銹。銹是鐵的一種氧化物。(動詞為 rust，形容詞為 rusty)

**使失光澤** (動) 使有光澤表面腐蝕。例如：銀的表面形成一層硫化銀，使銀失去光澤；生成硫化銀是一種腐蝕或使銀失去光澤 (名詞為 tarnish)。

**chemical reaction** a process (p.157) in which new substances (p.8) are formed, i.e. a chemical change takes place. Energy, usually heat, is needed to make the chemical reaction take place, or the chemical reaction produces (↓) heat.

**reaction** (n) in chemistry, a process (p.157) in which two substances (p.8) have an effect on each other and new substances are produced (↓). **react** (v).

**react** (v) to behave in such a way that a chemical reaction (↑) takes place, e.g. sodium metal reacts with water, a chemical reaction takes place and sodium hydroxide and hydrogen are formed (sodium and water react).

**reactant** (n) a substance which takes part in a chemical reaction (↑), e.g. sodium and water are reactants when these substances react.

**product** (n) a new substance formed from a chemical reaction (↑), e.g. hydrogen and sodium hydroxide are the products of the reaction between sodium metal and water. **produce** (v).

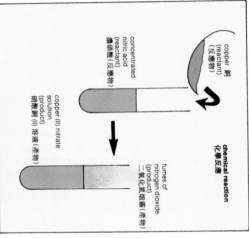

copper 銅
(reactant)
(反應物)

concentrated
nitric acid
(reactant)
濃硝酸 (反應物)

copper (II) nitrate
solution
(product)
硝酸銅 (II) 溶液 (產物)

chemical reaction
化學反應

fumes of
nitrogen dioxide
(product)
二氧化氮煙霧 (產物)

化學反應 生成新物質 (第 8 頁)，亦即發生化學變化的一種過程 (第 157 頁)。化學反應必須供于能量，才能發生；化學反應亦能產生 (↓) 熱量。

反應 (名) 化學上指兩種物質的過程 (第 157 頁)。(動詞為 react)

起反應 (動) 起作用 (動) 使發生化學反應 (↑) 的表現與方式。例如金屬鈉與水起作用，發生化學反應同時生成氫氧化鈉和氫氣 (鈉和水起作用)。

反應物 (名) 指參加化學反應 (↑) 的物質。例如，鈉和水反應時，鈉和水是反應物。

產物 (名) 化學反應 (↑) 時所生成的新物質。例如，氫和氫氧化鈉是金屬鈉和水之間反應的產物。(動詞為 produce)

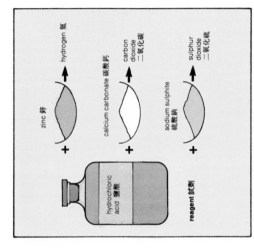

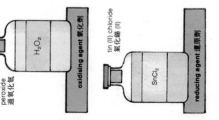

**reagent** (*n*) a substance which causes a chemical reaction (p.55) to take place. The mineral acids (p.55) are common reagents as they have known effects on many inorganic (p.55) substances.

試劑 (名) 能引致發生化學反應 (↑) 的一種物質。無機酸 (第 55 頁) 是常用試劑，對多種無機 (第 55 頁) 物質都有已知的效應。

**agent** (*n*) a substance, or a form of energy, which is used to produce a named effect, e.g. an oxidizing agent causes oxidation (p.70); light causes a photochemical (p.65) effect and so is an agent for the effect.

劑：介質：作用物 (名) 用於產生各名效應的一種物質或能量形式。例如：氧化劑產生氧化作用 (第 70 頁)；光引致光化學 (第 65 頁) 效應，因而是光化學效應的介質。

**take place** to come into being, when describing an event, i.e. an event takes place. When an event *happens*, it is not expected, when an event *takes place*, it is planned. *See occur*[1] (↓).

發生：進行 在敘述一事件而該事件業已實現，即發生一件事。"一件事" 這個詞，是指事件時如用 "happen" 這個詞，是指事件沒有預期發生。用 take place 時則指事件的發生是事先計劃好的。參見 "出現" (↓)。

**occur**[1] (*v*) to come into being at a certain time when describing an event, e.g. an event occurs, it is known when it will occur, although a person does not control it. To contrast *take place* and *occur*: a chemical reaction (↑) *takes place* because a person controls it, it cannot *occur* because it has not happened before and it is under the control of a person. A birthday occurs once a year, it does not take place.

出現 (動) 當敘述一事件時，指該事件在某一段時間內發生。例如，出現 (occur) 一件事，指雖然人們不能控制該事件，但卻知道該事件出現。take place：進行 (發生)。進行 與 occur (出現) 的比較：進行化學反應 (↑)，因為人們能控制此反應；不出現化學反應因為過去未發生過，而且這是受制人們控制的。生日每年出現不是發生一次。

**reversible reaction** 可逆反應
steam 蒸汽 / red-hot iron 赤熱的鐵 / heat 加熱 / hydrogen 氫 / oxide of iron 鐵的氧化物 / hydrogen 氫 / oxide of iron 鐵的氧化物 / heat 加熱 / steam 蒸汽 / iron 鐵

**reversible reaction** a reaction (p.62) in which reactants (p.62) form products (p.62) and the products can then react to form the reactants. For example, steam and iron form hydrogen and an oxide of iron. The oxide of iron and hydrogen react to form steam and iron. The reaction is written thus:

steam + iron ⇌ oxide of iron + hydrogen

**irreversible reaction** a reaction (p.62) in which the products will not react and the reaction stops when the products are formed.

**chain reaction** (n) a reaction (p.62) in which the first reaction between molecules (p.77) or atoms (p.110) forms products which then react with further molecules or atoms so that each reaction becomes stronger until the reaction is explosive.

**cycle** (n) a set of events which is repeated time after time. Each complete set is one cycle, and the cycle is repeated. For example, a carbon atom starts in a molecule of carbon dioxide in the air; it is used in photosynthesis (p.61) in a plant and becomes part of a carbohydrate (p.205) molecule; the carbohydrate is eaten by an animal and digested; the digested product takes part in respiration (p.61) and the carbon atom becomes part of a carbon dioxide molecule, and is breathed out; the carbon atom is once again in a carbon dioxide molecule in the air, the cycle has been completed, and starts all over again, **cyclic** (adj).

**combination** (n) the joining of elements (p.8) by chemical bonds (p.133) to form compounds (p.8). The joining of compounds by addition (p.188) to form new substances, e.g. the combination of iron and oxygen forms iron (III) oxide; the combination of zinc and chlorine forms zinc chloride; the addition of bromine to ethene (p.174) is a combination of the two compounds.

可逆反應 指產物與反應物又反應而生成原反應物（第62頁）之後，該產生反應物又生成產物的一種反應（第62頁）。例如，水、蒸汽和鐵生成氫和鐵的氧化物。鐵的氧化物又反應生成水蒸汽和鐵。因此反應可寫成：

水蒸汽 + 鐵 ⇌ 鐵的氧化物 + 氫氧

不可逆反應 指產物不再反應而生成之後反應即終止的一種反應（第62頁）。

連鎖反應 （名）分子（第77頁）或原子（第110頁）之間一級反應生成產物，該產物接着又和更多的分子或原子反應，使反應一次比一次更劇烈直至成為一種爆炸性反應（第62頁）。

循環 （名）指一次接一次重複出現的一組事件。每一個完整的一組事件就是一個循環，此循環周而復始。例如，一個碳原子自空氣中一氧化碳的一個分子開始，此碳原子為一株植物在光合作用中（第205頁）所利用，形成碳水化合物的一部分；此碳水化合物為一動物所吃下然後消化；消化的產物參與呼吸作用（第61頁）使該碳原子又成為二氧化碳分子的一部分而被呼出大氣中；至此，該碳原子又再處於空氣中的二氧化碳分子中，循環至此結束，然後重新開始另一次循環。（形容詞 cyclic）

化合作用 （名）藉化學鍵（第133頁）使元素（第8頁）結合而生成化合物（第8頁）的化合物藉加成作用（第188頁）結合而成新物質，例如，鐵和氧結合生成氧化鐵（III）。鋅和氯化合生成氧化鋅；乙烯（第174頁）中加入溴是這兩種化合物的化合作用。

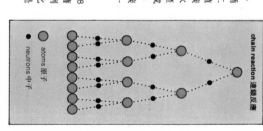

**chain reaction** 連鎖反應
○ atoms 原子
● neutrons 中子

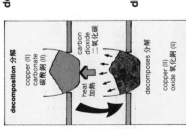

decomposition 分解

copper (II)
carbonate
碳酸銅 (II)

carbon
dioxide
二氧化碳

heat
加熱

decomposes 分解

copper (II)
oxide 氧化銅 (II)

**decomposition** (*n*) the changing of pure substances into simpler compounds. The process (p.157) is irreversible. Decomposition can be caused by temperature, light, electric current or micro-organisms, e.g. the decomposition of copper (II) carbonate when heated into copper (II) oxide and carbon dioxide; the decomposition of silver bromide by light into silver and bromine; the decomposition of copper (II) sulphate by an electric current. **decompose** (*v*).

**dissociation** (*n*) the separation of an ionic (p.123) or a covalent (p.8) compound (p.8) into simpler compounds; the process (p.157) is reversible, i.e. the simpler compounds can unite again. For example, sodium chloride, in water, undergoes (p.213) dissociation into ions (the ions can unite to form sodium chloride again); the dissociation of ammonium chloride into ammonia and hydrogen chloride (ammonia and hydrogen chloride can combine to form ammonium chloride). **dissociate** (*v*).

**thermal** (*adj*) caused by heat.
**photochemical** (*adj*) describes a chemical reaction caused by light. See *electrochemical (p.128)*.
**thermal decomposition** decomposition (↑) caused by heating a compound.

分解作用 (名) 將純淨物質變成較簡單化合物的過程。此過程 (第 157 頁) 是不可逆的。溫度、光、電流或微生物均可引起分解。例如，碳酸銅 (II) 被加熱分解成氧化銅 (II) 和二氧化碳；光使溴化銀分解成銀和溴；電流使硫酸銅 (II) 分解。(動詞爲 decompose)

離解作用 (名) 指離子 (第 123 頁) 化合物或共價化合物 (第 8 頁) 分離成較簡單化合物的過程；此過程 (第 157 頁) 是可逆的，亦即此種簡單的化合物可再次結合。例如，氯化鈉在水中遭受 (第 213 頁) 離解而成離子 (這些離子可再次結合成氯化鈉)；氯化銨離解成氨和氯化氫 (氨和氯化氫可再結合形成氯化銨)。(動詞爲 dissociate)

熱的 (形) 由熱引起的。
光化學的 (形) 描述一種由光引起的化學的。參見 "電化作用" (第 128 頁)
熱分解作用 加熱於一種化合物所引起的分解作用 (↑)。

**thermal dissociation** dissociation (↑) caused by heating a compound, e.g. the thermal dissociation of ammonium chloride.
**disintegrate** (*v*) to break into small pieces because of physical or chemical action, e.g. when hit hard a mineral (p.154) disintegrates into small pieces. **disintegration** (*n*).

熱離解作用 加熱於一種化合物所引起的離解作用 (↑)。例如氯化銨的熱離解。

分裂 (動) 因物理或化學作用而破裂成小塊。例如猛烈撞擊一種礦物 (第 154 頁) 使之破裂成小塊。(名詞爲 disintegration)

**hydrolysis**[1] (*n*) decomposition (p.65) caused by the chemical action of water. The salts of weak acids or weak bases undergo (p.213) hydrolysis, e.g. iron (III) chloride in water decomposes into iron (III) hydroxide and hydrochloric acid, as iron (III) hydroxide is a weak base.

**dehydrate** (*v*) (1) to take away water from a substance, to make it dry, e.g. to take water from ethanol (ethyl alcohol) to make it as dry as possible. (2) to take away the elements of water from a compound by a chemical action, e.g. concentrated sulphuric acid removes water from ethanol to form ethene:

$$C_2H_5OH \rightarrow C_2H_4 + (H_2O \text{ taken away})$$

Dehydrate is the opposite process of *hydrate* (p.90). **dehydration** (*n*).

**desiccate** (*v*) to take away all traces of water, i.e. a stronger action than dehydrate (↑). **desiccation** (*n*), **desiccator** (*n*), **desiccant** (*n*), **desiccant** (*n*) a substance which will desiccate (↑) compounds (p.8).

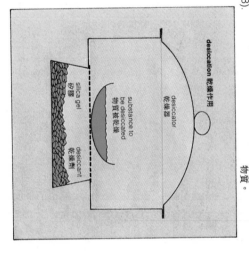

desiccation 乾燥作用

desiccator
乾燥器

silica gel
矽膠

substance to
be desiccated
物質經乾燥

desiccant
乾燥劑

dehydrating ethanol
乙醇脫水

ethanol 乙醇

fused calcium
chloride (takes
away water)
熔融的氯化鈣
(脫除水)

dehydrate 脫水

水解作用 (名) 藉水的化學作用而引起的分解作用 (第 65 頁)。弱酸或弱鹼的鹽遭受 (第 213 頁) 水解作用。例如氯化鐵 (III) 在水中分解成氫氧化鐵 (III) 和鹽酸。因為氫氧化鐵 (III) 是一種弱鹼。

脫水 (動) (1) 除去物質中之水分使之乾燥。例如去除乙醇中的水分使之盡可能無水。(2) 藉化學作用以去除化合物中的水的元素 (第 8 頁)。例如，用濃硫酸去除乙醇中的水使生成乙烯。

$$C_2H_5OH \rightarrow C_2H_4 + (去除的 H_2O)$$

脫水是水合 (第 90 頁) 的相反過程。(名詞為 dehydration)

乾燥 (動) 去除全部水迹。其作用比脫水 (↑) 更強。(名詞為 desiccation 乾燥作用，desiccator 乾燥器，desiccant 乾燥劑)

乾燥劑 (名) 指可使化合物 (第 8 頁) 乾燥 (↑) 的物質。

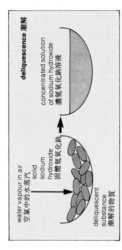

water vapour in air
空氣中的水蒸汽

solid
sodium
hydroxide
固體氫氧化鈉

deliquescent
substance
潮解的物質

**deliquescence 潮解**

concentrated solution
of sodium hydroxide
濃氫氧化鈉溶液

**deliquesce** (*v*) to absorb (p.35) water from the air by crystals (p.91) so that the crystals dissolve (p.30) and form a concentrated solution, e.g. sodium hydroxide flakes deliquesce in air and form a solution.
**deliquescent** (*adj*), **deliquescence** (*n*).

**effloresce** (*v*) to lose water of crystallization (p.91) from a crystal to the air, e.g. sodium carbonate crystals effloresce in air and form a powder. $Na_2CO_3.10H_2O \rightarrow Na_2CO_3.H_2O$. **efflorescent** (*adj*), **efflorescence** (*n*).

**hygroscopic** (*adj*) describes crystalline (p.15) and amorphous (p.15) substances which take water from the air and become damp, e.g. sodium chloride is hygroscopic and becomes damp when left in air. **hygroscopicity** (*n*).

**neutralization** (*n*) the reaction (p.62) between an acid (p.45) and a base (p.46) or alkali, during which they destroy each other's properties and form a salt and water. The solution of the salt is neutral, i.e. neither acidic nor alkaline. For example, the neutralization of hydrochloric acid by sodium hydroxide solution. **neutralize** (*v*), **neutral** (*adj*).

潮解 (動)　晶體 (第 91 頁) 由空氣中吸收 (第 35 頁) 水分，結果晶體溶解 (第 30 頁)，形成濃溶液。例如，氫氧化鈉片在空氣中潮解形成溶液。(形容詞爲 deliquescent，名詞爲 deliquescence)

風化 (動)　結晶 (第 91 頁) 水由晶體逸失到空氣中。例如碳酸鈉晶體在空氣中風化，形成粉末。
$Na_2CO_3.10H_2O \longrightarrow Na_2CO_3.H_2O$。
(形容詞爲 efflorescent，名詞爲 efflorescence)

吸濕的 (形)　形容晶狀 (第 15 頁) 和無定形 (第 15 頁) 物質吸收空氣中的水分變潮濕，例如氯化鈉是吸濕的，當露置於空氣中便會變潮。(名詞爲 hygroscopicity)

中和作用 (名)　指酸 (第 45 頁) 和鹼 (第 46 頁) 或強鹼之間的反應 (第 62 頁) 生成鹽和水，使酸和鹼的性質都消失的作用。鹽溶液是中性的；換言之，既非酸性亦非鹼性。例如，用氫氧化鈉溶液中和鹽酸。(動詞爲 neutralize，形容詞爲 neutral)

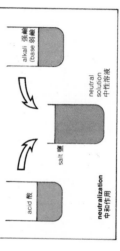

acid 酸

alkali 強鹼
(base 弱鹼)

salt 鹽

neutral
solution
中性溶液

**neutralization**
**中和作用**

**displacement** (*n*) a chemical reaction (p.62) in which one element takes the place of another element which is in a compound. For example, iron put into a solution of copper (II) sulphate causes the displacement of copper. The reaction is: Fe + CuSO₄→ Cu + FeSO₄ Copper metal appears in the solution as the iron disappears. **displace** (*v*).

**replace** (*v*) (1) to put one thing in place of another because the thing replaced is no longer wanted or no longer of use. (2) to put one atom (p.110), an ion or a functional group (p.185) in place of another atom, ion or group. Usually hydrogen in an acid is replaced by a metal, when the metal acts on (p.19) the acid to form a salt and hydrogen. **replacement** (*n*). **replaceable** (*adj*).

**replaceable** (*adj*) describes a hydrogen atom in an acid which can be replaced, directly or indirectly, by a metal atom.

**base exchange** a chemical reaction (p.62) in which two inorganic radicals (p.45) replace (↑) each other, e.g. as in the reaction:
calcium chloride + sodium carbonate ⟶ calcium carbonate + sodium chloride
In this example the chloride and carbonate radicals exchange with each other.

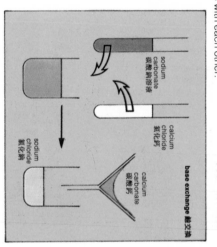

sodium
carbonate
碳酸鈉溶液

calcium
chloride
氯化鈣

sodium
chloride
氯化鈉

calcium
carbonate
碳酸鈣

**base exchange 鹼交換**

**置換作用**（名）　一種元素代化合物中另一種元素的化學反應（第 62 頁）。例如，鐵置於硫酸銅 (II) 溶液中產生銅的置換作用，其反應為：
Fe + CuSO₄ ⟶ Cu + FeSO₄，
隨著鐵消失，銅金屬出現在溶液中。（動詞為 displace）

**置換**（動）　(1) 將一物代替另一物，因原來被置換的物品不再要或已沒有用了；(2) 將一個原子（第 110 頁）、離子或官能團（第 185 頁）代替另一個原子、離子或官能團。一般而言，金屬中的氫為一種金屬所置換。（名詞為 replacement，形容詞為 replaceable）

**可置換的**（形）　指酸中的氫原子可直接或間接為金屬原子所置換。

**鹼交換作用**　兩個無機的根（第 45 頁）相互置換 (↑) 的一種化學反應（第 62 頁）。例如下列反應：
氯化鈣 + 碳酸鈉 ⟶ 碳酸鈣 + 氯化鈉
其中，氯和碳酸根相互置換。

copper (II) sulphate
solution
硫酸銅 (II) 溶液

iron (II) displaces copper
鐵置換銅

iron 鐵

iron (II)
solution
鐵置銅溶液

copper 銅

**displacement 置換**

放：釋放（動）放出受物理的束的某物或放出能量。一般都是由液體中放出一種氣體。例如，當一種細碎的（第 13 頁）固體放入溶有二氧化碳溶液中時，二氧化碳氣體便放入空氣中；當打開煤氣閥時，煤氣放入火爐中。（名詞爲 release）

釋放：放出（動）藉破壞化學鍵（第 133 頁）而放出一種氣體，例如將鋅屬放入鹽酸中，隨着酸的鍵被破壞而釋出氫氣。比較 given off（第 41 頁）與釋出（liberate）的比較：鋅放入鹽酸中，散發出氫氣，這是實際的觀察（第 42 頁）因爲並無現供有關反應的資料；鋅放入鹽酸中，釋出氫氣，提供有關化學反應的資料。（名詞爲 liberation）

親合力：化合力（名）衡量一種元素或化合物與另一種化合物起反應（第 62 頁）的能力。親合力越大，反應越強烈。例如，氫氧化鈉對二氧化碳有親合力；濃硫酸對水有很大的親合力，此反應很劇烈。

溫和的（形）描述介於強、弱之間的反應。許多種反應都是溫和的，因此通常都不用這個詞來描述，而只有用弱的、強的、劇烈的（↓）、爆炸性的（第 204 頁）這些詞來描述反應。

劇烈的（形）形容反應比強反應還要強，如不小心，其反應之强度會使儀器破壞。（名詞爲 violence）

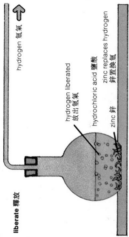

liberate 釋放

hydrogen 氫氣

hydrogen liberated
放出氫氣

hydrochloric acid 鹽酸

zinc replaces hydrogen
鋅置換氫

zinc 鋅

**release** (v) to set free something held physically or to set free energy. Usually a gas in a liquid is released, e.g. when a finely divided (p.13) solid is put into a solution of carbon dioxide, the carbon dioxide gas is released into the air; when a gas tap is opened, gas is released into a burner. **release** (n).

**liberate** (v) to set free a gas by breaking chemical bonds (p.133), e.g. when zinc metal is put in hydrochloric acid, hydrogen is liberated, as bonds in the acid have been broken. To compare *give off* (p.41) with *liberate*: hydrogen is *given off* when zinc is put in hydrochloric acid; this is a practical observation (p.42) as no information is given about the reaction; hydrogen is *liberated* when zinc is put in hydrochloric acid gives information about the chemical reaction. **liberation** (n).

**affinity** (n) a measure of the ability of one element or compound to react (p.62) with another compound. The greater the affinity, the stronger is the reaction, e.g. sodium hydroxide has an affinity for carbon dioxide; concentrated sulphuric acid has a great affinity for water, the action is violent.

**moderate** (adj) describes a reaction midway between weak and strong. Many reactions are moderate and so are not usually described as such; only weak, strong, violent (↓) and explosive (p.204) reactions are so described.

**violent** (adj) describes a reaction which is stronger than strong; the strength of the reaction can break apparatus unless care is taken. **violence** (n).

**oxidation** (n) (1) the addition of oxygen to an element or compound. (2) the removal of hydrogen from a compound. (3) the removal of electrons (p.110) from an atom (p.110) or ion (p.123). (4) an increase in the oxidation number (p.78) of an element. For example: (1) calcium + oxygen forms calcium oxide; (2) hydrogen chloride is oxidized to chlorine by the removal of hydrogen; (3) a Cu atom is oxidized to $Cu^{2+}$, or a $Fe^{2+}$ ion is oxidized to $Fe^{3+}$; (4) Mn(VI) is oxidized to Mn(VII). **oxidize** (v), **oxidizing** (adj), **oxidant** (n).

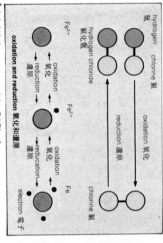

oxidation and reduction 氧化和還原

**reduction**[1] (n) (1) the removal of oxygen from a compound. (2) the addition of hydrogen to an element or compound. (3) the addition of electrons (p.110) to an atom (p.110) or ion (p.123). (4) a decrease (p.219) in the oxidation number (p.78) of an element. For example: (1) the reduction of zinc oxide to zinc; (2) the reduction of chlorine to hydrogen chloride; (3) the reduction of $Cu^{2+}$ to Cu or of $Fe^{3+}$ to $Fe^{2+}$; (4) Mn(VII) is reduced to Mn(IV). **reduce** (v), **reducing** (adj).

**redox process** if substance A oxidizes (↑) substance B, then substance B reduces (↑) substance A. Oxidation and reduction always occur together, so such a reaction is called a redox process. e.g. hydrogen reduces copper (II) oxide to copper, but copper (II) oxide oxidizes hydrogen to water (hydrogen oxide), the reaction is a redox process.

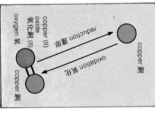

oxidation and reduction 氧化和還原

**氧化作用** (名) 指下列四種情況之一：(1)在一種元素或化合物中加入氧(第215頁)；(2)自一種化合物中除去氫(第215頁)；(3)自一種元素或化合物中除去電子(第110頁)或離子(第123頁)中減去一個原子(第110頁)；(4)增加一種元素的氧化值(第78頁)。例如：(1)鈣＋氧生成氧化鈣；(2)氫化氫藉脫除氫，氧化成氯；(3)Cu原子氧化成Cu²⁺，Fe²⁺離子氧化成Fe³⁺；(4)錳(VI)氧化成錳(VII)。(動詞為 oxidize，形容詞為 oxidize，名詞為 oxidant)

**還原作用** (名) 指下列四種情況之一：(1)除去化合物中的氧(第215頁)；(2)在元素或化合物中加入氫；(3)在原子(第110頁)或離子(第123頁)加入電子(第110頁)；(4)降低一種元素的氧化值(第78頁)。例如：(1)氧化鋅還原成鋅；(2)氯還原成氧化氫；(3)Cu²⁺還原成Cu或Fe³⁺還原成Fe²⁺；(4)Mn(VII)還原成Mn(IV)。(動詞為 reduce，形容詞為 reducing)

**氧化還原過程** 如A物質氧化(↑)B物質，則B物質促使還原(↑)A物質。氧化和還原總是在一起出現，因此這種反應稱為氧化還原過程。例如：氫將氧化銅(II)還原成銅，但氧化銅(II)將氫氧化成水(氧化氫)，此氧化反應就是氧化還原過程。

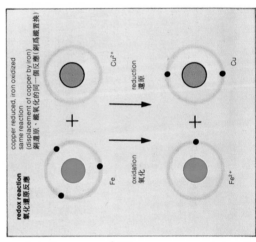

**redox reaction**
**氧化還原反應**

copper reduced, iron oxidized
same reaction
(displacement of copper by iron)
銅還原，鐵氧化的同一個反應（銅為鐵置換）

Fe

Cu²⁺

oxidation
氧化

reduction
還原

Fe²⁺

Cu

**autoxidation** (*n*) an oxidation by oxygen in the air at room temperature, e.g. the oxidation of iron in the air to form rust (p.61).

**disproportionation** (*n*) a process (p.157) in which a substance undergoes (p.213) oxidation (p.70) and reduction (p.70) at the same time, e.g. Cu⁺ undergoes disproportionation into Cu and Cu²⁺ One atom of Cu⁺ is oxidized to Cu²⁺ and one atom is reduced to Cu.

$$Cu_2{}^+SO_4{}^{2-} \longrightarrow Cu + Cu^{2+}SO_4{}^{2-}$$

**oxidizing agent** a substance which oxidizes (↑) elements or compounds, e.g. oxygen is an oxidizing agent, it oxidizes iron to iron (II) oxide; potassium manganate (VII) solution is an oxidizing agent, it oxidizes iron (II) sulphate to iron (III) sulphate.

**oxidant** (*n*) another name for oxidizing agent (↑).

**reducing agent** a substance which reduces (↑) compounds, e.g. hydrogen is a reducing agent, it reduces copper (II) oxide to copper; tin (II) chloride solution is a reducing agent, it reduces iron (III) sulphate to iron (II) sulphate.

**自動氧化作用**（名）在室溫下和空氣中的氧所發生的氧化作用。例如鐵在空氣中氧化生成鏽（第61頁）。

**自身氧化還原作用**（名）一種物質同時遭受（第213頁）氧化（第70頁）和還原（第70頁）的過程（第157頁）。例如Cu⁺自身氧化還原作用變成Cu和Cu²⁺。Cu⁺的一個原子被氧化變成Cu²⁺，同時又有一個原子被還原成Cu。

$$Cu_2{}^+SO_4{}^{2-} \longrightarrow Cu + Cu^{2+}SO_4{}^{2-}$$

**氧化劑** 能使元素或化合物氧化(↑)的物質。例如氧是一種氧化劑，它使鐵氧化成氧化鐵(III)；高錳酸鉀(VII)溶液是一種氧化劑，它使硫酸亞鐵(II)氧化成硫酸鐵(III)。

**氧化劑**的另一英文名為**oxidant**。

**還原劑** 能使化合物還原(↑)的一種物質。例如氫是一種還原劑，它使氧化銅(II)還原成銅；氯化亞錫(II)溶液是一種還原劑，它使硫酸鐵(III)還原成硫酸亞鐵(II)。

**catalyst** (n) a substance which increases the rate of chemical reaction (p.62) but itself remains chemically unchanged, e.g. platinum is a catalyst for the reaction between sulphur dioxide and oxygen, it increases the rate of reaction. Many finely divided (p. 13) metals act as catalysts. **catalyse** (v), **catalysis** (n).

**catalysis** (n) a process in which a catalyst (↑) increases the rate of reaction (p.149).

**negative catalyst** a substance which decreases (p.219) the rate of reaction (p.149); it reacts with and destroys the catalyst.

**inhibitor** (n) a substance which slows down a chemical reaction (p.62); it may be a negative catalyst (↑) or a retarder (↓).

**retarder** (n) a substance used to decrease the rate of reaction (p.149) by physical and other means, e.g. the addition of a colloid (p.98) to slow down an ionic (p.123) reaction.

**promoter** (n) a substance used to increase the action of a catalyst (↑), e.g. finely divided (p.13) iron catalyzes the reaction between nitrogen and hydrogen to form ammonia, the addition of iron oxide increases the effect of the catalyst, that is, iron oxide is a promoter.

**autocatalysis** (n) a process (p.157) which takes place when one of the products (p.62) of a chemical reaction (p.62) acts as a catalyst (↑) for the reaction, e.g. potassium manganate (VII) in acid solution oxidizes ethanedioic (oxalic) acid, the Mn²⁺ ions, a product of the reaction, catalyze the reaction so autocatalysis takes place.

**enzymatic** (adj) describes any effect caused by enzymes. An enzyme is a catalyst (↑) produced by living things and which takes part in chemical changes in living things.

**enzyme** (n).

**poison** (n) a substance which prevents a catalyst (↑) from acting, e.g. arsenic is a poison which prevents platinum from acting as a catalyst. Substances which are poisonous for catalysts are also poisonous for living things as they prevent enzymatic (↑) reactions in living things.

催化劑：觸媒劑（名） 指能增加化學反應（第 62 頁）速度（但本身不發生化學變化）的物質。例如，鉑是二氧化硫和氧發生反應的一種催化劑。鉑能增加反應之速度。許多細碎的（第 13 頁）金屬均能起催化劑作用。（動詞屬 catalyze，名詞屬 catalysis）

催化作用（名） 用催化劑（↑）以增加反應速度的過程。

負催化劑（名） 能降低（第 219 頁）反應速度（第 149 頁）之物質；它與催化劑反應並破壞催化劑。

抑制劑（名） 能延緩化學反應（第 62 頁）的一種物質，抑制劑可以是一種負催化劑（↑）也可以是一種阻滯劑（↓）。

阻滯劑：延緩劑（名） 能藉物理或其他方法降低反應速度（第 149 頁）之物質，例如加入膠體反應速度（第 149 頁）以延緩離子（第 123 頁）的反應。

促進劑（名） 用於增強催化劑（↑）作用的一種物質。例如細碎的（第 13 頁）鐵（粉）可以催化氮與氫之反應，加入氧化鐵鐵粉可以提高催化劑之效應，亦即氧化鐵是一種助催化劑。

助催化劑（名） 用於增強催化劑（↑）作用的一種物質，例如細碎的（第 13 頁）鐵（粉）可以催化氮與氫之反應，加入氧化鐵鐵粉可以提高催化劑之效應，亦即氧化鐵是一種助催化劑。

自動催化作用（名） 化學反應（第 62 頁）產物之一作爲催化劑（↑）使此反應得以進行的過程（第 157 頁）。例如酸溶液中的高錳酸鉀（VII）氧化乙二酸（即草酸）氧化。反應產物Mn²⁺離子催化此反應，從而發生自動催化作用。

酶的：酶作用的（形） 描述由酶引起的任何效應。酶是指生物體內產生的一種催化劑（↑），酶能催化生物體內的化學變化。（名詞屬 enzyme）

毒物（名） 能阻止催化劑（↑）起作用的一種物質。例如，砷是一種阻止鉑起催化作用的毒物。對催化劑有毒性的物質，也同樣對生物體有毒性；因這種物質會阻止生物體內的酶催化（↑）反應。

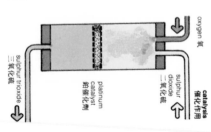

oxygen 氧
sulphur dioxide 二氧化硫
sulphur trioxide 三氧化硫
platinum catalyst 鉑催化劑
**catalysis** 催化作用

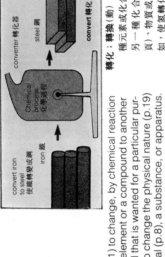

convert iron
to steel
使鐵轉變成鋼
iron 鐵

converter 轉化器

Chemical process
化學過程

steel 鋼

**convert** 轉化

**convert** (*v*) (1) to change, by chemical reaction (p.62), an element or a compound to another compound that is wanted for a particular purpose. (2) to change the physical nature (p.19) of a material (p.8), a substance, or apparatus. For example to convert nitrogen to ammonia; to convert iron to steel; to convert water to steam. *See transform (p.144).* **conversion** (*n*), **converter** (*n*).

轉化；轉換（動）　（1）藉化學反應（第 62 頁）使一種元素或化合物改變成某一特定用途所需之另一種化合物；（2）改變一種材料（第 8 頁）、物質或儀器之物理本性（第 19 頁）。例如，使氮轉化成氨；使鐵轉化成鋼；使水轉化成水蒸汽。參見"轉變"（第 144 頁）。（名詞為 conversion、converter）

**transmute** (*v*) to change one element (p.8) into another, using a radioactive (p.138) change or using bombardment (p 143) of subatomic particles (p.110). **transmutation** (*n*).

嬗變；蛻變（動）　利用放射性（第 138 頁）變化或採取次原子微粒（第 110 頁）轟擊法（第 143 頁）使一種元素（第 8 頁）轉變成另一種元素。（名詞為 transmutation）

**decolorize** (*v*) to take away colour so that a coloured substance becomes colourless (p.15), e.g. iodine solution is brown, but when iodine is converted to iodide ions, by adding potassium iodide, the iodine solution is decolorized and becomes colourless. **decolorization** (*n*).

脫色（動）　除去顏色，使一種有色的物質變成無色的（第 15 頁）物質。例如碘溶液是棕色的液體，但在加入碘化鉀時碘轉化成碘化物離子，碘溶液被脫色變成無色的液體。（名詞為 decolorization）

**bleach** (*v*) to take colour from a coloured substance, usually a pigment (p.162), leaving the substance white, e.g. to bleach cotton with chlorine, the natural yellow colour of cotton is changed to white; to bleach paper using sulphur dioxide. Chlorine and sulphur dioxide are the common bleaches. **bleach** (*n*), **bleaching** (*n*).

漂白（動）　由有色的物質中脫除有色物質，一般是顏料（第 162 頁）的顏色，使物質變白色。例如用用氯氣漂白棉花，使棉花之天然黃色變成白色；用二氧化硫漂白紙張。氯氣和二氧化硫是常用的漂白劑。（名詞為 bleach、bleaching）

neutron
中子

magnesium
atom 鎂原子

sodium
atom. 鈉原子

proton 質子

**transmute** 嬗變

transmutation of
magnesium by neutrons
用中子使鎂嬗變

white cloth
白布

coloured
cloth 有色布

bleach
(contains bleaching agent)
（含漂白劑）

**bleaching** 漂白

**detonate** (v) to initiate (↓) an explosion (p.58), using an electric spark, or similar agent, called a **detonator** (n).

**initiate** (v) to cause a process (p.157) to start when the agent of the cause takes no further part in the process, e.g. chlorine and methane do not react under room conditions, but heating the mixture decomposes (p.65) chlorine molecules (p.77) to chlorine atoms; these atoms react with methane and afterwards the reaction continues without heating, i.e. heat initiates the reaction. **initiation** (n).

**decrepitate** (v) to burst with small explosive (p.58) sounds, when crystals (p.91) are heated, e.g. lead (II) nitrate crystals decrepitate when heated and many small explosions can be heard. **decrepitation** (n).

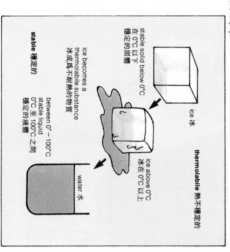

ice 冰

stable solid below 0°C
在 0°C 以下
穩定的固體

ice becomes a
thermoliable substance
冰成為不耐熱的物質

**stable** 穩定的

**thermoliable** 熱不穩定的

ice above 0°C
冰在 0°C 以上

water 水

between 0°– 100°C
stable liquid
0°C 至 100°C 之間
穩定的液體

**stable** (adj) describes a substance (p.8) which is not readily changed by heat, chemical reagents (p.63), light, or other forms of energy, e.g. calcium carbonate is stable at all temperatures except very high ones. Compare inert (p.19), inactive (p.19). **stability** (n), **stabilizer** (n), **stabilize** (v).

**stabile** (adj) another name for stable (↑).

引爆 (動) 利用電火花或雷管之類介質引發 (↓) 一次爆炸 (第 58 頁)。

引發 (動) 在起因介質不進一步參與過程下引致一個過程 (第 157 頁) 開始。例如氯和甲烷在室內條件下不起反應，但將其混合物加熱就可使氯分子 (第 77 頁) 分解 (第 65 頁) 成氯原子；這些氯原子又與甲烷反應，之後無需再加熱也可使反應繼續進行，亦即熱引發了反應。(名詞為 initiation)。

爆裂 (動) 當晶體 (第 91 頁) 被加熱時引起爆裂並伴有小爆炸 (第 58 頁) 聲。例如硝酸鉛 (II) 晶體故加熱時會發生爆裂並可聽到許多小爆炸聲。(名詞為 decrepitation)

穩定的 (形) 描述一種物質 (第 8 頁)，光或其他形式能量的作用下都是穩定的。例如，碳酸鈣在任何溫度下都是穩定的，除非溫度極高。可與 "不活潑的"（第 19 頁）"惰性的"（第 19 頁）作比較。(名詞為 stability、stabilizer，動詞為 stabilize)

穩定的 (形) 英文亦拼寫為 **stabile**。

**labile** (*adj*) describes a substance (p.8) that readily changes, either physically or chemically, i.e. it decomposes readily. Under normal conditions (p.103), a labile substance is stable (↑), but a change in conditions can easily produce a change in the substance.

**unstable** (*adj*) describes a substance which tends (p.216) to decompose without apparent cause, e.g. nitrogen triiodide is unstable when dry, an insect touching the dry powder can cause it to explode.

**thermostable** (*adj*) describes a substance which is stable (↑) when heated.

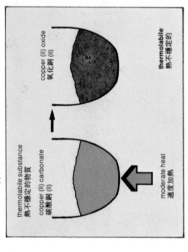

thermolabile substance
熱不穩定的物質

copper (II) carbonate
碳酸銅 (II)

moderate heat
適度加熱

copper (II) oxide
氧化銅 (II)

**thermolabile**
熱不穩定的

**thermolabile** (*adj*) describes a substance which loses its nature (p.19) or decomposes when heated.

**spontaneous** (*adj*) describes an event which takes place without any apparent (p.223) cause, e.g. an unstable (↑) compound can explode spontaneously. **spontaneity** (*n*).

**instantaneous** (*adj*) describes an event which lasts such a short time that the length of time cannot be measured, e.g. when a match is struck the appearance of the flame is instantaneous. **instant** (*n*).

**nascent** (*adj*) describes an element at the moment of its coming into being, e.g. when zinc acts on hydrochloric acid, the hydrogen evolved is nascent at the moment of formation.

易變的 (形) 描述一種物質 (第 8 頁) 在物理上或化學上都易變化，亦即它是易分解的。在正常條件下 (第 103 頁)，易變的物質是穩定的 (↑)。但在條件改變時，此物質就容易發生變化。

不穩定的 (形) 描述沒有明顯 (第 223 頁) 原因而傾向於 (第 216 頁) 分解的一種物質。例如，三碘化氮在乾燥時是不穩定的，昆蟲接觸此種乾粉末會引致發生爆炸。

熱穩定的 (形) 描述受熱時是穩定的 (↑) 一種物質。

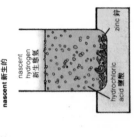

**nascent** 新生的

nascent hydrogen
新生態氫

hydrochloric acid 鹽酸

zinc 鋅

熱不穩定的 (形) 形容受熱時會失去其本性 (第 19 頁) 或分解的一種物質。

自發的 (形) 一種沒有任何明顯 (第 223 頁) 原因而發生的事件。例如一種不穩定的 (↑) 化合物自發爆炸。(名詞為 spontaneity)

瞬間的 (形) 描述事件只能在短暫時間內持續。其時間的長度不能測度，例如，擦一根火柴時火焰的出現是瞬時的。(名詞為 instant)

新生的 (形) 描述剛出現瞬間的一種元素。例如鋅作用於鹽酸時，在氫氣形成瞬間放出的氫是新生的氫。

**theory** (n) a description of the causes and effects of natural events and processes (p.157). The theory must rest on experimental observation (p.42). Certain parts of the theory may have to be assumed (p.222) as observation on these parts is not possible, e.g. an atom (p.110) cannot be observed, its properties are assumed. A theory can be used to describe, explain, classify (p.120) and predict (p.85) natural events or processes. **theoretical** (adj).

**atomic theory** the theory that all solids, liquids and gases are made up of atoms (p.110).

**Dalton's atomic theory** all elements (p.8) are made up of small particles (p.110) called atoms. An atom cannot be made, destroyed or divided. Atoms of the same element are alike. Compounds (p.8) are made by atoms combining chemically to form a molecule. A molecule has a small, whole number of atoms. All molecules of a compound are alike. Chemical change takes place when small, whole numbers of atoms combine or are separated.

**law of constant proportions** all pure specimens (p.43) of the same chemical compound (p.8) contain the same elements (p.8) combined in the same proportions (p.8) by mass.

**proportion** (n) if several ratios (p.79) are equal, then their figures are in proportion, e.g. 3/4 = 6/8, so 3 is to 4 as 6 is to 8, the figures are in proportion. **proportional** (adj).

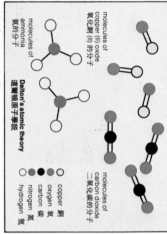

molecules of copper (II) oxide
氧化銅(II)的分子

molecules of ammonia
氨的分子

molecules of carbon dioxide
二氧化碳的分子

**Dalton's atomic theory**
道爾頓原子學說

| | |
|---|---|
| copper | 銅 |
| oxygen | 氧 |
| carbon | 碳 |
| nitrogen | 氮 |
| hydrogen | 氫 |

學說：理論（名） 對種種自然事件和過程（第 157 頁）起因和效應的敘述。學說必須以實驗觀察（第 42 頁）為依據。當學說的某些部分不可能觀察時，也許需對這些部分作出假設（第 222 頁）。例如，原子（第 110 頁）是不能觀察的，因而其性質是假定的。學說可用於闡述、解釋、分類（第 120 頁）和預示（第 85 頁）自然事件或過程。（形容詞為 theoretical）

原子論 此學說認為一切固體、液體和氣體都是由原子（第 110 頁）構成的。

道爾頓原子學說 此學說闡明一切元素（第 8 頁）都是由稱為原子的細小的微粒（第 110 頁）所構成的。原子是不能創生、不能消失也不能再分割。同一種元素的原子是相同的。所有化合物（第 8 頁）都是由原子化學結合產生的分子所構成。分子有小整數的原子。同一化合物的全部分子都相同。當小整數的原子結合或分離時就發生化學變化。

定比定律 同一種化合物（第 8 頁）的所有純樣品（第 43 頁）都含有相同的元素（第 8 頁），這些元素都以相同的質量比例（第 8 頁）結合。

比例（名） 如果有若干個比率（第 79 頁）是相等的，那麼其數字也是按比例的。例如 3/4 = 6/8，即 3 比 4 等於 6 比 8，其數字是按比例的。（形容詞為 proportional）

**combining weight** the weight of an element or radical (p.45) which will combine (p.64) with, or displace (p.68), 1 g of hydrogen or 8 g of oxygen. This name is not much used in modern chemistry.

**molecule** (*n*) the smallest particle (p.110) of an element or compound (p.8) that can exist (p.213) by itself. A molecule generally consists of a group of atoms (p.110) combined by covalent (p.136) bonds. Ionic (p.123) compounds do not form molecules. **molecular** (*adj*).

**symbol** (*n*) a letter or sign used to show a chemical element, a quantity, a mathematical operation, or a piece of apparatus, e.g. Na is used for sodium, Cl for chlorine, *V* for volume, *p* for pressure. See *diagram*.

**化合量** 指能化合（第 64 頁）或置換（第 68 頁）1 克氫或或 8 克氧的一個元素或根（第 45 頁）的重量。現代化學已很少用這個名稱。

**分子**（名） 指本身能存在（第 213 頁）的一種元素或化合物（第 8 頁）的最小微粒（第 110 頁）。分子通常都是由一組藉共價鍵（第 136 頁）結合的原子（第 110 頁）所構成。離子（第 123 頁）化合物不形成分子。（形容詞爲 molecular）

**符號**（名） 表示化學元素、數量、數學運算或儀器用的字母或記號。例如：Na 表示鈉，Cl 表示氯，*V* 表示體積，*p* 表示壓力。見下表：

| symbols in chemistry 化學符號 | | | |
|---|---|---|---|
| Na sodium 鈉 | H hydrogen 氫 | | |
| K potassium 鉀 | C carbon 碳 | | |
| Cu copper 銅 | O oxygen 氧 | | |
| Fe iron 鐵 | N nitrogen 氮 | | |
| Mn manganese 錳 | S sulphur 硫 | | |
| Ba barium 鋇 | Cl chlorine 氯 | | |
| Zn zinc 鋅 | I iodine 碘 | | |
| Mg magnesium 鎂 | P phosphorus 磷 | | |
| Ca calcium 鈣 | Ag silver 銀 | | |

| symbols for quantities 量的符號 | |
|---|---|
| *t* time 時間 | *p* pressure 壓力 |
| *m* mass 質量 | *T* temperature 溫度 |
| *l* length 長度 | *c* concentration 濃度 |
| *v* volume 體積 | *n* mole fraction 莫爾分數 |
| *p* density 密度 | *L* Avogadro constant 亞伏加德羅常數 |
| *R* molar gas constant 莫爾氣體常數 | *F* Faraday constant 法拉弟常數 |

**formula** (n) (formulae n.pl.) (1) a chemical formula shows the number of atoms (p.110) of each element in a molecule or the ions (p.123) in a compound. These are other kinds of formulae; see *structural formula* (p.181). (2) a physical formula shows the relation (p.232) between different quantities (p.81), e.g. $pV$ = constant; this is a formula showing the relation between the pressure and the volume of a gas.

**formula weight** this name has been replaced by *relative formula mass* (↓).

**relative formula mass** the mass of one mole (p.80) of molecules (p.77), or ions (p.123) of a compound (p.8). It is calculated from the formula (↑) of a compound and the *relative atomic masses* (p.113) of the elements in the compound.

**equation** (n) the formulae (↑) of compounds or elements are arranged in an equation to show the reactants (p.62) and products (p.62) of a chemical reaction. For example:

$$2KNO_3 \xrightarrow{\text{heat}} 2KNO_2 + O_2$$

This shows (1) two molecules of potassium nitrate decompose (p.65) on heating to form two molecules of potassium nitrite and one molecule of oxygen OR (2) two moles (p.80) of potassium nitrate, when heated, decompose to form two moles of potassium nitrite and one mole of oxygen. Ionic (p.123) equations show reactions between ions. For example:

$$Ba^{2+} + SO_4{}^{2-} \longrightarrow Ba^{2+} SO_4{}^{2-}$$

**oxidation number** a number showing the oxidation state (p.135) of a metal in a compound (p.8) with the number written in roman numerals, e.g. lead (II) nitrate, with lead having an oxidation number of II and an oxidation state of + 2; in potassium manganate (VI), maganese in the acid radical (p.45) has an oxidation number of VI and an oxidation state of + 6.

**balance²** (v) (1) to keep an object still, under equal and opposite forces. (2) to have equal numbers of atoms of each element on the opposite sides of an equation, e.g. $2KNO_3 = 2KNO_2 + O_2$ is balanced because there are two potassium atoms, two nitrogen atoms and six oxygen atoms on each side of the equation. **balanced** (adj).

式量 此名稱已為 "相對式量" (↓) 代替。

相對式量 (名) 將化合物或元素的化學式 (第 80 頁) 的化合物 (第 8 頁) 數目。可由化合物之化學式 (第 8 頁) 和化合物中各元素之相對原子質量 (第 113 頁) 計算得。

反應式：方程式 (名) 將化合物或元素的化學式排在一條方程式中表示化學反應之反應物 (第 62 頁) 和產物 (第 62 頁) 的公式 (↑)。例如：

$$2KNO_3 \xrightarrow{\text{加熱}} 2KNO_2 + O_2$$

這條反應式表示：(1) 兩個硝酸鉀分子在加熱時分解 (第 65 頁) 成兩個亞硝酸鉀和一個氧分子。或 (2) 兩摩爾 (第 80 頁) 硝酸鉀在加熱時分解成 2 摩爾亞硝酸鉀和 1 摩爾氧。離子 (第 123 頁) 反應式表示離子之間的反應。例如：

$$Ba^{2+} + SO_4{}^{2-} \longrightarrow Ba^{2+} SO_4{}^{2-}$$

氧化值 (第 135 頁) 的數值，該數值以羅馬數字寫出。例如鉛 (II)，表示鉛的氧化值為 II、氧化態為 + 2；錳 (VI) 酸鉀，表示錳在鉻酸根 (第 45 頁) 中的氧化值為 VI 及氧化態為 + 6。

平衡² (動) (1) 使一個物體在力的大小相等、方向相反下保持靜止。(2) 使方程式兩邊各元素的原子數目相等。例如，$2KNO_3 = 2KNO_2 + O_2$ 是平衡的，因兩邊等式兩邊各有兩個鉀原子、兩個氮原子和 6 個氧原子。(形容詞為 balanced)

化學式：公式 (名) (1) 化學式表示化合物分子或離子 (第 123 頁) 中各種元素的原子 (第 110 頁) 數目；化學式的其他型式見結構式 (第 181 頁)；(2) 物理式表示不同量 (第 81 頁) 之間的關係；例如 $pV$ = 常數；這公式表示一定質量的氣體的壓力與體積之間的關係。(複數為 formulae)

---

**chemical formulae**
化學式

**CuO**
copper oxide 氧化銅

**CO₂**
carbon dioxide 二氧化碳

**NH₃**
ammonia 氨

**BaSO₄**
barium sulphate 硫酸鋇

**physical formulae**
物理式

$$pV = nRT$$

$$p = \frac{m}{V}$$

**calculate** (v) to get a result (p.39) by using arithmetical processes on numbers or the values of quantities (p.81), e.g. to calculate density (p.12) from measurements on the mass and volume of a specimen (p.43). **calculation** (n).

**error** (n) a mistake in a calculation (↑) or in a statement (p.222). The error can be caused by incorrect calculation, readings (p.39), inferences etc.

**exact** (adj) describes a calculation (↑) or statement (p.222) without errors (↑).

**approximate** (adj) describes a measurement which is sufficiently correct to be used in a calculation, e.g. the relative atomic mass of oxygen is 15.99, an approximate value is 16, good enough for most calculations. **approximation** (n).

**ratio** (n) the ratio of two numbers, or of two quantities (p.81), is obtained by dividing one by the other and simplifying the fraction, e.g. the ratio of 28 to 40 is 28/40 = 7/10, i.e. the ratio is 7:10.

**multiple** (adj) describes (1) an object made up of two or more like parts. (2) a number which consists of a smaller number multiplied a number of times, e.g. 55 and 44 are multiples of 11.

**fixed** (adj) describes a quantity (p.81) that has been made unchanging instead of allowing it to change or to be changed, e.g. a metal cylinder contains a fixed volume of gas; the volume has been made unchanging because the volume of the cylinder does not change. **fix** (v).

**arbitrary** (adj) describes something that has been decided because it is the most suitable or the easiest thing to do and has no relation to any theory (p.76), e.g. the choice of the metre as the standard of length was an arbitrary choice.

**mean** (adj) describes a value of a quantity (p.81) or a number which is equally far from a high and a low value. If there are two numbers x and y, then (x + y)/2 is the arithmetic mean, and $\sqrt{x \times y}$ is the geometric mean, e.g. $6\frac{1}{2}$ is the arithmetic mean of 9 and 4; 6 is the geometric mean of 9 and 4. **mean** (n).

**average** (adj) describes a value (or a number) obtained by adding together all the values and dividing by the number of values, e.g. the average of 14, 16 and 18 is (14 + 16 + 18) ÷ 3 = 16.

---

calculation of
density 密度的計算

mass of metal
金屬之質量
= 296.25 g

volume of metal
金屬之體積
= 37.5 cm³

density of metal
金屬之密度
= 296.25 g
─────────
= 37.5 cm³
= 7.9 gcm⁻³

**calculation 計算結果**

**calculation 計算法**

```
       26.7
   ×      6
     ──────
     160.2
      31.3    +  191.5
              −   46.7
                 ─────
                 144.8
```

calculation with numbers
用數字計算

---

計算（動）以算術方法計算得出有關數字或數量（第 81 頁）值的結果（第 39 頁）。例如從樣品（第 43 頁）的質量和體積的測量值計算得出密度（第 12 頁）值。（名詞為 calculation）。

誤差（名）指計算結果（↑）或陳述（第 222 頁）中的錯誤。不正確的計算結果、讀數（第 39 頁）、推斷等都可產生差錯。

正確的（形）描述計算（↑）或陳述（第 222 頁）沒有誤差（↑）。

近似的（形）描述測量足夠準確、可供計算之用。例如氧的相對原子質量為 15.99，近似值為 16，對大多數計算結果已足夠準確了。（名詞為 aproximation）。

比率（名）由一個數除以另一個數並將分數化簡得出。例如 28 對 40 的比為 28/40 = 7/10，即比為 7:10。

多重的；成倍的（形）（1）描述一個物體由兩個或多個相同部分構成；（2）由一個較小的數乘以一個整數的數的數，例如 55 和 44 都是 11 的倍數。

固定的（形）描述一個數量（第 81 頁）已使之不可變以代其可變或被改變。例如金屬瓶內裝著固定體積的氣體；因氣瓶的體積不改變，所以氣體的體積也不改變。（動詞為 fix）

任意的（形）描述某些事情被決定為最適合做或最容易做，且與任何理論之（第 76 頁）都無聯繫。例如選擇 "米" 作為長度標準是一種任意的選擇。

均數的（形）描述一個數值及低數值等距。假定有 x 和 y 兩個數字，則 (x + y)/2 的算術平均為均數何平均的數。例如：$6\frac{1}{2}$ 為 9 和 4 的算術平均為均數，6、9 和 4 的幾何平均數。（名詞為 mean）

平均的（形）形容一個數值（或一個數）是由全部數值相加後除以數值之數目得出，例如 14、16 和 18 的平均值為 (14 + 16 + 18) ÷ 3 = 16。

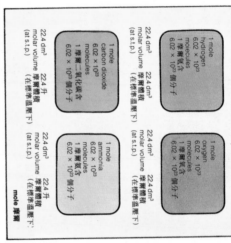

**mole** (*n*) the standard (p.229) for measurement of an amount (↓) of substance. One mole of a substance is that amount of substance which contains the same number of elementary particles (p.13) as there are atoms in 0.012kg (12 grams) of carbon-12. (The isotope (p.114) of carbon with a mass number (p.113) of 12). The particles can be ions (p.123), atoms (p.110), molecules (p.77), electrons (p.110) or any other named particle. If the particles are not named, they are assumed (p.222) to be atoms in elements, molecules in covalent (p.136) compounds or ions in an electrovalent (p.134) compound. **molar** (*adj*).

**mole fraction** the amount of a substance expressed as a fraction of a mole (↑), e.g. 0.2 mole is a mole fraction.

**molar volume** the volume of one mole (↑) of a substance in a named state of matter (p.9). The molar volume of any gas at s.t.p. (p.102) is always 22.4 dm³. The symbol $V_m$ is used for molar volume.

**Avogadro constant** the number of particles in 1 mole (↑). The approximate value is $6.02 \times 10^{23}$ mole. The symbol for the Avogadro constant is $L$.

摩爾（名）以 1 摩爾物質總量（↓）指所含基本粒子（第 229 頁）。1 摩爾物質總量是指所含基本粒子（第 13 頁）數目和 0.012 千克（12 克）碳 -12 同位素（第 114 頁）（其質量數（第 113 頁）為 12）的原子數目相同的物質總量。此基本粒子可以是離子（第 123 頁）、原子（第 110 頁）、分子（第 77 頁）、電子（第 110 頁）或任何其他已知的粒子。若為未知之粒子，則將之假定（第 222 頁）為元素的原子、共價（第 136 頁）化合物的分子或電價（第 134 頁）化合物的離子。（形容詞為 molar）

摩爾分數 以 1 摩爾（↑）的分數表示的物質的總量。例如，0.2 摩爾是摩爾分數。

摩爾體積 以 1 摩爾（↑）物質在一已知物態（第 9 頁）中佔有的體積。在標準狀態（第 102 頁）下任何氣體的摩爾體積總是 22.4 升。摩爾體積的符號為 $V_m$。

亞伏加德羅常數 1 摩爾（↑）所含的粒子數目，近似值是 $6.02 \times 10^{23}$。符號是 $L$。

**amount** (*n*) the amount of a substance is proportional (p.76) to the number of elementary particles (p.13) it contains. The amount of substance is a physical quantity (↓).

總量（名）一種物質的總量與其所含有的基本粒子（第 13 頁）數目成正比例（第 76 頁）。物質的總量是一個物理量（↓）。

**quantity** (*n*) (1) any measurement of materials, substances (p.8) or energy (p.135) is a quantity. Examples of quantities are mass, length, time, temperature, amount of matter, atomic number, mass number, wavelength, concentration, heat, density. *Compare quality (p.15)*. (2) a measurement of materials or substances that does not give a value, e.g. a quantity of lime; this could be measured in kilograms or moles (↑). **quantitative** (*adj*).

量：數量（名）（1）材料、物質（第 8 頁）或能量（第 135 頁）的量度就是它的量。質量、長度、時間、溫度、物質的總量、原子序、質量數、波長、濃度、熱量、密度都是量的例子。與「品質、質」（第 15 頁）作比較；（2）不指明其值的材料或物質的量可以用千克或摩爾（↑）未量的總量；而此量可以用千克或摩爾（↑）未量度。（形容詞爲 quantitative）

**concentration** (*n*) the amount (↑) of solute (p.86) dissolved in a solution (p.86). Concentration can be stated as: (a) grams of solute in 1 dm³ solution; (b) moles of solute in 1 dm³ solution; (c) as a percentage. For example, 80g of sodium hydroxide dissolved in 1 dm³ solution has a concentration of 80g dm⁻³, 2mol dm⁻³ or 8% **concentrate** (*v*), **concentrated** (*adj*).

濃度（名）指溶解於溶液（第 86 頁）中的溶質（第 86 頁）的量（↑）。濃度有以下三種表達方式：（a）以 1 升溶液中溶質的克數表示；（b）以 1 升溶液中溶質的摩爾數表示；（c）以百分數表示。例如，1 升溶液中溶有 80 克氫氧化鈉，其濃度爲 80 克／升、2 摩爾／升或 8%。（動詞爲 concentrate，形容詞式爲 concentrated）

**dilution** (*n*) (1) the process (p.157) of adding more solvent (p.86) to a solution (p.86); this lowers the concentration (↑) of the solute (p.86), e.g. the dilution of a concentrated solution of sodium hydroxide to make a dilute solution. (2) the volume in dm³ of a solution containing one mole of solute, e.g. a solution of sodium hydroxide containing 80 g dm⁻³ has a dilution of 0.5. *See diluent (p.56)*. **dilute** (*v*), **dilute** (*adj*).

稀釋；稀釋度（名）（1）在溶液（第 86 頁）中加入更多溶劑（第 86 頁）的過程（第 157 頁）；稀釋降低了溶質（第 86 頁）的濃度（↑）。例如，將濃的氫氧化鈉溶液稀釋製得稀溶液；（2）含 1 摩爾溶質的溶液的體積（以升爲單位），即每 80 克／升氫氧化鈉的溶液，其稀釋度爲 0.5。參見「稀釋劑」（第 56 頁）。（動詞爲 dilute，形容詞爲 dilute）

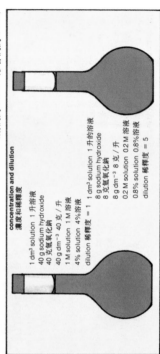

**concentration and dilution**
濃度和稀釋度

1 dm³ solution 1 升溶液
40 g sodium hydroxide
40 克氫氧化鈉
40 g dm⁻³ 40 克／升
1 M solution 1 M 溶液
4% solution 4% 溶液
dilution 稀釋度 = 1

1 dm³ solution 1 升溶液
8 g sodium hydroxide
8 克氫氧化鈉
8 g dm⁻³ 8 克／升
0.2 M solution 0.2 M 溶液
0.8% solution 0.8% 溶液
dilution 稀釋度 = 5

**analysis** (*n*) a process (p.157) to find the constitution (↓) of a compound (p.8), the composition (↓) of a mixture, the concentration (p.81) of a solution, or the identity (p.225) of a substance, e.g. the analysis of the constitution of ethanoic acid; the analysis of a mixture in order to name the substances present. See assay (p.155). **analyse** (*v*), **analytical** (*adj*).

**volumetric analysis** a way of analysis which uses solutions (p.86) of known concentrations (p.81) in titration (p.39). The reactions of the solutions may be acid-alkali, redox (p.70), or other kinds.

**gravimetric analysis** a way of analysis which uses reactions (p.62) forming precipitates (p.30). The precipitates are dried and weighed. Other reactions, such as the reduction (p.70) of oxides to metals may be used. In all cases, a solid is prepared (p.43) and weighed.

**composition** (*n*) the chemical composition of a substance gives the proportion of the elements united in the substance. The proportions can be given as a percentage composition of elements by mass, or as the proportion of atoms of each element in a covalent (p.136) molecule (p.77). The composition of a mixture gives the proportion of substances in it. **composed of** (*v*).

**stoichiometric** (*adj*) describes compounds which obey the law of constant proportions (p.76), or processes of analysis (↑) which measure mass or volume and assume (p.222) the law. **stoichiometry** (*n*).

**structure**[1] (*n*) the arrangement of connected parts of a whole with the parts depending on each other to form the structure, e.g. the structure of an atom (p.110) shows the arrangement of electrons, protons and neutrons; the structure of a crystal shows the arrangement of ions (p.123) and their dependence on each other. **structural** (*adj*).

**constitution** (*n*) the constitution of a compound (p.8) gives the number and position (p.211) in relation (p.232) to each of the other atoms forming a molecule (p.77), a radical (p.45) or an ion (p.123). The relative positions of the atoms can be shown by structural or graphic formulae (p.181). **constitute** (*v*), **constituent** (*n*).

分析（名）指找出化合物（第 8 頁）的構造（↓）、混合物的組成（↓）、溶液的濃度（第 81 頁）或物質的本質（第 225 頁）的一種方法（第 157 頁）。例如，分析乙酸的構造；分析一種混合物以說出混合物中存在的物質。參見“化驗”（第 155 頁）。（動詞為 analyse，形容詞為 analytical）。

容量分析 用已知濃度（第 81 頁）的各種溶液（第 86 頁）進行滴定（第 39 頁）的分析方法。溶液中的反應可為酸鹼反應、氧化還原反應（第 70 頁）或其他類型的反應。

重量分析 利用形成沉澱物（第 30 頁）的種種反應（第 62 頁）將沉澱物烘乾和稱重，進行分析的方法。也可以利用其他類型的反應，例如利用氧化物還原（第 70 頁）成金屬的反應，在所有情況下都要製備（第 43 頁）固體物並進行稱重。

組成：成分（名）一個物質的化學組成指出該物質中所結合的各元素之比例。此比例可以用各元素的質量百分組成表達，或者用共價（第 136 頁）分子（第 77 頁）中各元素的原子比例表達。混合物的組成表示出混合物中各個物質之比例。（動詞 composed of，意為“由……組成”。）

化學計量的（形）描述服從定比定律（第 76 頁）的化合物，或描述用於測量質量或體積以及用於假定（第 222 頁）定律的分析（↑）方法。（名詞為 stoichiometry）

結構（名）整體之連接部分與彼此相關部分的排列或結構。例如，原子（第 110 頁）的結構顯示電子、質子和中子的排列；晶體的結構顯示離子（第 123 頁）的排列以及其彼此間的依賴關係。（形容詞為 structural）

構造（名）化合物（第 8 頁）之構造表示出該化合物中形成一個離子—個分子（第 77 頁）的數目和位置（第 211 頁）的關係（第 232 頁）或一個其他原子—個根（第 45 頁）。原子的相對位置可由結構式和圖式（第 181 頁）表示。（動詞為 constitute，名詞為 constituent）

a structure of connected
and dependent parts
making a whole
一種連結構和構成
一個整體的相關部分

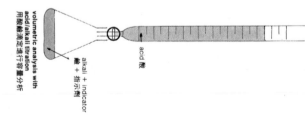

volumetric analysis with
acid/alkali titration
用酸鹼滴定法進行容量分析

acid 酸

alkali + indicator
鹼 + 指示劑

**molecular structure** a molecule consists of atoms combined by covalent bonds (p.136). A covalent bond has a direction in space, so each molecule has bonds directed in space which give the molecule a structure. The important molecular structures have the shape of a straight line, a tetrahedron (↓), an octahedron (↓) and a pyramid (p.84). Cations and anions (p.125) also have a molecular structure.

**分子結構** 分子係由一些原子藉共價鍵（第136頁）結合而成的。共價鍵在空間有方向，因而每一分子的鍵在空間都有方向，使分子具有一種結構。分子結構主要有直線型、四面體（↓）型、八面體（↓）型和棱錐體（第84頁）型。陽離子和陰離子（第125頁）也具有一種分子的結構。

**linear** (adj) in the shape of a straight line. A linear molecule has three atoms in a straight line, e.g. carbon dioxide, nitrogen monoxide. The central atom has no lone pairs (p.133) of electrons.

**直線型的(形)** 成直線形狀的。直線型分子具有排成一直線的三個原子。例如，二氧化碳和一氧化氮的原子，其中心原子沒有共用電子對的（第133頁）電子。

H₂O

non-linear 非直線型

**non-linear** (adj) in the shape of a bent line, see diagram. A non-linear molecule has three atoms joined by two bonds which are at an angle to each other, e.g. water has two hydrogen atoms combined with one oxygen atom and the angle between the covalent bonds is 104.5°. The central, oxygen atom has two lone pairs (p.133) of electrons.

**非直線型的(形)** 成曲折線形狀（見圖）的。非直線型分子具有由兩個鍵相連的三個原子，而兩鍵間彼此互成一角度。例如水的分子係由兩個氫原子和一個氧原子結合而成，其共價鍵的角度爲104.5°，中心氧原子有兩個共用的電子對（第133頁）電子。

**trigonal planar** (adj) describes a molecule with one central atom and three other atoms joined to it. All four atoms are in the same plane, e.g in the carbonate ion ($CO_3^{2-}$), see p.49, the angle between the bonds is.120° The central, carbon atom has no lone pairs (p.133) of electrons.

**平面三角形的(形)** 描述具一個中心原子和其他三個原子與之相連的分子，全部四個原子同在一個平面上。例如碳酸根離子（$CO_3^-$）（見第49頁），其鍵間角度爲120°，中心的碳原子沒有共用電子對（第133頁）電子。

**tetrahedral** (adj) in the shape of a tetrahedron, see diagram. A central atom is joined to four other atoms, and all the covalent bonds (p.136) have equal angles of 109.5° between each pair of bonds, e.g. methane ($CH_4$) has a tetrahedral structure. The central, carbon atom has no lone pairs (p.133) of electrons. **tetrahedron** (n).

**四面體的(形)** 成四面體形狀（見圖）的。一個中心原子與其他四個原子相連，所有共價鍵（第136頁）均具有109.5°的相同角度。例如甲烷（$CH_4$）具有四面體結構，其中心碳原子沒有共用電子對（第133頁）電子。（名詞爲 tetrahedron）

**octahedral** (adj) in the shape of an octahedron, see diagram. A central atom is joined to six other atoms, e.g. the hexacyanoferrate ions (p.53) have an octahedral structure. The central atom has no lone pairs (p.133) of electrons in an octahedral structure.

**八面體的(形)** 成八面體形狀（見圖）的。一個中心原子與其他6個原子相連。例如六氰合鐵酸根離子（第53頁）爲八面體結構，結構中的中心原子沒有共用電子對（第133頁）電子。

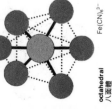

CO₂

linear 直線型

CO₃²⁻

trigonal planar
平面三角形

CH₄

tetrahedral
四面體

Fe(CN)₆³⁻

octahedral
八面體

**pyramidal** (*adj*) in the shape of a pyramid. A pyramid in molecular structure can have a triangular base or a square base. **pyramid** (*n*).

**trigonal pyramidal** in the shape of a pyramid with a triangular base. A central atom is joined to three other atoms, e.g. ammonia ($NH_3$) has a trigonal pyramidal structure. The central, nitrogen atom has one lone pair (p.133) of electrons, which completes a tetrahedral structure.

**square pyramidal** in the shape of a pyramid with a square base. A central atom is joined to five other atoms. This structure is not common. The central atom has one lone pair (p.133) of electrons.

**square planar** in the shape of a square. A central atom is joined to four other atoms and all five atoms are in the same plane, e.g. the tetrachloroplatinate (VI) ion ($PtCl_4^{2-}$). The central, platinum atom has two lone pairs (p.133) of electrons, which complete an octahedral (p.83) structure.

**trigonal bipyramidal** in the shape of two triangular pyramids joined at their bases. A central atom is joined to five other atoms, e.g. phosphorus pentachloride ($PCl_5$) has a trigonal bipyramidal structure. The central, phosphorus atom has no lone pairs (p.133) of electrons.

稜錐形的 (形) 成稜錐體形狀的。稜錐體分子結構中可有一個三角形的底或正方形的底。(名詞為 pyramid)。

三角錐形的 包含一個三角形底的稜錐體形狀的。其一中心原子與其他三個原子相連。例如氨 ($NH_3$) 為三角錐形結構。中心氮原子有一對不共用電子 (第133頁) 電子，構成四面體結構。

正方錐形的 包含一個正方形底的稜錐體形狀的。其一個中心原子與其他五個原子相連。這種結構不常見。中心原子有一對不共用電子 (第133頁) 電子。

正方平面的 成正方形形狀的。其一個中心原子與其他五個原子相連，而五個原子同在一平面上。如四氯鉑 (VI) 酸根離子 ($PtCl_4^{2-}$) 中心鉑原子有兩對不共用電子 (第133頁) 電子，構成八面體 (第83頁) 結構。

三角雙錐的 以兩個三角錐體的底連接成的形狀。其一個中心原子與其他五個原子相連。例如五氯化磷 ($PCl_5$) 為三角雙錐形的結構。中心磷原子沒有不共用電子 (第133頁) 電子。

| SHAPE 形狀 | NUMBER OF BONDS 鍵數目 | NUMBER OF LONE PAIRS 不共用電子對數目 | EXAMPLE 示例 |
| --- | --- | --- | --- |
| linear 直線型的 | 2 | 0 | $CO_2$ |
| trigonal planar 平面三角形的 | 3 | 0 | $NO_3$ |
| tetrahedral 四面體的 | 4 | 0 | $CH_4$ |
| trigonal pyramidal 三角錐形的 | 3 | 1 | $NH_3$ |
| non-linear 非直線形的 | 2 | 2 | $H_2O$ |
| trigonal bipyramidal 三角雙錐形的 | 5 | 0 | $PCl_5$ |
| octahedral 八面體的 | 6 | 0 | $FeCN_6^{3-}$ |
| square pyramidal 正方稜形的 | 5 | 1 | — |
| square planar 正方平面的 | 4 | 2 | $PtCl_4^{2-}$ |

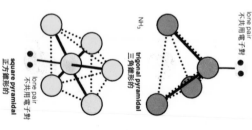

PCl₅
**trigonal bipyramidal**
三角雙錐形的

$PtCl_4^{2-}$
lone pair
不共用電子對
**square planar**
正方平面的
lone pair
孤對

$NH_3$
lone pair
不共用電子對
**square pyramidal**
正方稜形的
lone pair
不共用電子對
**trigonal pyramidal**
三角錐形的

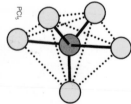

含量：容量（名）（1）指某種物質在混合物中佔的份量，尤指一種礦石中某種金屬佔的份量。例如礦石（第 154 頁）中銀的含量；（2）指容器能裝載的總量。通常作為該容器的容量。

可利用的：可得到的（形）描述在需要時可獲得的任何東西。例如，可從發生器（第 27 頁）得到的一種氣體；植物能利用空氣中的二氧化碳；酸中含的氫是可自一種化學反應中得到的。任何可利用的東西未必都以未化合態存在（第 213 頁）。

初始的（形）指一個過程的最初部分。指時間上的第一項事件，例如滴定管的第一次讀數，就是初始讀數，在滴液流入燒瓶之前由滴定管（第 213 頁）讀出；某種物質經受（第 213 頁）化學變化之前其初始質量；在一次實驗中第一次記錄的溫度是初始溫度。

中間的（形）描述任何部分或事件介於初始（↑）和最後（↓）部分或事件之間。

最後的（形）指一個過程的最末部分。指時間上的最後事件。例如，完成滴定之後在滴定管讀出的最後讀數；物質經受（第 213 頁）化學變化之後最後留下的質量；實驗結束時最後記錄下的溫度。

定性的（形）描述不作實際測量而其全（第 81 頁）只作比較（第 224 頁）的一種觀察（第 42 頁），例如溫、熱、紅熱，赤熱是對溫度等作的定性觀察（與溫度有關）。氫氣迅速放出（第 40 頁）是一種定性的觀察（與緩慢變放出作比較）。

定量的（形）描述包括量在內的一種觀察。例如一次量測量是一次化學反應中放出的熱量，以焦耳為量度單位（名詞為 quantity）。

預示：預測（動）說出將來會發生的事件並信它是正確的。例如利用波義耳定律（第 105 頁）預示壓力變化後的氣體體積。預示是利用化學定律作出的（名詞為 prediction，形容詞為 predictable）。

**content** (*n*) (1) the amount of a substance in a mixture, particularly the amount of a metal in a mineral, e.g. the silver content of an ore (p.154). (2) the amount a vessel contains. Usually given as the contents of the vessel.

**available** (*adj*) describes anything that can be obtained if it is needed, e.g. a gas is available from a generator (p.27); carbon dioxide in the air is available to plants; hydrogen in an acid is available from a chemical reaction. Anything available does not necessarily exist (p.213)

**initial** (*adj*) the first part of a process, the first event in time, e.g. the first reading on a burette is the initial reading, taken before the contents (↑) are run into a flask; the mass of a substance before it undergoes (p.213) a chemical change is its initial mass; the first temperature recorded in an experiment is the initial temperature.

**intermediate** (*adj*) describes any part, or event, between the initial (↑) and final (↓) part or event.

**final** (*adj*) the last part of a process, the last event in time, e.g. the final reading on a burette after titration is finished; the final mass left after a substance has undergone (p.213) chemical change; the final temperature recorded at the end of an experiment.

**qualitative** (*adj*) describes an observation (p.42) in which no actual measurement is made, but quantities (p.81) are compared (p.224), e.g. warm, hot, red-hot, are qualitative observations on temperature; a rapid evolution (p.40) of hydrogen is a qualitative observation (by comparison with a slow evolution). **quality** (*n*).

**quantitative** (*adj*) describes an observation with a measurement, e.g. a quantitative measurement of the heat given out in a chemical reaction, measured in joules. **quantity** (*n*).

**predict** (*v*) to say what is going to happen in the future and to be sure it is correct, e.g. to predict the volume of a gas after a change in pressure, using Boyle's law (p.105). Predictions are made by using chemical laws. **prediction** (*n*). **predictable** (*adj*).

start 開始
initial reading 初始讀數
**initial** 初始的
final reading 最後讀數
end 終點
**final** 最後的
contents of flask 燒瓶的內容物

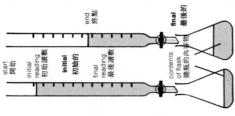

**initial** 初始的
initial pressure 初始壓力
intermediate pressure 中間壓力
**intermediate** 中間的

intermediate pressure 中間壓力
**intermediate** 中間的
**final** 最後的
final pressure 最後的壓力

**solute** (n) a solid or a gas which dissolves (p.30) in a liquid.

**solvent** (n) the liquid in which a solute (↑) is dissolved (p.30).

**solution** (n) the result of dissolving a solute (↑) in a solvent (↑). If it is a homogenous (p.54) mixture of two or more substances it is called a *true solution*. Liquids other than water can be used as solvents (↑) to make a solution, but water is the usual solvent. For example, iodine can be dissolved in trichloromethane (chloroform). If a solvent other than water is used, it must be stated (p.222). A solution can also be made by dissolving one liquid in another; the liquid in greater quantity is the solvent, e.g. a solution of ethanol (ethyl alcohol) in water.

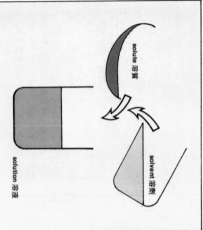

solute 溶質

solvent 溶劑

solution 溶液

**suspension** (n) finely divided (p.13) particles of an insoluble (p.17) substance suspended (p.31) in a liquid forming an homogenous (p.54) mixture. For example, clay shaken up with water forms a suspension; lime added to water forms a solution (↑) and the excess solid forms a suspension. When a suspension is filtered, the solid substance is collected as a residue (p.31).

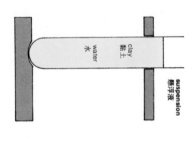

Clay 黏土

water 水

suspension 懸浮液

溶質 (名) 指溶解 (第 30 頁) 於一種液體中的固體或氣體。

溶劑 (名) 指溶質 (↑) 溶解 (第 30 頁) 於其中的液體。

溶液 (名) 溶質 (↑) 溶解於一種溶劑 (↑) 中之結果。若此溶液為兩種或多種物質之均勻之 (第 54 頁) 混合物，則稱之為 "真溶液"。除水外還可用其他液體作溶劑，例如，碘可溶於三氯甲烷 (氯仿) 之中；如用水之外的其他溶劑，則應說明 (第 222 頁)。溶液也可由一種液體溶解於另一種液體而製得，用量較多的液體為溶劑，例如酒精 (乙醇) 在水中所成的溶液。

懸浮液 (名) 指一種不溶解的 (第 17 頁) 物質之細碎 (第 13 頁) 粒子，懸浮 (第 31 頁) 於一種液體中形成均勻的 (第 54 頁) 混合物。例如，黏土用水搖混形成懸浮液；石灰加入水形成溶液 (↑) 而過量的固體形成懸浮液。懸浮液過濾時，固體物質就成為濾渣 (第 31 頁) 而被收集。

**solubility** (*n*) (1) the property of being soluble (p.17); or a qualitative (p.85) observation on how soluble a substance is. (2) the mass in grams of a solid which will dissolve in 100 grams of solvent (↑) at a stated (p.222) temperature in the presence of excess solute (↑). (3) the volume of gas, measured in cm³, which will saturate (↓) 100 g of solvent at a stated temperature. Water is considered to be the usual **solvent**, any other solvent must be named. **soluble** (*adj*).

溶解性：溶解度（名） (1) 指可溶解的（第 17 頁）的性質；或有關一種物質可溶解了多少的定性（第 85 頁）觀察；(2) 在指定（第 222 頁）溫度下在過量溶質 (↑) 存在時，每 100 克溶劑 (↑) 中可溶解的固體質量克數；(3) 在指定溫度下能使 100 克溶劑飽和 (↓) 的氣體之體積，以 cm³ 爲量度單位。水是常用之溶劑，若爲其他溶劑則需指明。(形 容詞爲 soluble)

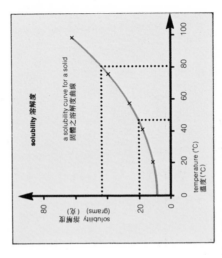

solubility 溶解度

a solubility curve for a solid
固體之溶解度曲線

solubility 溶解度 (grams) (克)

temperature (°C)
溫度 (°C)

saturated 飽和的

saturated solution
飽和的溶液

excess solute
(does not dissolve)
過量的溶質 (不溶解)

**saturated**[1] (*adj*) (1) describes a solution (↑) which will not dissolve any more of a solute (↑). (2) describes air containing water vapour when the air will not hold any more water vapour at a particular temperature. **saturation** (*n*). **saturate** (*v*).

飽和的（形） (1) 描述已再不能溶解更多溶質 (↑) 的一種溶液 (↑)；(2) 描述含水蒸汽的空氣在一特定溫度下，已不能再含有更多水蒸汽。(名詞爲 saturation，動詞爲 saturate)

**unsaturated**[1] (*adj*) describes any solution (↑), or air, which is not saturated (↑) as more solute (↑) can be dissolved (p.30).

未飽和的（形） 描述任何溶液 (↑) 或空氣沒有飽和 (↑) 因爲尚有更多溶質 (↑) 可以被溶解（第 30 頁）。

**supersaturated** (*adj*) an unstable (p.75) state of a solution (↑) which contains more solute (↑) at a given temperature than it should contain. This can happen when a warm, saturated (↑) solution is slowly cooled. **supersaturate** (*v*). **supersaturation** (*n*).

過飽和的（形） 指一種溶液 (↑) 的不穩定（第 75 頁）狀態，在指定溫度下，此溶液中所含之溶質 (↑) 比它應含者爲多。將溫暖的飽和溶液緩慢冷卻即會出現過飽和溶液。(動詞爲 supersaturate，名詞爲 supersaturation)

**aqueous** (*adj*) describes a solution (p.86) with water as the solvent (p.86), e.g. an aqueous solution of potassium hydroxide is made by dissolving potassium hydroxide in water.

**non-aqueous** (*adj*) describes a solution (p.86) with a solvent (p.86) other than water; the solvent is usually named, e.g. an alcoholic (ethanolic) solution of potassium hydroxide.

**concentrated** (*adj*) describes a solution (p.86) containing a high proportion of solute (p.86). **concentration** (*n*), **concentrate** (*v*).

**dilute** (*adj*) describes a solution (p.86) containing a low proportion of solute (p.86). **dilute** (*v*).

**molar concentration** a term used incorrectly, but often used to mean the concentration in moles (p.80) in 1 dm³ of solution (p.86). It is better to use the word *concentration* and say mol dm⁻³

**molarity** (*n*) the number of moles (p.80) of solute (p.86) in one cubic decimetre of solution (p.86), e.g. a solution of 1 mole of sodium hydroxide is 40 g. A solution of sodium hydroxide containing 4 g dm⁻³ contains 0.1 mol dm⁻³, so the molarity is 0.1.

**molality** (*n*) a method of measuring concentration, giving the number of moles (p.80) of solute (p.86) in one kilogram of solvent (p.86).

**M-value** (*n*) a method of describing concentration of a solution (p.86) giving the number of moles of solute (p.86) in one cubic decimetre of solution, i.e. the value of the molarity (↑). If a solution has a molarity of 0.1 it is described as 0.1 M.

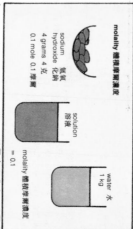

molality 體積摩爾濃度

sodium hydroxide 氫氧化鈉
4 grams 4 克
0.1 mole 0.1 摩爾

solution 溶液

water 水 1 kg

molality 體積摩爾濃度 = 0.1

---

水的（形）　描述用水作為溶劑（第 86 頁）的一種溶液（第 86 頁）。例如氫氧化鉀水溶液是將氫氧化鉀溶解在水中製成的。

非水的（形）　描述含有除水之外的溶劑（第 86 頁）的一種溶液（第 86 頁）。通常都指明所用之溶劑。例如氫氧化鉀之酒精（乙醇）溶液。

濃縮的（形）　描述含高比例的溶質（第 86 頁）的一種溶液（第 86 頁）。（名詞為 concen-tration，動詞為 concentrate）

稀釋的（形）　描述含低比例的溶質（第 86 頁）的一種溶液（第 86 頁）。（名詞為 dilution，動詞為 dilute）

摩爾濃度　克分子濃度　用此術語不正確，但往往用於指 1 升溶液（第 86 頁）中所含溶質之摩爾數（第 80 頁）。用 "濃度"這一詞述以 mol dm⁻³ 表示更恰當。

體積摩爾濃度　體積克分子濃度（名）　指 1 立方分米（1 升）溶液（第 86 頁）中所含溶質之摩爾數（第 80 頁）。例如 1 摩爾氫氧化鈉的質量為 40 克。1 升含 4 克氫氧化鈉的溶液為每升含 0.1 摩爾，所以體積摩爾濃度為 0.1。

重量摩爾濃度（名）　計量濃度的一種方法，表示在 1 千克溶劑（第 86 頁）中溶解的溶質的摩爾數（第 80 頁）。

M 值（名）　描述溶液（第 86 頁）濃度的一種方法，表示在 1 立方分米（升）溶液中所含溶質（第 86 頁）的摩爾數，即體積摩爾濃度（↑）的數值。如果溶液之摩爾濃度為 0.1，即稱之為 0.1 M。

---

dissolved in 溶於溶液中
solution
1 dm³ solution 1 升溶液
4 g dm⁻³ 4 克 / 升
0.1 M solution 0.1 M 溶液
molarity 摩爾濃度
= 0.1

sodium hydroxide 氫氧化鈉
4 grams 4 克
0.1 mole 0.1 摩爾

molarity / 摩爾濃度

**gram molecular weight 克分子量**

$$NaOH$$
$$23 + 16 + 1$$
$$= 40$$

40 g is gram
molecular weight
克分子量為 40 克

**gram molecule** the molecular weight of a
compound stated (p.222) in grams. It is found
by adding the atomic weights (p.114) of the
elements in the compound. This measurement
is no longer used; instead the amount of a
compound is given in moles (p.80).

克分子 以克數表示（第 222 頁）的化合物之分子
量。以化合物中各元素原子量（第 114 頁）之
總和而求得。現已不用此計算法，而以摩爾
數（第 80 頁）表示化合物之總量。

**equivalent weight 當量**

$$Na + H_2O \rightarrow NaOH + (\overset{\uparrow}{H})$$
23 g    40 g    1 g

23 g is equivalent weight of sodium
鈉之當量為 23 克

$$Mg + 2HCl \rightarrow MgCl_2 + (\overset{\uparrow}{H_2})$$
24 g          2 g

12 g is equivalent weight of magnesium
(direct replacement of hydrogen by
a metal)
鎂之當量爲 12 克（用一種金屬直接置換氫）

**equivalent weight** the mass of an element which
will combine (p.64) with or displace (p.68),
directly or indirectly, 1 gram of hydrogen. The
equivalent weight of an acid is the mass, in
grams, of an acid that contains 1 g of replaceable (p.68) hydrogen. The equivalent weight of
an alkali, or a base, is that mass that neutralizes
(p.67) the equivalent weight of an acid. The
equivalent weight of an oxidizing agent is the
mass of the agent which provides 8 g oxygen.
This measurement is no longer used.

當量 指能直接或間接化合（第 64 頁）或置換（第
68 頁）1 克氫原子的一種元素的質量。酸的當量
是含 1 克可置換（第 68 頁）氫的酸的質量克
數。強鹼或鹼的當量是指它中和（第 67 頁）
一種酸之當量的質量。氧化劑之當量是指能
提供 8 克氧的此種氧化劑之質量。現已不用
此種算法。

**normal solution** a solution (p.86) containing the
equivalent weight (↑) of a substance dissolved
in one cubic decimetre of solution, e.g. 98 g
sulphuric acid contains 2 g replaceable
hydrogen, hence the equivalent weight of
sulphuric acid is 49. A normal solution of
sulphuric acid contains 49 g dm⁻³

當量溶液 指在一立方分米（1 升）溶液中含有所
溶解物質之當量（↑）的溶液（第 86 頁）。例如
98 克硫酸中含有 2 克可置換的氫，故硫酸之
當量爲 49。硫酸之當量溶液爲每升硫酸溶液
中含 49 克硫酸。

**normality** ($n$) the fraction of the equivalent weight
(↑) of a solute (p.86) dissolved in one cubic
decimetre of solution (p.86), e.g. if 1 dm³ of
solution contains 98 g sulphuric acid, then it
contains 2 equivalent weights and its normality
is 2. This is usually written as 2N.

當量濃度（名）指在一立方分米（1 升）溶液（第
86 頁）中所溶解的溶質（第 86 頁）之當量（↑）
的分數。例如，如 1 升溶液中含 98 克硫酸，
則該溶液含 2 個當量，即其當量濃度爲 2。
通常表示成 2N。

**standard solution** a solution (p.86) whose concentration is known accurately. The concentration can be given in grams per cubic decimetre, as a molarity (↑) or as a normality (↑).

標準溶液 指已準確知其濃度的一種溶液（第 86
頁）。其濃度可按照摩爾濃度（↑）或當量濃度
(↑)以每立方分米（升）所含的克數表示。

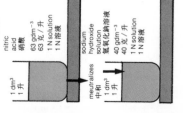

**normal solution 當量溶液**

nitric
acid
硝酸
63 gdm⁻³
63 克/升
1 N solution
1 N 溶液

1 dm³
1 升

sodium
hydroxide
solution
氫氧化鈉溶液
40 gdm⁻³
40 克/升
1 N solution
1 N 溶液

neutralizes
中和
1 dm³
1 升

**solvation** (*n*) a process in which molecules of a solvent (p.86) become attached to ions (p.123), or molecules (p.77), of solutes (p.86). **solvate** (*v*).

**hydration** (*n*) solvation (↑) when water is the solvent (p.86). *See aqua-ion (p.132).* **hydrate** (*v*). **dehydrate** (*v*).

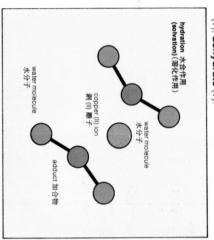

hydration (solvation) 水合作用 (溶化作用)

water molecule 水分子

copper (II) ion 銅(II) 離子

water molecule 水分子

adduct 加合物

**adduct** (*n*) a compound (p.8) with some molecules (p.77), usually of solvent (p.86), combined with metal ions through weak bonds, e.g. hydrogen bonds.

**hydrate** (*n*) a crystal (↓) with molecules of water combined with ions (p.123) in its crystal structure, e.g. zinc sulphate has seven water molecules in its crystal lattice (p.92): ZnSO₄. 7H₂O. **hydrated** (*adj*).

**water of crystallization** water molecules in hydrates (↑). A particular crystalline substance always has the same number of molecules of water of crystallization combined with the ions (p.123) of a molecule of that substance, e.g. CuSO₄.5H₂O; CuCl₂.2H₂O; Na₂SO₄.10H₂O; Na₂CO₃.10H₂O; Na₂CO₃.H₂O.

**mother liquor** the solution left after crystals (↓) have formed.

**supernatant** (*adj*) describes the liquid above a precipitate (p.30) or a sediment (p.31). A supernatant liquid is separated from a sediment by decantation (p.31).

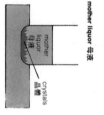

mother liquor 母液

mother liquor 母液

crystals 晶體

溶劑化作用（名）　指溶劑（第 86 頁）之分子與溶質（第 86 頁）之離子（第 123 頁）或分子（第 77 頁）相連結的過程。（動詞爲 solvate）

水合作用（名）　用水作溶劑（第 86 頁）時之溶劑化作用（↑）。參見“水合離子”（第 132 頁）。（動詞爲 hydrate 水合、dehydrate 脫水）

加合物（名）　含有一些溶劑（第 86 頁）分子（第 77 頁），靠弱鍵（如氫鍵）與金屬離子結合的化合物（第 8 頁）。

水合物（名）　在晶體結構中，含有與離子（第 123 頁）相結合的水分子的晶體（↓）。例如硫酸鋅的晶格（第 92 頁）中含有七個水分子，即 ZnSO₄.7H₂O（形容詞爲 hydrated）

結晶水　指水合物（↑）中的水分子。一種特定的結晶物質總是含有相同數目的結晶水分子，與該物質分子之離子（第 123 頁）相結合。例如 CuSO₄.5H₂O；CuCl₂.2H₂O；Na₂SO₄.10H₂O；Na₂CO₃.10H₂O；Na₂CO₃.H₂O。

母液　指生成晶體（↓）之後留下的溶液。

上澄清的　描述液體在沉澱物（第 30 頁）或沉積物（第 31 頁）上面，可藉傾析作用（第 31 頁）將上澄清液與沉積物分離。

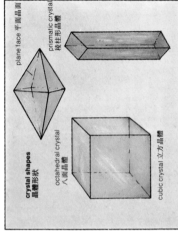

crystal shapes
晶體形狀

plane face 平面晶面

prismatic crystal
稜柱形晶體

octahedral crystal
八面晶體

cubic crystal 立方晶體

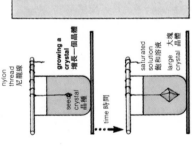

nylon
thread
尼龍線

growing a
crystal
增長一個晶體

seed
crystal
晶種

time 時間

saturated
solution
飽和溶液

large 大塊
crystal 晶體

**crystal** (*n*) a solid substance (p.8) with a regular shape. It has plane (p.92) faces, which are always at the same angle for similar sides in all crystals of the substance. The crystal shape is a property (p.9) of a crystalline substance. **crystalline** (*adj*), **crystalloid** (*n*), **crystallize** (*v*).

**crystalloid** (*n*) a substance which forms crystals (↑) and forms a true solution (p.86) in water. A crystalloid passes through a permeable membrane (p.99). *See colloid (p.98)*.

**crystallization** (*n*) the process of crystals (↑) forming in a solution of a crystalline (↑) substance; also the production of crystals in an experiment (p.42).

**recrystallization** (*n*) to form crystals (↑) of a substance, then to dissolve the crystals in a solvent and to crystallize (↑) again. This makes sure the crystals consist only of the pure substance.

**fractional crystallization** a process for separating two crystalline (↑) solids which have solubilities of nearly the same value. The substances undergo recrystallization (↑) many times. After each crystallization the crystals will be richer in one of the substances and the mother liquor (p.90) richer in the other. At the end of separation (p.34) pure specimens (p.43) of both substances can be obtained.

**晶體** (名) 具有規則形狀的固體物質(第 8 頁)。晶體有平面(第 92 頁)晶面,在物質面的所有晶體中,這些晶面總是與相類似邊保持同一角度。晶體形狀是結晶物質的一種屬性(第 9 頁)。形容詞為 crystalline,名詞 crystalloid 意為晶質,動詞為 crystallize。

**晶質** (名) 能形成晶體(↑)並可在水中形成真溶液(第 86 頁)的一種物質。晶質可透過滲透性膜(第 99 頁)。參見" 膠體"(第 98 頁)。

**結晶作用** (名) 在結晶(↑)的溶液中生晶(↑)的過程;也指在實驗(第 42 頁)中產生晶體之過程。

**再結晶作用** (名) 使一種物質形成晶體(↑)之後,使之溶解於一種溶劑中,及重新結晶(↑)。這可確保晶體所含物質很純。

**分級結晶:分步結晶** 使兩種溶解度相近的結晶(↑)固體分離的過程。這類物質可經受多次再結晶(↑)。每經一次結晶之後,晶體更富集其中一種物質,母液(第 90 頁)則富集另一種物質。分離(第 34 頁)之後,可取得兩種物質的純樣本(第 43 頁)。

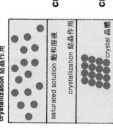

crystallization 結晶作用

saturated solution 飽和溶液

crystallization 結晶作用

crystal 晶體

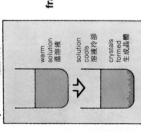

crystallization 結晶作用

warm
solution
溫暖液

solution
cools
溶液冷卻

crystals
formed
生成晶體

**polymorphism** (*n*) the state of a solid substance (p.8) existing (p.213) in two, or more, crystalline (p.15) forms. Elements and compounds can be polymorphic. See *allotropy* (p. 118). Examples are: mercury (II) oxide which has a red and a yellow crystalline form; sodium carbonate which has several hydrates, the decahydrate ($10H_2O$) forms monoclinic (p.96), the decahydrate ($7H_2O$) forms rhombic (p.96) crystals and the transition point (↓) between the two forms is 32°C. **polymorphic** (*adj*).

**enantiotropy** (*n*) polymorphism (↑) in which two stable (p.74) crystalline forms exist, one below a transition point (↓) and one above, the change between the two forms is reversible (p.216), e.g. sulphur has two crystalline forms, rhombic $S_\alpha$ and monoclinic $S_\beta$. $S_\alpha$ is stable below 96°C while $S_\beta$ is stable above 96°C. 96°C is the transition point (↓). **enantiotropic** (*adj*).

**monotropy** (*n*) polymorphism (↑) in which there is only one stable (p.74) crystalline form and other forms are unstable, e.g. red phosphorus is the stable form of phosphorus and white phosphorus is unstable. There is no transition point (↓) between the two forms. **monotropic** (*adj*).

**transition point** (*n*) the temperature at which one crystalline form (p.15) form of a substance changes to another crystalline form in a reversible (p.216) change. At the transition temperature, both crystalline forms can exist (p.213) together.

**lattice** (*n*) a regular arrangement of points in space with a pattern (↓) that can be recognized.

**crystal lattice** (*n*) a lattice (↑) with atoms, molecules, or ions at the points of the lattice. A lattice has a pattern in three dimensions and reaches to the faces of the crystal (p.91).

**isomorphism** (*n*) the state of crystals of two different substances having the same crystal lattice (↑), hence the same crystal shape. **isomorphic** (*adj*).

**plane** (*n*) a flat surface. Two planes can meet in a straight line. Three planes can meet in the same straight line, or at a point. **planar** (*adj*).

**axis** (*n*) (*n.pl. axes*) (1) a line drawn through an object which divides it into two equal halves. (2) a line about which an object turns. **axial** (*adj*).

同質多晶現象 (名) 一種固體物質 (第 213 頁) 兩種或多種結晶 (第 15 頁) 形式的狀態。元素和化合物可以是同質多晶的。參見 "同素異形現象" (第 118 頁)。例如：氧化汞 (II) 有紅色和黃色兩種晶體；碳酸鈉有若干種水合物：十水合物 ($10H_2O$) 形成單斜 (第 96 頁) 晶體，七水合物 ($7H_2O$) 形成斜方 (第 96 頁) 晶體；兩種晶形之間的轉變點 (↓) 為其 32°C。(形容詞為 polymorphic)

鏡像異構體 (名) 指存在兩種穩定 (第 74 頁) 晶形的同質多晶型 (↑) 現象。其中一種在轉變點以下穩定，另一種在轉變點以上穩定，兩種晶形之間的轉變是可逆的 (第 216 頁)。例如：硫有斜方晶體 $S_\alpha$ 和單斜晶體 $S_\beta$ 兩種晶形。96°C 以下時 $S_\alpha$ 穩定，$S_\beta$ 在 96°C 以上穩定。96°C 為其轉變點 (↓)。(形容詞為 enantiotropic)

單變現象 (名) 只存在一種穩定 (第 74 頁) 晶形，而另一種晶形不穩定的多晶型 (↑) 現象。例如紅磷是穩定形的碳，白磷則不穩定。兩種晶形之間無轉變點 (↓)。(形容詞為 monotropic)

轉變點 (名) 一種晶形在可逆 (第 216 頁) 變化中變成另一種結晶形式時的溫度。在轉變溫度時，兩種晶形可同時存在 (第 213 頁)。

格子 (名) 點在空間中以一種可識別的圖像 (↓) 作規則排列。

晶格 (名) 在各點上排列有原子、分子或離子的一種格子 (↑)。晶格具有三度空間的圖像並延伸到晶體 (第 91 頁) 的各個晶面。

異質同晶 (名) 指兩種不同物質的晶體形狀相同的兩種不同物質的晶體形狀，因而晶體形狀也相同的狀態。(形容詞為 isomorphic)

平面 (名) 一個平坦的表面。兩個平面在一條直線相交，三個平面可以在同一條直線或同一個點相交。(形容詞為 planar)

晶軸 (名) (1) 一條畫經物體而將之分為兩等半的線。(2) 指物體繞之轉動的一條線。(名詞複數為 axes，形容詞為 axial)

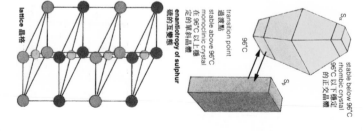

$S_\alpha$

96°C

$S_\beta$

stable above 96°C
monoclinic crystal
在 96°C 以上穩
定的單斜晶體

stable below 96°C
rhombic crystal
在 96°C 以下穩定
的正交結晶

transition point
過渡點

**enantiotropy of sulphur**
硫的互變態

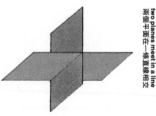

**lattice 晶格**

**two planes meet in a line**
兩個平面在一條直線相交

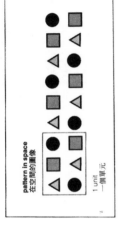

symmetry 對稱
axis of symmetry 對稱的軸
point of symmetry 對稱的點
plane of symmetry 對稱的平面

pattern in space 在空間的圖像
1 unit 一個單元

**pattern** (n) a regular arrangement of points, objects, or shapes which recur (p.217) in space; each arrangement is called a unit of the pattern. Processes or events which occur regularly in time form a dynamic pattern.

圖像：圖案（名）在空間重複出現（第217頁）的點、物體或形狀的規則排列；每一種排列稱爲一個圖像單元。有規則地按時出現的過程或事件構成一個動態的圖像。

**symmetry** (n) the state of having a regular shape such that a line or plane can be drawn which divides the shape into two equal halves, e.g. a circle has symmetry as any line through the centre divides it into two equal halves. **symmetrical** (adj), **asymmetrical** (adj).

對稱（名）有規則形狀的狀態，因而可畫一條線將此形狀平分成相等的兩半部。例如，圓是對稱的，因爲通過圓心的任何直線都將圓分成相等的兩個半圓。形容詞爲 symmetrical 對稱的，asymmetrical 不對稱的。

**crystal symmetry** symmetry (↑) of the lattice (↑) or shape of a crystal (p.91). Crystals can have planes, lines or centres of symmetry.

晶體對稱 指晶體的晶格（第91頁）的對稱（↑）或形狀的對稱（↑）。晶體可具有對稱的平面、線或中心。

**crystal face** (n) a plane surface of a crystal.

晶面（名）指晶體的平面。

**plane of symmetry** a plane which divides the lattice (↑) or the shape of a crystal into two equal halves.

對稱面 使晶體（↑）或晶體形狀分成相等兩半的一個面。

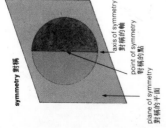

orientation 取向
90°
orientation of 90° to other line 與另一條線成90°的取向

**orientation** (n) the direction in which an object points in relation to its surroundings (p.103), e.g. the orientation of one line in relation to another line which it meets at a point.

取向：方位（名）物體所指方向與其環境（第103頁）的關係。例如，一直線的取向與相交於一點之另一線的關係。

**perfect crystal** (n) a crystal (p.91) which has no breaks in the lattice (↑), i.e. no atoms, ions or molecules missing from the lattice.

完美晶體（名）指晶格（↑）沒有損壞的一種晶體（第91頁），即晶格中沒有損失原子、離子或分子的晶體。

**irregular** (adj) describes any arrangement, in space or in time, which is not regular. The opposite of regular.

不規則的（形）描述任何一種排列在空間或時間上沒有規則，是規則的相反詞。

**interstice** (n) a small space between the ions, atoms or molecules in a crystal lattice (↑), e.g. in steel, there are atoms of carbon in the interstices of the iron atoms forming the lattice of the metal crystals. See metal crystals (p.95). **interstitial** (adj).

間隙（名）晶體晶格（↑）中離子、原子或分子之間的細小空間。例如，在鋼鐵中的數個原子處於形成金屬晶體晶格的鐵原子空隙中，參見"金屬晶體"（第95頁）。（形容詞爲 interstitial）。

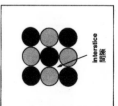

interstice 間隙

**cleavage plane** a plane in a crystal (p.91) along which a blow will separate the crystal into two parts with a clean cut. The crystal is *cleaved* by the blow. Along any other plane the crystal will be shattered (↓). **cleave** (v), **cleavage** (n).

**shatter** (v) to break a hard, or brittle (p.14) object into very many small pieces by a hard blow. e.g. a cup dropped on the ground is shattered.

**slip plane** a plane in a crystal (p.91), particularly a metal crystal (↓), along which one part of the crystal will move relative to the other part. When a force acts on the crystal. One part of the crystal lattice *slips* over the other part. Slip planes in metal crystals cause metals to be *ductile* (p.14).

**giant structure** a crystal lattice (p.92) with no separate molecules (p.77). The ions (p.123) or atoms (p.110) forming the lattice are all joined, one to another, by bonds (p.133), so that the crystal appears to be one very large molecule. Many giant structures are held together by ionic bonds (p.134).

**molecular crystal** a crystal consisting (p.55) of separate molecules held together by weak van der Waals' (p.137) forces. The crystal is very soft and has a low melting point (p.12). Many organic compounds form molecular crystals.

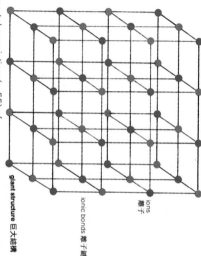

giant structure 巨大結構

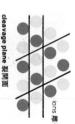

cleavage plane 裂開面

ions 離子

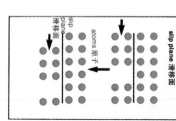

slip plane 滑移面

slip plane 滑移面

atoms 原子

裂開面 晶體 (第 91 頁) 中的一個面，沿此面衝擊可將晶體分裂成輪廓清晰的兩部分。晶體就是藉衝擊力的。如沿任何其他晶面衝擊，則此晶體便會破碎 (↓)。 (動詞局cleave，名詞局 cleavage)。

破碎 (動) 藉一次重擊將硬或脆的物體打碎成許多小塊。例如杯子跌落地面時被碎了。

滑移面 指晶體 (第 91 頁)，尤其是金屬晶體 (↓) 中的一個面，當力作用在晶體上時，晶體的某一部分可沿此面相對於另一部分移動。晶體晶格之一部分"滑"過其他部分，金屬晶體的滑移面是使金屬具有"延性的" (第 14 頁) 原因。

巨大結構 指不具有獨立分子 (第 77 頁) 的晶格 (第 92 頁)。構成晶格之離子 (第 123 頁) 或原子 (第 110 頁)，一個接一個靠鍵 (第 133 頁) 全部相連，因而使晶體看來像一個很大的分子。許多巨大結構都是靠離子鍵 (第 134 頁) 連在一起的。

分子晶體 靠弱范德華耳斯 (第 137 頁) 力連在一起的獨立分子組成的晶體。此種晶體很軟，熔點 (第 12 頁) 低。許多種有機化合物能形成分子晶體。

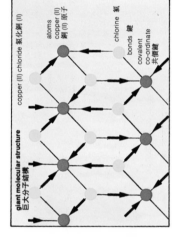

giant molecular structure
巨大分子結構

copper (II) chloride 氯化銅 (II)

atoms
copper (II)
銅 (II) 原子

chlorine
氯

bonds 鍵
covalent
co-ordinate
共價鍵

**giant molecular crystal** a crystal with a giant structure (↑) composed of atoms joined by covalent bonds (p.136). A giant molecular crystal is very hard and has a very high melting point (p.12). Aluminium oxide and anhydrous copper (II) chloride form giant molecular crystals.

**metal crystal** a giant structure (↑) formed by metals. The crystal lattice consists (p.55) of positive ions of metals. A cloud of electrons can move throughout the lattice and holds the positive ions together. The cloud of freely moving electrons accounts for the ability of metals to conduct electric current (p.122) and heat.

**close packing** an arrangement of atoms put as close to each other as possible. This forms a close packed lattice of one layer (p.18). Two arrangements of close packing are found in metal crystals (↑), hexagonal (↓) and cubic (↓).

**hexagonal close packing** a second layer of atoms lies in the hollows formed by groups of three atoms in the first layer of close packing (↑). A third layer of atoms similarly lies in the hollows formed by groups of three atoms in the second layer; each atom in the third layer is directly below an atom in the first layer.

**cubic close packing** the atoms in the first two layers are arranged in the same way as the atoms in hexagonal close packing (↑). The third layer of atoms lies in the hollows formed by groups of three atoms, but these atoms are not below the atoms in the first layer.

巨大分子晶體　含有巨大結構 (↑) 的一種晶體，巨大結構由原子靠共價鍵 (第 136 頁) 連接組成。巨大分子晶體極硬，熔點 (第 12 頁) 極高。氧化鋁和無水氯化銅 (II) 形成巨大分子晶體。

金屬晶體　由金屬所形成的一種巨大結構 (↑)。晶體晶格由金屬之陽離子組成。有一電子雲可在整個晶格游移使陽離子結合在一起，自由移動的電子雲是金屬能夠傳導電流 (第 122 頁) 和熱的原因。

緊密堆積　使原子彼此間盡可能靠近的一種排列。結果形成一層 (第 18 頁) 緊密堆積的晶格。金屬晶體 (↑) 有六方緊密堆積 (↓) 和立方緊密堆積 (↓) 兩種排列。

六方緊密堆積　第二層原子處於緊密堆積的第一層中由三個原子一組的原子組形成的空隙之中。第三層原子同樣地處於第二層中由三個原子一組的原子組形成的空隙之中；第三層中的每一個原子都直接排在第一層中的每一個原子之下。

立方緊密堆積　首兩層中的原子排列與六方緊密堆積 (↑) 中原子排列方式相同。第三層的原子處於由三個原子一組的原子組形成的空隙之中，但這些原子不是排於第一層中各原子之下。

close packing
緊密堆積

close packing of atoms
原子的緊密堆積
**hexagonal close packing**
**六方緊密堆積**

● top layer 頂層
● middle layer 中層
○ bottom layer 底層

**cubic close packing**
**立方緊密堆積**

● top layer 頂層
● middle layer 中層
○ bottom layer 底層

**crystal systems** crystals are grouped into systems according to their symmetry (p.93).

**cubic system** a system with three equal axes of symmetry (a = b = c). The angle between any two axes is 90°.

**tetragonal system** a system with two equal axes of symmetry and a third axis which is longer or shorter (a = b ≠ c). The angle between any two axes is 90°.

**orthorhombic system** a system with three unequal axes of symmetry (a ≠ b ≠ c). The angle between any two axes is 90°.

**monoclinic system** a system which has three unequal axes of symmetry. The angle between two pairs of the axes is 90°, the angle between the third pair of axes is not 90° but either greater or smaller.

**triclinic system** a system which has three unequal axes of symmetry. The angle between any two axes is not 90°.

**hexagonal system** a system which has four equal axes of symmetry at an angle of 120° to each other with a fourth axis of symmetry at an angle of 90°, and of a different length.

晶系 晶體按其對稱（第 93 頁）分類成若干系統。

立方晶系 三對稱軸等長（a = b = c）的一種晶系。任何兩軸間的夾角均為 90°。

四方晶系 兩對稱軸等長而第三軸較長或較短（a = b≠c）的一種晶系。任何兩軸間的夾角均為 90°。

斜方晶系 三對稱軸不等長（a≠b≠c）的一種晶系。任何兩軸間的夾角均為 90°。

單斜晶系 三對稱軸不等長的一種晶系。兩對稱軸間的夾角為 90°，而第三對稱軸之間的夾角不是 90°，可大於或小於 90°。

三斜晶系 三對稱軸不等長的一種晶系。任何兩軸間的夾角都不是 90°。

六方晶系 三對稱軸等長的一種晶系，等長對稱軸彼此間之夾角為 120°，長度不同的第四對稱軸與等長對稱軸間的夾角為 90°。

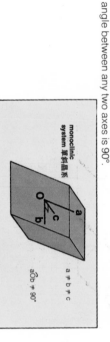
monoclinic system 單斜晶系
aÔb ≠ 90°
a ≠ b ≠ c

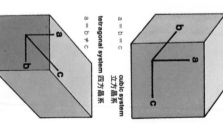
orthorhombic system 斜方晶系
a ≠ b ≠ c

tetragonal system 四方晶系
a = b ≠ c

cubic system 立方晶系
a = b = c

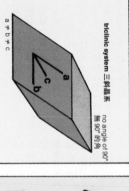
triclinic system 三斜晶系
no angle of 90°
無 90° 的角
a ≠ b ≠ c

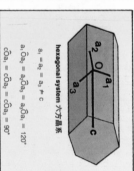
hexagonal system 六方晶系
$a_1 = a_2 = a_3 ≠ c$
$a_1Ôa_2 = a_2Ôa_3 = a_3Ôa_1 = 120°$
$cÔa_1 = cÔa_2 = cÔa_3 = 90°$

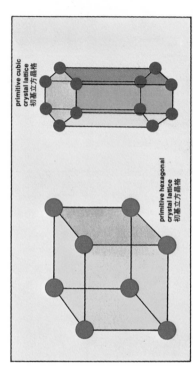

**primitive cubic crystal lattice**
初基立方晶格

**primitive hexagonal crystal lattice**
初基立方晶格

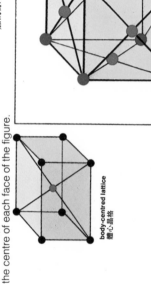

**face-centred lattice**
面心晶格

**body-centred lattice**
體心晶格

**primitive structure** this is the simplest crystal structure with an atom (p.110) or ion (p.123) at each corner of the figure for each of the crystal structures. See *diagram*, which shows a primitive cubic lattice and a primitive hexagonal lattice.

**body-centred lattice** this structure is a primitive structure (↑) with an additional atom or ion at the centre of the figure.

**face-centred lattice** this structure is a primitive structure (↑) with additional atoms or ions at the centre of each face of the figure.

**初基結構** 為最簡單的晶體結構，晶體結構的各個角上都有一個原子（第110頁）或離子（第123頁）。（見圖，圖示初基立方晶格和初基六方晶格。

**體心晶格** 這種結構是在圖形的中心增加一個原子或離子的初基結構（↑）。

**面心晶格** 這種結構是在圖形的各面的中心增加的原子或離子的初基結構（↑）。

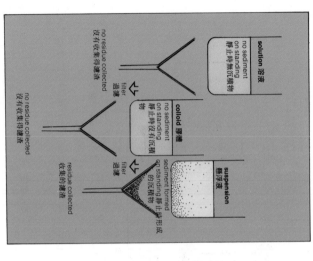

no sediment
on standing
靜止時無沉積物

**solution** 溶液

no residue collected
沒有收集得濾渣

filter
過濾

no sediment
on standing
靜止時沒有沉
物

**colloid** 膠體

no residue collected
沒有收集得濾渣

filter
過濾

sediment formed
on standing
靜止時形成
的沉積物

**suspension**
懸浮液

residue collected
收集得濾渣

**colloids** 膠體
effect of a light beam on
true solutions and colloids
光束對真溶液和膠體的效應

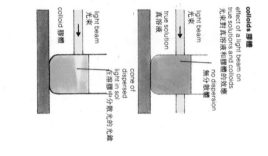

no dispersion
無分散物

true solution
真溶液

light beam
光束

cone of
dispersed
light in sol
在溶膠中分散光的光錐

light beam
光束

colloid 膠體

**colloid** (n) a constituent (p.54) of a disperse (↓) system (p.212) in which that constituent is dispersed throughout another constituent, e.g. a constituent (clay) is dispersed in water (the other constituent). The colloidal constituent is present (p.217) in particles which are between 1 nm and 100 nm in size. If the colloid is a solid and it is dispersed in a liquid, the size of the particles (p.13) is greater than those in a solution (p.86), but smaller than those in a suspension (p.86). Solid colloidal particles dispersed in a liquid pass through a filter (p.30) but not through a permeable mem-brane (↓). Examples of colloids include: milk (a dispersion of colloidal particles of fat in water); foam (p.100); aerosol (p.100); sol (p.100); froth (p.100); smoke (p.100). **colloidal** (adj).

膠體（名）指分散（↓）體系（第 212 頁）中之一種成分（第 54 頁）充分地分散於另一成分中。例如一種成分（黏土）分散於水（另一成分）之中。膠態成分以粒子形式存在（第 217 頁）。粒子直徑大於 1 nm 至 100 nm 之間。如果該膠體屬固體並分散於一種液體中，則其粒子（第 13 頁）直徑大於真溶液（第 86 頁）中之粒子而小於懸浮液（第 86 頁）中之粒子。分散於液體中之固體膠體粒子可通過濾器（第 30 頁）但不能透過滲透性膜（↓）。例如：奶（脂肪粒在水中的膠態分散體）、泡沫（第 100 頁）、氣溶膠（第 100 頁）、溶膠（第 100 頁）、浮泡沫（第 100 頁）、煙（第 100 頁）都是膠體。（形容詞為 colloidal）

**disperse** (*v*) to spread particles (p.13), or similar small objects, over a large area or volume. The action points to the agent causing it, e.g. a liquid disperses a solid colloid (↑) over all its volume. **dispersion** (*n*), **disperse** (*adj*), **dispersed** (*adj*).

**dispersion medium** the liquid, or gas, that disperses (↑) a colloid (↑) over all its volume. It has a continuous phase whereas the disperse phase is discontinuous.

**disperse phase** the colloidal (↑) substance that is dispersed (↑) over all the dispersion medium (↑).

**membrane** (*n*) a thin piece of material such as parchment paper, cellophane, a bladder. A liquid can pass slowly through a membrane if the membrane is permeable (↓). A crystalloid (p.91) solid can also pass through a permeable membrane, but a colloidal (↑) solid and a suspension cannot pass through it.

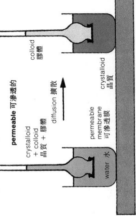

crystalloid + colloid
晶質 + 膠體

water 水

permeable membrane 可滲透膜

diffusion 擴散

permeable 可滲透的

crystalloid 晶質

colloid 膠體

**permeable** (*adj*) describes any object which allows fluids (p.11) to pass through by diffusion (p.35). Permeable membranes (↑) allow molecules (p.77) of fluids to diffuse. Other substances which are finely divided, e.g. sand, also allow particles (p.13) to diffuse. **permeability** (*n*).

**coagulate** (*v*) to cause small particles to join together to form bigger particles or a mass of particles, e.g. a colloidal sol (p.100) can be coagulated to form a suspension (p.86); an iron (III) oxide sol in water is coagulated to a suspension by a solution of aluminium chloride. **coagulation** (*n*), **coagulated** (*adj*).

aluminium chloride solution 氯化鋁溶液

coagulate 凝結

iron (III) oxide sol 氧化鐵 (III) 溶膠

coagulates 凝集劑

iron (III) oxide suspension 氧化鐵 (III) 懸浮液

分散 (動) 使粒子 (第 13 頁) 或類似的小物體散佈在一個大面積或大容積上。其作用指向引起其分散的介質。例如液體將固體膠體 (↑) 分散在其全部容積上。(名詞爲 dispersion，形容詞爲 disperse，dispersed)

分散介質 使膠體 (↑) 分散於其整個容積內的液體或氣體。分散介質爲連續相，分散相則爲不連續相。

分散相 指散於所有全部分散介質 (↑) 中的膠態 (↑) 物質。

膜 (名) 諸如羊皮紙、玻璃紙、膀胱之類薄片材料。如爲可滲透 (↓) 膜，則液體可慢慢通過此膜。晶質 (第 91 頁) 固體也可通過可滲透膜、膠態 (↑) 固體和懸浮體則不能讓通過可滲透膜。

可滲透的 (形) 描述能讓流體 (第 11 頁) 藉擴散作用 (第 35 頁) 通過的任何物體。可滲透性膜 (↑) 能讓流體之分子 (第 77 頁) 滲出。其他細碎的物質，如砂子也能讓粒子 (第 13 頁) 擴散。(名詞爲 permeability)

凝集：凝聚 (動) 使細小粒子聯結一起而成更大粒子或粒子團。例如小粒子的溶膠 (第 100 頁) 可以凝集形成懸浮液 (第 86 頁)；氯化鋁溶液滲於水中之氧化鐵 (III) 之溶膠凝集成懸浮液。(名詞爲 coagulation，形容詞爲 coagulated)

**sol** (*n*) a sol is formed by a colloidal (p.98) solid dispersed (p.99) in a liquid. The colloidal solid passes through a permeable membrane (p.99). Crystalloids (p.91) are separated from colloids by dialysis (p.91) but does not pass through a filter paper (p.30) but does not pass through a permeable membrane (p.99). Crystalloids (p.91) are separated from colloids by dialysis (p.91). Sols are described as lyophilic or lyophobic (↓).

**hydrosol** (*n*) a sol (↑) in which the dispersion medium (p.99) is water.

**emulsion** (*n*) an emulsion is formed from two liquids, one of which is dispersed (p.99) in the other liquid in the colloidal state. If the emulsion is not stabilized (↓) the liquids separate out, e.g. coconut oil and water form an emulsion when shaken together; the two liquids separate on standing. Soap acts as an emulsifying agent (↓) to stabilize the emulsion. Milk is an emulsion with butter fat dispersed in a dilute sugar solution. **emulsify** (*v*), **emulsified** (*adj*).

**foam** (*n*) a foam is formed from a gas dispersed (p.99) in a liquid. It consists of small bubbles of gas with a thin film (p.18) of liquid round the bubbles. A foam is usually stable. Foams are formed when a gas and a liquid are forced through a jet under pressure. **foam** (*v*), **foaming** (*adj*).

**froth** (*n*) a foam (↑) which has large bubbles and is less stable. Mixtures of ethanol and water form a froth, with carbon dioxide as the gas. **froth** (*v*).

**aerosol** (*n*) in an aerosol, a liquid is dispersed in a gas, usually air. It is formed by atomization (↓). A mist is an example of an aerosol. Many insecticides are used as aerosols.

**smoke** (*n*) smoke is formed by solids dispersed in a gas, usually air, e.g. small particles (p.13) of carbon dispersed in air in wood smoke.

**gel** (*n*) an intermediate (p.85) stage (p.159) between a sol (↑) and a suspension. The colloidal particles (p.13) form long thin threads round the liquid dispersion medium (p.99), the result appears to be solid but it is easily deformed. A jelly is a gel. **gel** (*v*), **gelation** (*n*).

---

溶膠（名） 溶膠是在膠體（第 98 頁）固體分散（第 99 頁）在液體中形成的。膠體固體可以透過可透性膜（第 99 頁）。晶質（第 91 頁）藉滲析（第 91 頁）自膠體中分離出來。溶膠可分成親液溶膠和疏液（↓）溶膠。

水溶膠（名） 分散介質（第 99 頁）為水的溶膠（↑）。

乳狀液（名） 乳狀液係由兩種液體所形成，其中一種液體分散（第 99 頁）在另一種處於膠體狀態的液體之中。如果乳狀液是不穩定的（↓），則液體可以分離出來。例如，將椰子油和水一起搖混可形成一種乳狀液，在靜止時兩種液體復又分離出來，肥皂作用於乳化劑（↓）可使乳狀液液穩定。牛奶是乳脂分散於稀糖液中形成的一種乳狀液。（動詞為 emulsify，形容詞為 emulsified）

泡沫（名） 氣體分散（第 99 頁）在液體中形成泡沫。泡沫係由細小的氣體泡泡以及包著氣泡的一層液體薄膜（第 18 頁）所組成。泡沫通常都是穩定的。當在壓力下迫使氣體和液體通過一個噴嘴時可形成泡沫。（動詞為 foaming）

浮泡（名） 指大而較不穩定的泡沫（↑）。當其具有二氧化碳氣體時酒精和水的混合物組成為浮泡。（動詞為 froth）

氣溶膠（名） 液體分散於氣體（一般是空氣）中形成氣溶膠。氣溶膠是由氣化作用（↓）所形成的。霧是氣溶膠的例子。許多種殺蟲劑都是作為氣溶膠（噴霧劑）使用。

煙（名） 固體分散在氣體（一般是空氣）中形成煙。例如，炭的小粒子（第 13 頁）在碳的固體煙中。

凝膠（名） 指介於溶膠（↑）和懸浮質（第 99 頁）間一個中間（第 13 頁）階段（第 159 頁）。膠體粒子形成細長的絲線。繞著液體固體但卻易變形。凝膠是一種凝膠（但其固態好像是固體）。（動詞為 gel，名詞 gelation 意為膠凝作用）

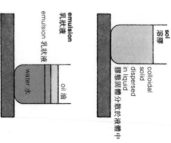

smoke 煙

smoke
solid
dispersed
by gas
固體屬氣
體所分散

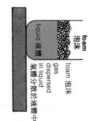

aerosol 噴霧劑

liquid
dispersed
in air
液體分散於氣空中

foam 泡沫

liquid
dispersed
in liquid
液體屬氣

gas
dispersed
in liquid
氣體分散於液體中

foam 泡沫

sol
溶膠

colloidal
solid
in liquid
膠體固體分散於液體中

emulsion 乳狀液

emulsion
乳狀液

water 水

oil 油

**lyophilic** (*adj*) describes a sol (↑) in which the disperse phase (p.99), a solid, has an attraction (p.124) for the dispersion medium (p.99), a liquid. The colloid (p.98) goes readily into solution; after coagulation (p.99), it is readily dispersed (p.99) again by adding more liquid, e.g. a starch sol is lyophilic.

**lyophobic** (*adj*) describes a sol (↑) in which the disperse phase (p.99), a solid, has a repulsion (p.124) for the dispersion medium (p.99), a liquid. The colloid (p.98) readily comes out of solution; after coagulation (p.99) it does not become dispersed (p.99) again, e.g. a gold sol.

**hydrophilic** (*adj*) describes a lyophilic (↑) sol with water as the dispersion medium (p.99).

**hydrophobic** (*adj*) describes a lyophobic (↑) sol with water as the dispersion medium (p.99).

**stabilize** (*v*) to make a sol, emulsion, or gel, stable, by preventing them from changing to a suspension, or from coagulating or from separating. **stable** (*adj*), **stabilizer** (*n*), **stabilization** (*n*).

**emulsifying agent** an agent (p.63) which stabilizes an emulsion, e.g. soap stabilizes a coconut oil and water emulsion.

**atomize** (*v*) to blow a liquid through a fine (p.13) jet (p.29). This forms an aerosol (↑) with the liquid dispersed (p.99) in air. **atomizer** (*n*).

**cataphoresis** (*n*) the movement of colloidal (p.98) particles (p.13) towards electrodes in a lyophobic (↑) sol, or in a smoke (↑). Lyophilic (↑) sols may undergo (p.213) cataphoresis if the colloid has an electric charge. The colloidal particles may move either to the anode (p.123) or the cathode (p.123) depending on their charge, e.g. clay and metallic particles are negatively charged; iron (III) chloride and aluminium hydroxide are positively charged in aqueous (p.88) sols. Platinum electrodes are usually used in cataphoresis.

**electrophoresis** (*n*) another name for cataphoresis (↑).

**thixotropy** (*n*) a property of some colloidal liquids by which the liquid has a decreased viscosity at an increased rate of flow. **thixotropic** (*adj*).

親液的 (形) 指固體分散相 (第 99 頁) 對液體分散介質 (第 99 頁) 有吸引力 (第 124 頁) 的一種溶體 (↑)。這種膠體 (第 98 頁) 很易溶入溶液中、凝聚、凝體 (第 99 頁) 之後、可藉加入更多的液體而使之重新分散 (第 99 頁)。例如，澱粉溶膠為親液的。

疏液的 (形) 指固體的分散相 (第 99 頁) 對液體的分散介質 (第 99 頁) 有排斥力 (第 124 頁) 的一種溶膠 (↑)。這種膠體 (第 98 頁) 容易從溶液中分離出來，凝聚 (第 99 頁) 之後不能重新分散 (第 99 頁)。例如，金的溶膠。

親水的 (形) 指一種親液 (↑) 溶膠以水為分散介質 (第 99 頁)。

疏水的 (形) 指一種疏液 (↑) 溶膠以水為分散介質 (第 99 頁)。

使穩定 (動) 為防止溶膠、乳狀液或凝膠轉變成懸浮液，或使之凝聚生凝聚或分離而使之穩定。(形容詞為 stable、名詞為 stabilizer 穩定劑、stabilization 穩定作用)

乳化劑 (名) 使乳狀液穩定化 (↑) 的一種化學劑 (第 63 頁)。例如，肥皂使椰子油和水的乳狀液穩定。

使霧化：噴成霧狀 (動) (使液體吹過一個細的分散 (第 29 頁) 噴嘴 (第 13 頁) 至空氣中的氣溶膠 (↑)。(名詞為 atomizer)

電泳：陽離子電泳 (名) (第 98 頁) 粒子 (第 13 頁) 在流液 (↑) 溶膠 (↑) 中、膠體帶電荷。若膠體帶電荷，則溶液 (↑) 溶膠會遷受 (第 213 頁) 電泳。膠體粒子可能向陰極 (p.123) 移動、也可能向陽極 (第 123 頁) 移動，視其電荷而定。例如黏土粒子和金屬粒子都帶負電荷；氯化鐵 (III) 和氫氧化鋁在水的 (第 88 頁) 溶膠中帶正電荷。電泳一般使用鉑電極。

電泳：電粒泳 (↑) 的別稱。

觸變性：遞流性 (名) 某些膠態液體所具有的一種性質。這類液體的黏度隨流動速率增加而下降。(形容詞為 thixotropic)

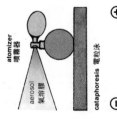

atomizer
噴霧器

aerosol
氣溶膠

cataphoresis 電粒泳

platinum electrodes
鉑電極

cathode
陰極

anode
陽極

water
水

moves upwards
向上移動

gold sol
金的溶膠

**pressure** (n) the force per unit area acting on a surface. It is the force in newtons acting on an area measured in square metres. A force of 200 newtons acting on an area of 5 sq metres gives a pressure of $200 N \div 5 m^2 = 40 N m^{-2}$. The unit of pressure is the pascal (Pa); a pressure of $1 N m^{-2} = 1 Pa$. Pressure in gases is also measured by the height of a column of mercury supported by the pressure; the height is given in millimetres (mm) of mercury. 1 mm of mercury = $133.322 N m^{-2}$. **press** (v).

**atmospheric pressure** the pressure (↑) exerted (p.106) by the air in the atmosphere on the Earth's surface. It is measured by a barometer. The everyday variations of atmospheric pressure are above and below this value.

**standard atmospheric pressure** an atmospheric pressure of $101\,325 N m^{-2}$ or 760 mm of mercury.

**s.t.p.** (abbr) an abbreviation for standard temperature (↓) and pressure (↑), i.e. 0°C or 273 K and $101\,325 N m^{-2}$

**temperature** (n) a physical property which determines the direction of the flow of heat between materials (p.8) in thermal (p.65) contact (p.217). Heat flows from a higher to a lower temperature.

**Celsius scale** a temperature scale (p.26) with 0° as the temperature of melting ice and 100° as the temperature of steam at standard atmospheric pressure (↑). One Celsius degree (1°C) is 1/100 of the temperature interval (p.220) between the ice point and the steam point. The symbol for Celsius temperature is θ.

**absolute scale** a temperature scale (p.26) with 0 as the absolute zero of temperature and a temperature interval of 1 kelvin. On this scale the temperature of melting ice is 273 K and of steam is 373 K. Absolute temperatures are changed to Celsius temperatures by: 273 + Celsius temperature = absolute temperature. The symbol for Celsius temperature is T.

**kelvin** (n) the S.I. unit of temperature, measured from the triple point of water at which ice, water and water vapour all exist (p.213) together. One kelvin is 1/273.16 of the value of the triple point temperature of 1 K = 1°C (symbol = K).

壓力（名） 每單位面積表面所受之作用力。此作用力以牛頓（N）爲單位。面積以平方米計量單位爲平方米。200 牛頓力作用在 5 平方米之面積，單位爲牛頓的斯卡（帕）。$200 N \div 5 m^2 = 40 N m^{-2}$。壓力單位爲帕斯卡（Pa）。1 牛頓·米⁻² 的壓力也以該壓力所支持 = 1帕（Pa）。氣體的壓力也以該壓力所支持之水柱高度來量度，以毫米（mm）數表示。1 mm 水柱高亦 = $133.322 N m^{-2}$。（動詞爲 press）

大氣壓 指大氣中之空氣施加（第 106 頁）在地球表面上的壓力（↑），可藉氣壓計測得。

標準大氣壓 指 $101325 N m^{-2}$ 或 760 mm 末柱之大氣壓（↑），大氣壓的每日變化在此值上或下。

標準溫壓（縮） 標準溫度（↑）和壓力（↑）的縮寫，即溫度 0°C（273 K）和壓力 $101325 N m^{-2}$。

溫度（名） 確定兩（第 8 頁）之間熱流方向之物理性質。熱量從高溫流向低溫。材料（第 8 頁）之間在熱（第 65 頁）接觸（第 217 頁）時。

攝氏溫標 指用 0° 作爲冰化之溫度，100° 作爲在標準大氣壓（↑）下水之蒸汽之溫度的一種溫度標度（第 26 頁）。攝氏一度（1°C）等於水之沸至沸點之間溫度間隔（第 220 頁）的 1/100。攝氏溫度之符號爲 θ。

絕對溫標（名） 以絕對零度爲溫度之絕對零度、溫度間隔爲 1 開爾文度（第 26 頁）的一種溫標，在此溫度上，冰之溶化溫度爲 273 K，水之沸點爲 373 K。絕對溫度與攝氏溫度之換算如下：絕對溫度 = 273 + 攝氏溫度。絕對溫度之符號爲 T。

開爾文（名） 爲溫度國際制單位（第 213 頁）的水的三相點溫度測量，自冰、水和水汽都一起存在（第 213 頁）的水的三相點溫度數值的 1/273.16。溫度 1 開爾文度之間隔爲 1°C。（其符號爲 K）。1 K 之溫度間隔 = 1°C。（其符號爲 K）。

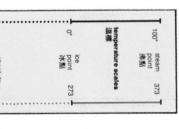

temperature scales 溫標

ice point 冰點
steam point 沸點
absolute zero 絕對零度
Celsius scale 攝氏溫標
absolute scale 絕對溫標

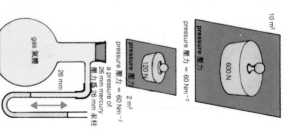

gas 氣體
mercury 水柱
pressure 壓力

pressure 壓力
a pressure of 26 mm mercury 壓力爲 26 mm 末柱

pressure 壓力 = 60 N m⁻²
a pressure of 60 N m⁻²

**surroundings** (n.pl.) all the objects and materials near to and around an object, or the objects and materials for an experiment, are the surroundings of that object or experiment. The surroundings may, or may not, have an effect on the object or experiment. **surround** (v).

**factor** (n) a possible cause of an effect or change. Some of the surroundings (↑) are factors, i.e. those that have an effect on an object or experiment. Factors for chemical reactions are temperature (p.81), pressure (↑), concentration (p.81), catalysts (p.72), surface nature (p.19).

**conditions** (n.pl.) the magnitude of factors (↑) of a particular object or experiment. To contrast factors and conditions: (1) temperature is a factor for the rate of reaction (p.149) between substances; a temperature of 60°C is an actual condition in an experiment. (2) water is a factor in the rusting (p.61) of iron; the presence of water is a necessary condition. Conditions can be necessary, i.e. without them, there is no reaction or effect, or conditions may be adverse, under which the reaction or effect does not take place or takes place very slowly, or conditions may be suitable, under which the reaction or effect is normal (p.229).

**vapour pressure** the pressure (↑) exerted (p.106) by a saturated (p.87) vapour (p.11). Vapour pressure increases with temperature. A liquid boils when its vapour pressure is equal to the atmospheric pressure (↑).

**ambient** (adj) describes conditions of the surroundings (↑), e.g. the ambient temperature is the temperature of the surrounding air.

環境：周圍事物（名，複）某一物體附近或周圍的一切物體和材料，或某一實驗所用的材料和材料都是該物體或實驗的環境。環境對物體或實驗可能有或沒有影響。（動詞為 surround）

因子：因素（名）指可能引起效應或變化的一種原因。某些環境（↑）是因素，即是對物體或實驗有影響的環境。溫度（↑）、壓力（↑）、濃度（第81頁）、催化劑（第72頁）、表面本性（第19頁）都是化學反應的因素。

條件（名，複）某一特定物體或實驗的各因素為（1）之量值。因素與條件之比較：(1)溫度是因素之間反應（第149頁）速度之條件；(2)水為導致溫度60°C為實驗的實際條件。水的存在是鐵生鏽（第61頁）因素之一；水的存在是個必要條件，就不會發生反應或沒有效應。條件有這些條件可以是必要的，即沒有，條件也可能是不必的，或者發生反應慢。條件也可能或沒有效應，或者發生得極慢。條件也可能是適宜的，在此條件下，反應或效應正常進行。（第229頁）

蒸汽壓（名）指飽和（第87頁）蒸汽（第11頁）所施加（第106頁）之壓力（↑）。蒸汽壓隨溫度升高而增加。當液體的蒸汽壓等於大氣壓（↑）時，液體便沸騰。

外界的：周圍的（形）描述周圍空氣之條件。如外界溫度係指周圍空氣之溫度。

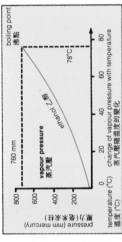

change of vapour pressure with temperature 蒸汽壓隨溫度的變化

boiling point 沸點 / 760 mm / 78°C / ethanol 乙醇 / vapour pressure 蒸汽壓 / pressure (mm mercury) 壓力（毫米汞柱） / temperature (°C) 溫度（°C）

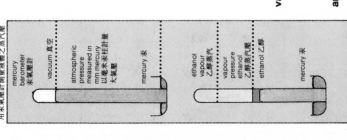

mercury barometer 汞氣壓計
using a mercury barometer to measure vapour pressure of a liquid 用汞氣壓計測量液體之蒸汽壓

vacuum 真空 / mercury barometer 汞氣壓計 / atmospheric pressure measured in mm mercury 以毫米汞柱計量大氣壓 / mercury 汞 / ethanol vapour 乙醇蒸汽 / vapour pressure ethanol 乙醇蒸汽壓 / ethanol 乙醇 / mercury 汞 / vapour pressure 蒸汽壓

**condensation²** (n) a change of state (p.9) from vapour, or gas, to liquid caused by a decrease in temperature. **condense** (v).

**liquefaction** (n) a change of state (p.9) from vapour to liquid caused by an increase in pressure. **liquefy** (v).

**permanent gas** a gas which is not easily liquefied because it has to be first cooled to a very low temperature, e.g. hydrogen, nitrogen and oxygen are permanent gases.

**noble gas** one of the following gases: helium, neon, argon, krypton, xenon, and radon. They do not take part in chemical reactions and are usually considered to be inert (p.19). They are permanent gases (↑) and also monatomic (↓).

**critical temperature** a particular temperature for each gas, above which the gas cannot be liquefied (↑) by pressure (p.102) alone. The critical temperature for oxygen is −119°C (154 K), so oxygen has to be cooled below −119°C before it can be liquefied by pressure; for this reason, oxygen is a permanent gas (↑). Critical temperature has the symbol $T_c$.

**critical pressure** (n) the pressure sufficient to liquefy a gas at its critical temperature (↑).

**monatomic** (adj) describes a gas with molecules (p.77) consisting of one atom only, e.g. neon and helium are monatomic gases.

**diatomic** (adj) describes a gas with molecules (p.77) consisting of two atoms (p.77).

**polyatomic** (adj) describes a gas with molecules (p.77) consisting of several atoms (p.110). Such gases are usually compounds, e.g. carbon dioxide.

**atomicity** (n) the number of atoms in a molecule of an element or a compound, when the element or compound consists of separate molecules. In most cases, the substances will be gaseous, e.g. helium has an atomicity of 1; hydrogen of 2; sulphur dioxide of 3.

**gram-molecular volume** the volume of one gram-molecule (the molecular weight in grams, calculated from the sum of the atomic weights in grams) of a substance. Molar volume is now used instead of gram-molecular volume.

冷凝作用（名）蒸汽或氣體在溫度降低時變成液體的物態變化（第 9 頁）。（動）同爲 condense。

液化作用（名）蒸汽在壓力升高時變成液體的物態變化（第 9 頁）。（動）同爲 liquefy。

永久氣體 必須先冷卻到一個極低溫度否則不易液化的一種氣體。例如氫、氮和氧都是永久氣體。

稀有氣體：惰性氣體 指下列氣體之一：氦、氖、氬、氪、氙和氡。這些氣體不參與化學反應，通常被看成是惰性的（第 19 頁）氣體。稀有氣體都是永久氣體（↑）。其分子都是單原子的（↓）。

臨界溫度 爲每一種氣體的特定溫度。高於此溫度時，不能只靠壓力（第 102 頁）使氣體液化。氧之臨界溫度爲 −119°C（154 K），所以氧必須冷卻到 −119°C 以下才能靠壓力液化。因此氧是一種永久氣體（↑）。臨界溫度之符號爲 $T_c$。

臨界壓力（名）在臨界溫度（↑）時足使氣體液化所需的壓力。

單原子的（形）描述只由一個原子（第 110 頁）組成其分子的某種氣體。例如氖和氦都是單原子的氣體。

雙原子的（形）描述由兩個原子（第 110 頁）組成其分子的某種氣體。

多原子的（形）指由若干個原子（第 110 頁）組成其分子的某種氣體。此種氣體通常都是化合物，如二氧化碳。

原子數（名）當元素或化合物由分離的分子所組成時，即分子中所含有之原子數。在多數情況下，這些物質都爲氣態。例如：氦的原子數爲 1；氫的原子數爲 2；二氧化硫爲 3。

克分子體積 指一克分子（以克爲單位的分子量，即分子中所有原子之體積量之總和，以克爲單位）的物質所佔有之體積。現已代之以摩爾體積克分子體積。

**diatomic** 雙原子的

hydrogen 氫 ($H_2$)

**monatomic** 單原子的

neon 氖 (Ne)

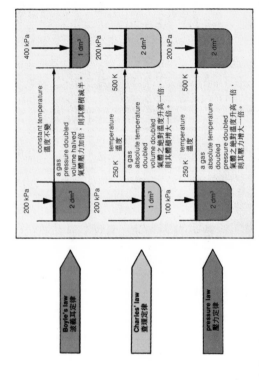

**Boyle's law** the volume of a fixed (p.79) mass of gas is inversely proportional to the pressure (p.102) at a constant (p.106) temperature; the relation can also be written as pressure times volume equals a constant. In symbols $pV = k$, where $k$ is a constant. The law is obeyed (p.107) at low pressures, but real (p.107) gases do not obey the law at high pressures.

**Charles' law** the volume of a given mass of gas increases by 1/273 of its volume at 0°C for each degree Celsius rise in temperature, if the pressure (p.102) remains constant. In symbols
$$V_\theta = V_0 (1 + \theta/273)$$

The volume of the gas is proportional to the absolute temperature (p.102), so $V \propto T$ and hence $V_1/V_2 = T_1/T_2$.

**pressure law** the pressure (p.102) of a given mass of gas increases by 1/273 of its pressure at 0°C for each degree Celsius rise in temperature, if the volume remains constant. In symbols $p_\theta = p_0 (1 + \theta/273)$.

The pressure of the gas is proportional to the absolute temperature (p.102) so $p \propto T$ and hence $p_1/p_2 = T_1/T_2$.

**波義耳定律** 在溫度不變（第 106 頁）時，一定（第 79 頁）質量之氣體體積與壓力（第 102 頁）成反比；此關係也可寫成壓力乘體積等於常數，以符號表示為 $pV = k$。氣體在低壓時遵循（第 107 頁）此定律，但在高壓時實際氣體（第 107 頁）不遵循此定律。

**查理定律** 在壓力（第 102 頁）保持不變、溫度每升高一攝氏度，一定質量的氣體體積增加其在 0°C 時體積的 $1/273$。以符號表示為：
$$V_\theta = V_0 (1 + \theta/273)$$
氣體之體積與絕對溫度（第 102 頁）成正比，即 $V \propto T$，所以 $V_1/V_2 = T_1/T_2$。

**壓力定律** 在體積保持不變、溫度每升高一攝氏度，一定質量的氣體壓力（第 102 頁）增加其在 0°C 時壓力的 $1/273$。以符號表示為：
$$P_\theta = P_0 (1 + \theta/273)$$
氣體之壓力與絕對溫度（第 102 頁）成正比，即 $P \propto T$，所以 $p_1/p_2 = T_1/T_2$。

**gas equation** the three laws, Boyle's, Charles' and pressure, can be combined into one equation which is $pV = kT$, where $k$ is a constant ($\downarrow$) depending on the mass of gas. The equation is usually written as $p_1V_1/p_2V_2 = T_1/T_2$. If the amount of gas is given as a mole fraction (p.80) with symbol $n$, then the equation becomes $pV = nRT$, where $R$ is the same constant for all gases.

**constant¹** (n) (1) the value of a physical quantity (p.81) which cannot be changed, e.g. the speed of light is a constant; the Avogadro constant (p.80), (2) the unchanging value of a physical quantity under experimental conditions (p.103), e.g. the boiling point of a liquid at s.t.p. (p.102).

**constant²** (adj) describes an unchanging value of a physical quantity which is controlled (p.221) by an observer.

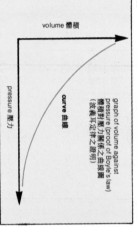

volume 體積

pressure 壓力

curve 曲線

graph of volume against pressure (proof of Boyle's law)
體積對壓力關係之曲線圖
（波義耳定律之證明）

**curve** (n) the line drawn in a graph showing the relation (p.232) between two quantities (p.81).

**derive** (v) to obtain a value or a statement (p.222) by several steps of experimental work or deduction (p.222). The work must start from something initially (p.222) known. For example, Boyle's law is derived from experiments on gases, measuring volumes and pressures, plotting (p.39) graphs and then deducing the relation between pressure and volume. The statement of the law is then derived. **derivation** (n), **derivative** (n), **derived** (adj).

**exert** (v) to bring into effect, e.g. a gas exerts a pressure on the walls of a vessel (p.25) so the pressure has an effect on the walls.

氣體方程式 波義耳定律、查理定律和壓力定律這三條定律可以合併寫成一條方程式 $pV = kT$，其中 K 是依賴於氣體質量的常數 (↓)。此方程式一般寫成 $p_1V_1/p_2V_2 = T_1/T_2$。符號為 n，則該方程式可寫成 $pV = nRT$，式中之 R 對於一切氣體都是相同的常數。

常數（名）（1）指不可改變之物理量（第 81 頁）的數值。例如，光速為常數；亞佛加德羅常數（第 80 頁）；(2)指在實驗條件（第 103 頁）下不變之物理量的值。例如，液體在標準狀態下（第 102 頁）之沸點。

不變的（形）描述受觀察者控制的（第 221 頁）物理量的不變值。

曲線（名）繪於圖上表示兩個量（第 81 頁）間關係（第 232 頁）的一條曲線。

推導（動）由若干實驗工作步驟或推論（第 222 頁）取得數值或報告（第 222 頁）。此工作一定要從一開始（第 85 頁）就知道的某些事情着手。例如波義耳定律係從對氣體的實驗，測量其體積和壓力、繪製（第 39 頁）曲線圖，然後推斷得出壓力和體積之間的關係，進而再推導出定律之陳述。（名詞為 derivation 推論、derivative 衍生物，形容詞為 derived）

施加（動）使生效應。例如氣體對容器壁施加壓力，因此壓力對器壁有影響。

理想的 (形) (1) 描述在任何壓力下都遵守 (↓) 波
義耳定律 (第 105 頁) 的氣體。此種氣體實際
上並不存在 (第 213 頁)。實際 (↓) 氣體在低
壓力下是理想氣體；(2) 描述一種物體、排
列或或理論 (第 76 頁) 是完美的，此種物體、
排列或理論一般都是假說 (第 108 頁) 的一部
分。

非理想的 (形) (1) 描述一種氣體在任何壓力下
都不遵守波義耳定律；(2) 描述 (↓) 任何物體
或排列不是理想的。

實際的；真實的 (形) 描述實際存在 (第 213 頁)
的任何物體，尤其是實際存在的任何氣體。
例如氫、氧和氮之類氣體都是實際氣體；
(2) 描述任何物件是如其所述的。例如一件
真皮袋子就不是人造皮做的袋子。

使符合 (動) 使遵循行為的模式 (第 93 頁)。例
如氣體符合波義耳定律，但在壓力高時符合
的程度差 (第 227 頁) 較差。一種材料或物質
(第 8 頁) 可以符合一條定律 (第 109 頁) 或者
符合其性質的任何其他論點 (第 222 頁)，但
實驗 (第 42 頁) 結果 (第 39 頁) 與預言 (第 85
頁) 未必完全一致。(名詞為 conformity)

遵守；依從 (動) 使與一條定律完全一致，因而
實驗 (第 42 頁) 結果 (第 39 頁) 與預言 (第 85
頁) 完全一致。例如一理想 (↑) 氣體遵守波義
耳定律。化合物定律比比較 (第 76 頁)
conform (符合) 與 obey (遵守) 之比較：如果
實驗結果與預言近似 (第 79 頁) 一致，則稱
此物質 "符合" 定律。如實驗結果完全一
致，則稱該物質 "遵循" 定律。

格雷姆定律 在溫度保持不變時，氣體擴散 (第
35 頁) 或溢透 (第 35 頁) 速率與其密度 (第 12
頁) 之平方根成反 (第 233 頁) 比 (第 76 頁)。
氣體的相對分子量 (第 114 頁) 越大，則其擴
散或溢透速率越低。

ideal (*adj*) (1) describes a gas which obeys (↓)
Boyle's law (p.105) at all pressures. Such
gases do not exist (p.213). Real (↓) gases are
ideal gases at low pressures. (2) describes an
object, arrangement or theory (p.76) that is
perfect; such an object, arrangement or theory
is usually part of a hypothesis (p.108).

non-ideal (*adj*) (1) describes a gas which does
not obey Boyle's law at all pressures. (2)
describes any object or arrangement which is
not ideal (↑).

real (*adj*) (1) describes any object that exists
(p.213), in particular any gas that exists, e.g.
gases such as hydrogen, oxygen and nitrogen
are all real gases. (2) describes any object that
is what it is said to be, e.g. a bag made of real
leather not *imitation* leather.

conform (*v*) to follow a pattern (p.93) of
behaviour, e.g. gases conform with Boyle's
law, but the degree (p.227) of conformity
becomes less at high pressures. A material or
substance (p.8) can conform with a law
(p.109), or any other statement (p.222) of its
properties, but the agreement between
experimental (p.42) results (p.39) and the
prediction (p.85) is not necessarily exact.
conformity (*n*).

obey (*v*) to be in exact agreement with a law, so
that experimental (p.42) results (p.39) are in
exact agreement with predictions (p.85), e.g.
an ideal (↑) gas obeys Boyle's law; chemical
compounds obey the law of constant
proportions (p.76). To contrast *conform* and
*obey*: if experimental results are approximately
(p.79) in agreement with predictions, then a
substance *conforms* to the law, if the results
are in exact agreement, a substance *obeys*
the law.

Graham's law the rate of diffusion (p.35), or
effusion (p.35), of a gas at a constant
temperature is inversely (p.233) proportional
(p.76) to the square root of its density (p.12).
The greater the relative molecular mass
(p.114) of a gas, the slower is its rate of
diffusion, or effusion.

porous
plug 多孔塞
gas 氣體

**Graham's law
格雷姆定律**

20 seconds
later
20 秒後

gas
effuses
氣體溢透

rate of effusion =
volume AB
20 seconds
溢透率 =
體積 AB
20 秒

**Avogadro's hypothesis** equal volumes of all gases at the same temperature and pressure contain the same number of molecules. See *Avogadro constant (p.80)*. The molar volume (p.80) of a gas at s.t.p. (p.102) contains the Avogadro number of molecules.

**Avogadro's principle** another name of Avogadro's hypothesis.

**hypothesis** (n) a statement which cannot be proved experimentally (p.42). From it, however, laws can be deduced (p.222) and proved.

gas A
pressure P_A
A 氣體之壓力

gas B
pressure P_B
B 氣體之壓力

gas C
pressure P_C
C 氣體之壓力

water
水

water vapour
水汽

gas
氣體

**Dalton's law**
道爾敦定律

total pressure
總壓力
= P_A + P_B + P_C

**Dalton's law** in a mixture of gases which do not react (p.62), the total pressure exerted (p.106) by the mixture of gases is equal to the sum of the partial pressures (↓) of each gas. Also called **Dalton's law of partial pressures.**

**partial pressure** the partial pressure of a gas in a mixture of gases is equal to the pressure which each gas would exert (p.106) if it alone filled the whole volume of the mixture.

**kinetic theory** a theory (p.76) which explains the behaviour of liquids and gases by considering them to consist (p.55) of molecules (p.77) in random (p.223) motion; the molecules hit each other and the walls of the containing vessel (p.25). The molecules are considered to be completely elastic (p.14).

atmospheric
pressure
大氣壓力
= pressure

wet gas
濕氣體壓力
= pressure

dry gas
乾氣體壓力
= pressure
+ pressure
water vapour
水汽壓力

亞伏加德羅假說 在相同的溫度和壓力下，等體積之一切氣體都含有同數目的分子。參見 “亞伏加德羅常數”（第 80 頁）。在標準狀態下（第 102 頁），一克分子體積（第 80 頁）之氣體含有亞伏加德羅數目的分子。

亞伏加德羅原理 為亞伏加德羅假說之別稱。

假說（名）指未經實驗（第 42 頁）證實之論點。但人們可由此論點來推論（第 222 頁）和證實所提之定律。

遵爾敦定律 在不發生反應（第 62 頁）時，混合氣體所施加（第 106 頁）總壓力等於每種氣體分壓（↓）的總和。此定律又稱為遵爾敦分壓定律。

分壓 某一氣體在混合氣體中的分壓，等於該氣體單獨充滿全部混合氣體積時所施加（第 106 頁）之壓力。

分子運動學說 闡明液體和氣體行為的學說。該學說（第 76 頁）認為液體和氣體都是由處於隨機（第 223 頁）運動的分子（第 77 頁）所組成（第 55 頁）。分子互相碰撞並碰撞所盛載之容器（第 25 頁）壁。這些分子被認為是完全彈性的（第 14 頁）。

s.t.p. 標準溫度
22.4 dm³ 22.4 升
6.02 × 10²³ molecules
6.02 × 10²³ 個分子

s.t.p. 標準溫度
22.4 dm³ 22.4 升
6.02 × 10²³ molecules
6.02 × 10²³ 個分子

hydrogen 氫

oxygen 氧

**Avogadro's hypothesis**
亞伏加德羅假說

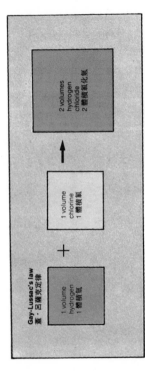

**Gay-Lussac's law** → **呂薩克定律** 在壓力和溫度保持不變下，參
**Gay-Lussac's law** the volumes of gases which
react, and the volumes of the products if
gaseous, are in the ratio (p.79) of small whole
numbers if pressure and temperature are kept
constant. For example, one volume of
hydrogen combines (p.64) with one volume of
chlorine to form two volumes of hydrogen
chloride. One and two are simple whole
numbers. *See* Charles' Law (p.105).

**law** (*n*) a statement about the properties of
materials and substances (p.8) that is
accepted as true. From it, predictions (p.85)
about the behaviour of substances can be
made. A law is universal (p.212) or limited (↓),
e.g. the law of constant proportions (p.76) is a
universal law as no substances are excepted
from it; Boyle's law is a limited law as it
depends on the condition of temperature
being constant.

**limited**[1] (*adj*) describes a law (↑), or theory
(p.76), which is true only under particular
conditions. **limitations** (*n*), **limit** (*v*).

**gas laws** statements about the behaviour of
gases under particular conditions. The
important laws are those of Boyle (p.105),
Charles (p.105) and Dalton (↑).

**Brownian motion** the random (p.223) motion of
molecules in liquids (p.10) and gases. The
motion is seen when pollen grains are
suspended (p.31) in water, or when smoke
particles are seen in air. In both cases, the
motion of molecules causes the motion of the
pollen grains or smoke particles.

蓋·呂薩克定律 在壓力和溫度保持不變下，參
加反應之各氣體的體積和各氣態產物的體積
成小的整數比（第79頁）。例如：1體積的氫
和1體積的氯化合（第64頁）生成2體積的
氯化氫。1和2都是簡單的整數。參見"查
理定律"（第105頁）。

定律（名） 被人們認為正確的對材料或物質（第8
頁）之性質或物質的論點。人們可自定律作出有關
物質性狀的預測（第85頁）。定律或是普遍
（第212頁）適用的或是有局限的（↓）。例如
定比定律（第76頁）是一條普遍適用的定
律，因為沒有哪一種物質不依此定律；波義
耳定律是一條有局限的定律，因為此定律依
賴於溫度不變這一條件。

有局限的（形） 描述某一定律（↑）或理論（第76
頁）只在特定條件下才真確。（名詞為
limitations，動詞為 limit）

氣體定律 闡明氣體在特定條件下的行為的定
律。波義耳定律（第105頁）、查理定律（第
105頁）和道爾敦定律（↑）是重要的氣體定
律。

布朗運動 指液體（第10頁）和氣體中分子的隨
機（第223頁）運動。花粉粒懸浮（第31頁）
於水中，或空氣中的煙霧微粒，都可見到這
種運動。這兩種情況都是分子運動引起花粉
粒或煙霧微粒的運動。

**atom** (n) the smallest particle (↓) of an element (p.8) that has the properties of that element. Atoms combine chemically to form molecules (p.77). An atom consists of a nucleus (↓) with electrons (↓) around it. **atomic** (adj).

原子 (名) 元素 (第 8 頁) 中具有該元素性質的最小粒子 (↓)。原子以化學結合形成分子 (第 77 頁)。原子由原子核及繞原子核運動的電子 (↓) 組成。(形容詞爲 atomic)

**subatomic** (adj) describes particles (↓) smaller than an atom (↑).

次原子的 (形) 亞原子的(形) 描述小於原子 (↑) 的粒子 (↓)。

**electron** (n) a subatomic (↑) particle with a negative electric charge (p.138). It is the smallest subatomic particle with a mass of $9.109 \times 10^{-31}$ kg.

電子 (名) 帶負電荷 (第 138 頁) 的次原子 (↑) 粒子。電子是最小的次原子粒子,其質量爲 $9.109 \times 10^{-31}$ 千克。

**proton** (n) a subatomic (↑) particle with a positive charge (p.138). The charge is equal and opposite to that of an electron (↑). The mass of a proton is 1840 times that of an electron.

質子 (名) 帶正電荷 (第 138 頁) 的次原子 (↑) 粒子。電荷與電子 (↑) 相等但符號相反。質子質量爲電子的 1840 倍。

**neutron** (n) a subatomic (↑) particle with a mass almost equal to the mass of a proton (↑). It has no electric charge (p.138).

中子 (名) 質量與質子 (↑) 幾乎相等的次原子 (↑) 粒子。中子不帶電荷 (第 138 頁)。

**nucleus** (n) the nucleus of an atom (↑) consists (p.55) of protons (↑) and neutrons (↑), except for hydrogen, which has a nucleus consisting of one proton. The nucleus has a positive electric charge, the size of which depends upon the number of protons it contains. The nucleus provides the mass of an atom. **nuclear** (adj).

原子核 (名) 原子核由質子 (↑) 和中子 (↑) 組成 (第 55 頁);但氫原子例外,其原子核只具有一個質子。原子核帶正電荷,核之大小取決於所含的質子數目。原子核提供原子之質量。(形容詞爲 nuclear)

**nuclear** (adj) describes anything to do with the nucleus (↑) of an atom.

原子核的 (形) 描述與原子核 (↑) 有關的任何事物。

**particle²** (n) a piece of substance so small that it is considered to have mass but not size or volume, that is, it is like a point. Atoms (↑) and molecules (p.77) are particles in this sense.

粒子 (名) 指物質的一部分小得被認爲是只具有質量但無大小或體積,亦即像一個點。就此意義而言,原子 (↑) 和分子 (第 77 頁) 都是粒子。

**positron** (n) a subatomic (↑) particle with a mass the same as the mass of an electron (↑), but with a positive charge equal and opposite to the charge on an electron.

正電子 (名) 質量與電子 (↑) 相同、帶正電荷,電荷量與電子相等但符號相反之次原子 (↑) 粒子。

**meson** (n) subatomic (↑) particles with a mass about half that of a proton (↑) and with zero spin; they appear to bind nuclear particles together.

介子 (名) 質量約爲質子 (↑) 的一半、具有零旋轉的次原子 (↑) 粒子;介子似用以結合原子核的粒子。

**orbit** (n) in the first models (p.223) of atoms, the electrons (↑) were considered to be in orbits round the nucleus (↑). An orbit is a circular path round a nucleus. This idea is no longer considered to be true.

軌道 (名) 在原子的最初模型 (第 223 頁) 中,電子 (↑) 被認爲是繞繞原子核的各軌道中。軌道是一條繞原子核的圓形路徑。現在這個概念現已被認爲是不正確的了。

**structure of an atom**
原子的結構

electrons around
nucleus
繞原子核運動的電子

nucleus
原子核

not to scale
不按比例

orbit 軌道

orbit
軌道

**shell** (*n*) a spherical (↓) space around a nucleus (↑); it contains extranuclear (p.113) electrons with energy levels (p.152) of that shell. An atom possesses different shells of increasing radii (↓) around its nucleus (↑). The radius of a shell is 100 000 times greater than the radius of the nucleus.

**sphere** (*n*) a solid in the shape of a ball.

**spherical** (*adj*).

**radius** (*n*) (*radii n.pl.*) (1) the line joining the centre of a circle to the line of the circle; the line drawn from the centre of a sphere to its surface. (2) the length of the line described above, e.g. a circle with a radius of 5 cm.

**orbital** (*n*) a space in which there can be one or two electrons (↑) but not more. The space is where there is a probability (p.223) of finding one or both electrons. The probability varies (p.218) over the space, and can be shown in various ways by a diagram (p.29). The position and motion of an electron in an orbital cannot be described, only the probability can be described. One way of showing probability is by shading, *see diagram*. Another way is by drawing the limits (p.211) of the orbital. Two electrons in an orbital do not interfere (p.216) with each other; two electrons form a stable (p.74) orbital, while one electron in an orbital is active in forming chemical bonds (p.133). The orbitals commonly taking part in bonds are s-, p- and d-orbitals (p.112).

**殼層** (名) 指環繞原子核 (↑) 之球形 (↓) 空間；殼層包含核外的 (第113頁) 電子，這些電子具有該殼層之能階 (第152頁)。一個原子具有不同的殼層，各殼層圍繞着原子核 (↑)，半徑 (↓) 依次增大。殼層之半徑比原子核半徑大 100000 倍以上。

**球** (名) 具球形的一個固體。(形容詞為 spherical)

**半徑** (名) (1) 指自圓心連接圓之直線：指自球心向球面畫出之直線；(2) 指上直線之長度。例如半徑爲 5 cm 之圓。(複數詞爲 radii)

**軌態；軌道** (名) 指可以含一個或最多兩個電子 (↑) 的一個空間。這是出現一個電子或兩個電子概率的空間，概率在此空間範圍內改變 (第218頁)，並可用簡圖 (第29頁) 以不同方式表示。您不能描述電子在軌道中出現的位置和運動，只能描述其出現的一種概率。描影法是表示概率的一種方法 (圖見第29頁)。另一種方法是畫出軌道的範圍 (第211頁)。處於一個軌道中的兩個電子互不干擾 (第216頁) 軌道；而在一個軌道中只有一個電子時，對形成化學鍵則很活躍 (第133頁)。通常參加鍵的是 s、p、d 軌道 (第112頁)。

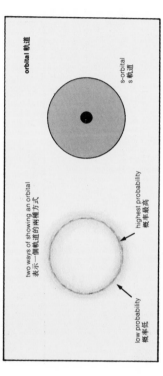

**orbital 軌道**

two ways of showing an orbital
表示一個軌道的兩種方式

highest probability
概率最高

low probability
概率最低

s-orbital
s 軌道

**s-orbital** (n) an orbital (p.111) in an atomic shell (p.111) containing one or two electrons (p.110). There is only one s-orbital in a shell. An s-orbital is spherically (p.111) symmetrical (p.93) about a nucleus. It has no directional properties.

**p-orbital** (n) an orbital (p.111) containing one or two electrons (p.110). There are three p-orbitals in an atomic shell (p.111). The three orbitals are at right angles to each other along three axes (p.92) and have directional properties.

**d-orbital** (n) an orbital (p.111) containing one or two electrons (p.110). There are five d-orbitals in an atomic shell (p.111). Four of the orbitals have the same shape, and one has a different shape. All five orbitals have directional properties.

**s-electron** (n) an electron (p.110) in an s-orbital (p.111). s-electrons have the least energy in a shell (p.111). Two s-electrons fill the s-orbital before an electron goes into a p-orbital.

**p-electron** (n) an electron (p.110) in a p-orbital (p.111). One p-electron goes into each of the three p-orbitals before the fourth electron goes into one p-orbital to fill it. A total of six p-electrons fill the three p-orbitals before an electron goes into the next higher energy level.

**d-electron** (n) an electron (p.110) in a d-orbital (p.111). One d-electron goes into each of the five d-orbitals before the sixth electron goes into one d-orbital to fill it.

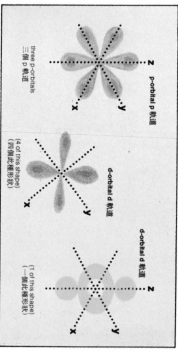

three p-orbitals
三個 p 軌道

p-orbital p 軌道

(4 of this shape)
(四個此種形狀)

d-orbital d 軌道

(1 of this shape)
(一個此種形狀)

d-orbital d 軌道

s 軌道 (名) 原子殼層 (第 111 頁) 中容納一個或兩個電子 (第 110 頁) 的一個軌道 (第 111 頁)。一個殼層中只有一個 s 軌道。s 軌道圍繞原子核呈球形 (第 111 頁) 對稱 (第 93 頁)。s 軌道無方向性。

p 軌道 (名) 容納一個或兩個電子 (第 110 頁) 的一個軌道 (第 111 頁)。一個原子殼層 (第 111 頁) 中有三個 p 軌道，分別沿三個軸 (第 92 頁) 互成直角，且有方向性。

d 軌道 (名) 容納一個或兩個電子 (第 110 頁) 的一個軌道 (第 111 頁)。一個原子殼層 (第 111 頁) 中有五個 d 軌道，其中四個軌道具有相同形狀，只有一個軌道具有不同之形狀，全部五個軌道都有方向性。

s 電子 (名) 指 s 軌道 (第 111 頁) 中之電子 (第 110 頁)。s 電子在一個殼層 (第 111 頁) 的能量最低。只有兩個電子填滿 s 軌道之後，第三個電子才能進入一個 p 軌道。

p 電子 (名) 指 p 軌道 (第 111 頁) 中的電子 (第 110 頁)。三個 p 軌道都各有一個電子進入之後，第 4 個電子才能進入而填滿一個 p 軌道。全部 6 個 p 電子都填滿三個 p 軌道之後，其他電子才可進入下一個更高的能級。

d 電子 (名) 指 d 軌道 (第 110 頁) 中的電子 (第 111 頁)。五個 d 軌道都各有一個電子進入之後，第六個電子才進入填滿一個 d 軌道。

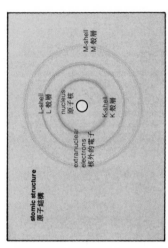

**atomic structure**
**原子結構**

L-shell
L 殼層

nucleus
原子核

extranuclear
electrons
核外的電子

M-shell
M 殼層

K-shell
K 殼層

**extranuclear** (*adj*) describes electrons (p.110) outside the nucleus (p.110) of an atom.

**atomic structure** the structure of the nucleus (p.110) and of the extranuclear (↑) electrons. The extranuclear electrons are arranged in atomic shells (p.111).

**K-shell** (*n*) the innermost atomic shell (p.111). It contains only one s-orbital. This arrangement completes the shell.

**L-shell** (*n*) the next atomic shell to the K-shell (↑). It contains one s-orbital and three p-orbitals. This arrangement completes the shell.

**M-shell** (*n*) the next atomic shell to the L-shell (↑). It contains one s-orbital, three p-orbitals and five d-orbitals. This completes the M-shell. There are further shells beyond the M-shell.

**atomic number** a number equal to the number of protons (p.110) in the nucleus (p.110) of an atom of an element. It has the symbol Z.

**mass number** a number equal to the sum of the number of protons (p.110) and number of neutrons (p.110) in the nucleus (p.110) of an atom of an element. It is the mass number of an isotope (p.114). The symbol for mass number is A.

**relative atomic mass** the ratio (p.79) of the mass of one atom of an element to one-twelfth of the mass of one atom of carbon-12. Most elements consist of isotopes (p.114) of its atoms, hence the relative atomic mass measures the average mass per atom of the normal isotopic composition of an element.

核外的（形）形容電子（第 110 頁）在原子核（第 110 頁）之外。

**原子結構**（名）指原子核（第 110 頁）和核外（↑）電子的結構。核外的電子排佈在原子殼層（第 111 頁）中。

**K 殼層**（名）指最靠近核的原子殼層（第 111 頁）。K 殼層只有一個 s 軌道，此排佈完成此殼層。

**L 殼層**（名）指 K 殼層的次一個原子殼層。L 殼層有一個 s 軌道和三個 p 軌道，這種排佈完成此殼層。

**M 殼層**（名）指 L 殼層的次一個原子殼層。M 殼層有一個 s 軌道、三個 p 軌道和 5 個 d 軌道。此排佈完成此殼層。M 殼層之外還有其他殼層。

**原子序** 等於元素原子核（第 110 頁）中質子（第 110 頁）數的一個數目，符號爲 Z。

**質量數** 等於某一元素的一個原子的原子核（第 110 頁）中質子（第 110 頁）數與中子（第 110 頁）數總和的數目。這是一個同位素（第 114 頁）的質量數。質量數的符號爲 A。

**相對原子質量** 某一元素的原子之質量與碳 -12 原子質量的 1/12 之比。大多數元素都由其原子的同位素（第 114 頁）所組成，因此相對原子質量作爲計算組成一元素的正常組成每一個原子的平均質量。

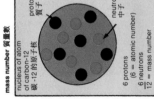

**mass number** 質量數

proton
質子

neutron
中子

nucleus of atom
of carbon-12
碳 -12 的原子核

6 protons (6 = atomic number)
6 neutrons
12 = mass number

6 個質子 (6 = 原子序數)
6 個中子
12 = 質量數

**isotope** (*n*) atoms of the same element (p.116) which have different mass numbers (p.113) are isotopes of the element. As all atoms of an element have the same number of protons (p.110), isotopes differ in having different numbers of neutrons (p.110) in the nucleus (p.110) of the atoms, e.g. there are two isotopes of carbon, one with a mass number of 12 and one with a mass number of 13. Both isotopes have six protons in their nuclei, so one isotope has six neutrons and the other seven neutrons. Isotopes are shown by writing the mass number after the element, e.g. carbon-12 is the isotope of carbon with a mass number of 12. **isotopic** (*adj*).

**isotopic ratio** (*n*) in a specimen (p.43) of an element. For elements obtained from natural sources (p.138) the isotopic ratio is always the same, e.g. the isotopic ratio for carbon is 98.9% C-12 and 1.1% C-13. The relative atomic mass of carbon is thus slightly greater than 12.

**relative isotopic mass** (*n*) (of an isotope) the ratio of the mass of one atom of the isotope to 1/12 of the mass of one atom of carbon-12.

**isotopic weight** (*n*) another name for relative isotopic mass (↑), no longer used.

**atomic mass unit** a mass equal to one-twelfth of the mass of an atom of the isotope (↑) carbon-12. Its value is 1.66043 × 10⁻²⁷ kg. The symbol for atomic mass unit is *amu*.

**atomic weight** (*n*) the ratio of the mass of one atom of an element to one-sixteenth the mass of an atom of oxygen. Relative atomic mass is now used instead of atomic weight, and the method of measuring is different with a standard of carbon-12 instead of oxygen.

**relative molecular mass** (*n*) the ratio (p.79) of the mass of one molecule of a substance (p.8) to the mass of one atom of carbon-12. The relative molecular mass of a compound is calculated by adding the relative atomic masses of all the atoms in a molecule of the compound. Relative molecular mass is used with covalent compounds, but not with ionic compounds. *See relative formula mass (p.78).*

同位素 (名) 同一種元素 (第116頁) 中，具有不同質量數 (第113頁) 的各種原子為該元素的同位素。由於同一種元素的所有原子，其質子 (第110頁) 數目都相同，故同位素的差別僅在於原子核 (第110頁) 的中子 (第110頁) 數目不同。例如碳有兩種同位素，兩種同位素的原子核中都有6個質子，所以其中一種有6個中子，另一種則有7個中子。元素名稱之後類上質量數即表示該元素的同位素。例如碳-12表示質量數為12的碳的同位素。(形容詞為 isotopic)

同位素比 (名) 在一種元素的一個標本 (第43頁) 中，不同同位素 (↑) 之比 (第79頁)。從天然源 (第138頁) 取得的元素，其同位素比永遠相同。例如，碳的同位素比為98.9%的C-12和1.1%的C-13。因此碳的相對原子質量稍大於12。

相對同位素質量 (名) （對同位素而言）指一個同位素的原子質量與一個碳-12原子之質量的 1/12 之比。

同位素重量 (名) 相對同位素質量 (↑) 之別稱，現已廢除不用此名稱。

原子質量單位 質量等於碳-12同位素 (↑) 之原子質量之 1/12。數值為 1.66043 × 10⁻²⁷ 千克。原子質量單位的符號為 *amu*。

原子量 (名) 一種元素的一個原子質量與一個氧原子質量的 1/16 之比。現已用“相對原子質量”，這一術語代替原子量，相對原子質量的不同是用碳-12代替氧作為標準。

相對分子質量 (名) 一個物質 (第8頁) 分子的質量與一個碳-12原子的質量之比 (第79頁)。化合物的相對分子質量等於化合物分子中全部原子的相對原子質量之和。相對分子質量使用於共價化合物而不使用於離子化合物。見“相對化學式質量”（第78頁）。

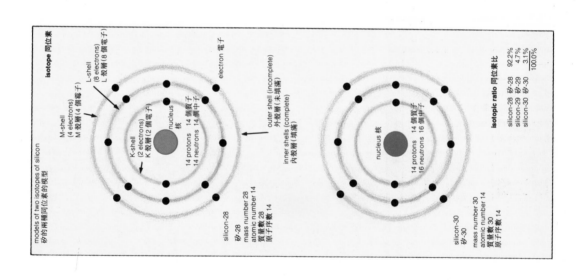

**isotope 同位素**

models of two isotopes of silicon
矽的兩種同位素的模型

M-shell
(4 electrons)
M 殼層 (4 個電子)

L-shell
(8 electrons)
L 殼層 (8 個電子)

K-shell
(2 electrons)
K 殼層 (2 個電子)

nucleus
核

14 protons
14 neutrons
14 個質子
14 個中子

electron 電子

outer shell (incomplete)
外殼層 (未填滿)

inner shells (complete)
內殼層 (填滿)

silicon-28
矽-28
mass number 28
atomic number 14
質量數 28
原子序數 14

nucleus
核

14 protons
16 neutrons
14 個質子
16 個中子

silicon-30
矽-30
mass number 30
atomic number 14
質量數 30
原子序數 14

**isotopic ratio 同位素比**

| | |
|---|---|
| silicon-28 矽-28 | 92.2% |
| silicon-29 矽-29 | 4.7% |
| silicon-30 矽-30 | 3.1% |
| | 100.0% |

**element²** (n) a substance (p.8) with all its atoms (p.110) having the same positive charge on the nucleus (p.110), i.e. the nucleus of all atoms has the same number of protons (p.110) and thus the same atomic number, and this determines the chemical nature of the element. **elementary** (adj).

**metal** (n) an element (↑) which forms positive ions (p.123) in chemical reactions. The general physical properties of metals are: (a) they conduct (p.122) electric current and heat; (b) they are lustrous (p.16), ductile (p.14) and malleable (p.14). The general chemical properties of metals are: (a) they form basic (p.46) oxides; (b) they form compounds with non-metals which are salts. Metals possess these properties in varying (p.218) degrees (p.227). All metals, except mercury, are solids. **metallic** (adj).

**non-metal** (n) an element (↑) which is not a metal (↑). Non-metals are solids or gases, except bromine which is a liquid. Their physical properties depend upon their structure (p.82). Solid non-metals are neither ductile (p.14) nor malleable (p.14), but are usually brittle (p.14). They do not conduct electric current, except for graphite (p.118), nor heat. They generally form negative ions (p.123) in chemical reactions, except for hydrogen which forms positive ions. Their oxides are generally acidic, but some are neutral, e.g. carbon monoxide.

元素（名） 指其所有全部原子（第110頁）都帶相同正電荷數的一類物質的（第110頁），即同種元素的所有原子的原子核都含有相同的質子（第110頁）數，因而原子序亦相同，而這確定該元素的化學本質。（形容詞爲 elementary）

金屬（名） 能在化學反應中形成正離子（第123頁）的一類元素（↑）。金屬的一般物理性質為：(a) 金屬能傳導（第122頁）電流和熱；(b) 金屬具有光澤（第16頁）、延性（第14頁）和展性（第14頁）。金屬的一般化學性質為：(a) 金屬能生成鹼性（第46頁）氧化物；(b) 金屬可與非金屬形成鹽的化合物。每一種金屬均在不同（第218頁）程度（第227頁）上具有這些性質。除水銀外，所有一切金屬都是固體。（形容詞為 metallic）。

非金屬（名） 指不屬於金屬（↑）的一類元素（↑）。除溴是液體外，其他非金屬不是固體就是氣體。非金屬的物理性質取決於其結構（第82頁）。固體非金屬既非延性的（第14頁）亦非展性的（第14頁），通常都是脆的（第14頁）。除了石墨（第118頁）外，非金屬不導電，也不導熱。在化學反應中，非金屬一般都生成負離子（第123頁），但氫在反應中形成正離子。非金屬的氧化物通常都是酸性的，但亦有一些是中性的，例如一氧化碳。

| | metals 金屬 | non-metals 非金屬 |
|---|---|---|
| elements 元素 | | |
| solids 固體 | iron 鐵　gold 金　sodium 鈉　copper 銅 | sulphur 硫　phosphorus 磷 |
| liquids 液體 | mercury 汞 | bromine 溴 |
| gases 氣體 | none 無 | oxygen 氧　hydrogen 氫　nitrogen 氮 |

**metalloid** (*n*) an element with some of the properties of metals and some of the properties of non-metals, e.g. antimony and arsenic are metalloids.

**alkali metal** a metal (↑) which forms an ion (p.123) with an oxidation state (p.135) of +1 only. The oxide is very soluble in water forming a hydroxide (p.48). Alkali metals are chemically very reactive (p.62); they are the elements of group I of the periodic table (p.119) and have one s-electron (p.112) in their outer shells (p.111).

**alkaline earth metal** a metal (↑) which forms an ion (p.123) with an oxidation state (p.135) of +2 only. The oxide is sparingly soluble in water forming a hydroxide (p.48). Alkaline earth metals are chemically reactive (p.62); they are the elements of group II of the periodic table (p.119) and have two s-electrons (p.112) in their outer shell (p.111).

**coinage metal** a metal (↑) which which does not oxidize (p.70) readily in the air, e.g. copper, silver, gold. The coinage metals are not chemically reactive (p.62); they form positive ions with variable (p.218) oxidation states (p.135); they are all transitional elements (p.121).

**base metal** a metal (↑) which is oxidized (p.70) by heating in air and is acted upon by mineral acids (p.55). Base metals are used for making articles for everyday use. For example, iron, lead, tin, zinc; some are transitional metals, e.g. iron, tin, lead.

**halogen** (*n*) a non-metal (↑), which forms a negative ion (p.123) with an electrovalency (p.134) of −1 and is in group VII of the periodic table (p.119). The halogens are fluorine, chlorine, bromine and iodine.

**類金屬** (名) 既具有某些金屬性質又具有某些非金屬性質的元素。例如銻和砷就是類金屬。

**鹼金屬** 能形成氧化態 (第135頁) 只為 +1 的離子 (第123頁) 的一種金屬 (↑)。其氧化物極易溶於水，生成氫氧化物 (第48頁)。鹼金屬在化學上是極活潑 (第62頁) 的，在週期表 (第119頁) 中列為 I 族的元素，其外殼層 (第111頁) 中有一個 s 電子 (第112頁)。

**鹼土金屬** 能形成氧化態 (第135頁) 只具有 +2 的離子 (第123頁) 的一種金屬 (↑)。其氧化物難溶於水但形成水作用也形成氫氧化物 (第48頁)。鹼土金屬是化學上活潑 (第62頁) 的；在週期表中 (第119頁) 中列為 II 族元素，其外殼層 (第111頁) 中有 2 個 s 電子 (第112頁)。

**鑄幣金屬** 在空氣中不易氧化 (第70頁) 的一種金屬 (↑)。例如銅、銀和金。鑄幣金屬在化學上不活潑 (第62頁) 這些金屬可形成具有變化 (第218頁) 的正離子。這些金屬都是過渡元素 (第121頁)。

**賤金屬** 在空氣中受熱會氧化 (第70頁) 並可與無機酸 (第55頁) 起作用的一種金屬 (↑)。賤金屬用於製造日用品。例如鐵、鉛、錫和鋅是賤金屬，其中有一些是過渡金屬，例如鐵、錫和鉛。

**鹵素** (名) 一種非金屬 (↑)，可形成電化價 (第134頁) 為 −1 的負離子 (第123頁)，在週期表 (第119頁) 中列為 VII 族的元素。鹵素包括氟、氯、溴和碘。

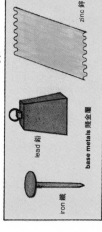

iron 鐵    lead 鉛

zinc 鋅

base metals 賤金屬

**coinage metal 鑄幣金屬**

copper 銅

silver 銀

gold 金

**allotropy** (*n*) the existence of two or more different forms of an element without a change of state (p.9). If the forms are crystalline (p.15) they are polymorphic (p.92) as well as allotropic, e.g. sulphur exists in five allotropes (↓). carbon has two crystalline allotropes. **allotropic** (*adj*), **allotrope** (*n*).

**dynamic allotropy** a kind of allotropy in which the allotropes are in a dynamic equilibrium (p.150) with each other; e.g. liquid sulphur has three allotropes (↓) which show dynamic allotropy.

**allotrope** (*n*) one of the forms of an element showing allotropy (↑).

**carbon** (*n*) a non-metal (p.116) with atomic number (p.113) 6, relative atomic mass (p.113) of 12.01 and in group IV of the periodic table (↓). See *isotope* (p.114). Carbon is one of the most important elements as it is a constituent (p.54) of all living things. It occurs (p.63) in two crystalline forms, diamond (↓) and graphite (↓). **carbonic** (*adj*), **carbonaceous** (*adj*).

**diamond** (*n*) a crystalline form of carbon with a tetrahedral (p.83) lattice (p.92). It is the hardest natural substance; it does not conduct electric current.

**graphite** (*n*) a crystalline form of carbon with a hexagonal (p.83) lattice (p.92). It is a soft substance that marks paper. It is the only non-metal that conducts electric current.

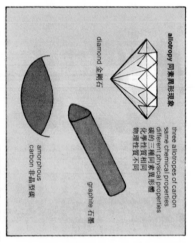

allotropy 同素異形現象

three allotropes of carbon same chemical properties different physical properties
碳的三種同素異形體化學性質相同物理性質不同

diamond 金剛石

amorphous carbon 非晶型碳

graphite 石墨

同素異形現象（名）同一種元素存在在兩種或多種不同形狀而沒有物態變化（第15頁）的現象。如果是晶體（第15頁）形狀，則這些晶體不僅是同素異形而且是多晶型的（第92頁）。例如硫有五種同素異形體（↓）；碳有兩種晶體的同素異形體。（形容詞爲 allotropic，名詞爲 allotrope）

動態同素異形現象（名）同素異形體間相互處於動態平衡（第150頁）的一種同素異形現象。例如液態硫有三種同素異形體（↓），呈現動態同素異形現象。

同素異形體（名）呈現同素異形現象（↑）的一種元素的形狀之一。

碳（名）原子序（第113頁）爲6，相對原子質量（第113頁）爲12.01，在週期表（↓）中排在第IV族的一種非金屬（第116頁）。多見"同位素"（第114頁）。碳是最重要的元素之一，因爲它是一切生物體內含有的一種成分（第54頁）。碳以兩種晶體形式出現（第63頁）。（形容詞爲 carbonic、carbonaceous）

金剛石：鑽石（名）具四面體（第83頁）晶格的一種碳的晶形。金剛石是最硬的天然物質，它不導電。

石墨（名）具六方的（第96頁）晶格（第92頁）的一種碳晶形。石墨是一種軟的物質，可在紙上留下痕迹。它是唯一一能導電的非金屬。

**periodic system** if the elements are arranged in order of increasing atomic number (p.113), a periodicity (p.120) is seen in their properties. For example, the elements with atomic numbers 3, 11, 19 are all chemically active metals; atomic numbers 9, 17, 35, 53 are all chemically active non-metals with an electrovalency (p.134) of − 1: atomic numbers 4, 12, 20, 38, 56 are all metals with an oxidation state (p.135) of II. The arithmetical difference between any pair of these numbers is 8 or 18. This suggests a relation (p.232) between the extranuclear (p.113) structure of the electrons and the periodicity of properties. The periodic system is shown by a periodic table (↓).

**periodic table** the periodic table is shown on p.120 and endpapers. The atomic number (p.113) of each element is shown and the table is arranged in periods (p.120) and groups (↓).

**group** (n) a vertical column of the periodic table (↑). Group I contains the alkali metals (p.117), group II the alkaline earth metals (p.117) and group VII the halogens (p.117). The elements in a group have very similar properties. For metals, chemical reactivity increases from elements with low atomic numbers (p.113) to those with high atomic numbers. For non-metals, chemical reactivity decreases (p.219) from elements with low atomic numbers. The elements in group I have one s-electron (p.111) in their outer shell (p.111), and those in group II have two s-electrons; elements in group VII have two s-electrons and five p-electrons (p.112).

週期系 將元素按原子序(第113頁)遞增順序排列，可看出元素性質的週期性(第120頁)變化。例如，原子序為3、11、19 的元素都是化學活潑的金屬；原子序數為9、17、35、53 的元素都是電價(第134頁)為 −1 的化學活潑非金屬；原子序為4、12、20、38、56 的元素都是氧化態(第135頁)II 的金屬。這些數字中，任何一對數字之間的算術差都是8或18。這就顯示出核外(第113頁)電子結構與性質週期性變化之間的一種關係(第232頁)。週期系是由週期表(↓)顯示出來。

週期表 週期表見本書第120頁和襯頁。週期表顯示各元素的原子序(第113頁)並按週期(第120頁)和族(↓)排列。

族 (名) 指週期表(↑)的每一縱行。I 族包含鹼金屬(第117頁)、II 族包含鹼土金屬(第117頁)、VII 族包含鹵素(第117頁)。同一族之元素，其性質極相似。金屬的化學反應性自原子序(第113頁)低之元素向原子序高之元素遞增。非金屬的化學反應性自原子序低之元素向原子序高之元素遞降(第219頁)。I 族中之元素，其外殼層(第111頁)有一個 s 電子(第112頁)，II 族中之元素有2個 s 電子，VII 族中之元素有2個 s 電子和5個 p 電子(第112頁)。

| group (I) alkali metals I族鹼金屬 | | | |
|---|---|---|---|
| atomic 原子序 number | relative atomic mass 相對原子質量 | element 元素 | chemical 化學活 activity 潑性 |
| 3 | 6.94 | lithium 鋰 | |
| 11 | 22.98 | sodium 鈉 | increases 增加 → |
| 19 | 39.10 | potassium 鉀 | |
| 37 | 85.47 | rubidium 銣 | |
| 55 | 132.91 | caesium 銫 | |
| **group (VII) halogens VII 族鹵素** | | | |
| 9 | 18.99 | fluorine 氟 | increases 增加 |
| 17 | 35.45 | chlorine 氯 | |
| 35 | 79.91 | bromine 溴 | |
| 53 | 126.90 | iodine 碘 | |

| period 週期 | group 族 1 | group 族 2 | transition elements 過渡元素 | group 族 3 | group 族 4 | group 族 5 | group 族 6 | group 族 7 | group 族 8 |
|---|---|---|---|---|---|---|---|---|---|
| 1 | 1<br>H<br>1.01 | | | | | | | | 2<br>He<br>4.00 |
| 2 | 3<br>Li<br>6.94 | 4<br>Be<br>9.01 | | 5<br>B<br>10.81 | 6<br>C<br>12.01 | 7<br>N<br>14.01 | 8<br>O<br>16.00 | 9<br>F<br>19.00 | 10<br>Ne<br>20.18 |
| 3 | 11<br>Na<br>22.99 | 12<br>Mg<br>24.31 | | 13<br>Al<br>26.98 | 14<br>Si<br>28.09 | 15<br>P<br>30.97 | 16<br>S<br>32.06 | 17<br>Cl<br>35.45 | 18<br>Ar<br>39.45 |
| 4 | 19<br>K<br>39.10 | 20<br>Ca<br>40.08 | | 31<br>Ga<br>69.72 | 32<br>Ge<br>72.59 | 33<br>As<br>74.92 | 34<br>Se<br>78.96 | 35<br>Br<br>79.90 | 36<br>Kr<br>83.80 |
| 5 | 37<br>Rb<br>85.47 | 38<br>Sr<br>87.62 | | 49<br>In<br>114.82 | 50<br>Sn<br>118.69 | 51<br>Sb<br>121.75 | 52<br>Te<br>127.90 | 53<br>I<br>126.90 | 54<br>Xe<br>131.30 |
| 6 | 55<br>Cs<br>132.91 | 56<br>Ba<br>137.33 | | 81<br>Tl<br>204.37 | 82<br>Pb<br>207.2 | 83<br>Bi<br>208.98 | 84<br>Po<br>(209) | 85<br>At<br>(210) | 86<br>Rn<br>(222) |
| 7 | 87<br>Fr<br>(223) | 88<br>Ra<br>226.03 | | | | | | | |

atomic number 原子序
symbol 符號
relative atomic mass 相對原子質量

**periodic table 週期表**

**period** (n) a horizontal row of the periodic table (p.119). It represents the gradual filling of s-orbitals (p.112) and p-orbitals (p.112), and for the transitional elements, the d-orbitals (p.112). The elements in a period change gradually from a characteristic (p.9) metal on the left of the period to a characteristic non-metal on the right. **periodic** (adj), **periodicity** (n).

**periodicity** (n) the regular occurrence (p.63) of similar chemical properties with increasing atomic number (p.113). This is chemical periodicity.

**classify** (v) to put materials, substances (p.8), objects, processes (p.157) into classes. In chemistry, to put elements into groups (p.119) of the periodic system; to put compounds into classes such as acids, alkalis, etc. **classification** (n).

週期 (名) 指週期表 (第 119 頁) 的每一橫排。週期表示 s 軌道 (第 112 頁) 和 p 軌道 (第 112 頁) 以及對於過渡元素而言還有 d 軌道 (第 112 頁) 的漸次填入的電子。同一個週期的各個元素，由週期左邊的金屬特徵 (第 9 頁) 逐步變向右邊的非金屬特徵。(形容詞屬 periodic，名詞屬 periodicity)

週期性 (名) 隨原子序數 (第 113 頁) 增加而有規律地出現 (第 63 頁) 相似的化學性質。此為化學的週期性。

分類 (動) 將材料、物質 (第 8 頁) 、物體、過程 (第 157 頁) 等歸類、化學上指將各種元素歸入週期系的族 (第 119 頁) 中或將化合物分成諸如酸、鹼等類別。(名詞屬 classification)

transition elements
過渡元素

| period 週期 | | | | transition elements 過渡元素 | | | | | | |
|---|---|---|---|---|---|---|---|---|---|---|
| 4 | 21 Sc 44.96 | 22 Ti 47.90 | 23 V 50.94 | 24 Cr 52.00 | 25 Mn 54.94 | 26 Fe 55.85 | 27 Co 58.99 | 28 Ni 58.70 | 29 Cu 63.55 | 30 Zn 65.38 |
| 5 | 39 Y 88.91 | 40 Zr 91.22 | 41 Nb 92.91 | 42 Mo 95.94 | 43 Tc 98.91 | 44 Ru 101.07 | 45 Rh 102.91 | 46 Pd 106.4 | 47 Ag 107.87 | 48 Cd 112.41 |
| 6 | 57* La 138.91 | 72 Hf 178.49 | 73 Ta 180.95 | 74 W 183.85 | 75 Re 186.2 | 76 Os 190.2 | 77 Ir 192.22 | 78 Pt 195.09 | 79 Au 196.97 | 80 Hg 200.59 |
| 7 | 89** Ac 227.03 | 104 Rf (261) | 105 Hn (260) | 106 (263) | | | | | | |

atomic number 原子序
symbol 元素符號
relative atomic mass 相對原子質量

*Lanthanide elements 57-71    *57-71 為鑭系元素
**Actinide elements 89-92    **89-92 為錒系元素

**transition element** an element in one of three periods (↑), 4, 5 and 6, each period containing 10 transitional elements. The first three periods of the periodic table (p.119) contain no transition elements. The transition elements are formed by the five d-orbitals (p.112) being filled in turn. The extranuclear (p.113) electrons cause the transitional elements to possess variable (p.218) electrovalencies (p.134) and the characteristics (p.9) of metals. These elements form complex ions (p.132) and many coloured compounds.

**s-block elements** the elements of groups I and II. They have 1 or 2 s-electrons (p.112) in their outer shell (p.111).

**p-block elements** the elements of groups III to VIII inclusive. They have 1 – 6 p-electrons (p.112) in their outer shell (p.111).

**d-block elements** the transitional elements. They have s-electrons (p.112) in their outer shell (p.111) and 1–10 d-electrons (p.112) in the next inner shell. The number and arrangement of the d-electrons determine the nature of the element.

**f-block elements** an inner transition series. They have s-electrons in the outer orbitals and fill the f-orbitals, allowing for 14 elements in all.

**過渡元素** 第 4、5、6 三個週期 (↑) 中任一個週期內的一種元素，各個週期都包含 10 個過渡元素。週期表 (第 119 頁) 中的前三個週期不包含過渡元素。過渡元素是由核外依次被填入的 5 個 d 軌道 (第 112 頁) 構成的。核外 (第 113 頁) 電子 (使過渡元素具有可變的 (第 218 頁) 的價 (第 134 頁) 和具有金屬特徵 (第 9 頁)。過渡元素可形成錯離子 (第 132 頁) 和多種有色的化合物。

**s 區元素** 包括第 I 族和第 II 族中的元素，其外殼層 (第 111 頁) 中有 1 或 2 個 s 電子 (第 112 頁)。

**p 區元素** 包括第 III 至第 VIII 族中的元素，其外殼層 (第 111 頁) 中有 1 至 6 個 p 電子 (第 112 頁)。

**d 區元素** 為過渡元素 (第 111 頁) 中有 s 電子 (第 112 頁)，次外殼層中有 1 至 10 個 d 電子 (第 112 頁)。d 電子的數目和排列決定元素的本質。

**f 區元素** 為一種內過渡序，其外殼層有 s 電子並填滿共 14 個電子 f 軌道。

**transition element**
extranuclear electrons
of manganese
過渡元素
錳的核外的電子

| shell 殼層 | s-electron s電子 | p-electrons p電子 | d-electrons d電子 |
|---|---|---|---|
| K | 2 | | |
| L | 2 | 2 2 2 / 6 | shell 殼 |
| M | 2 | 2 2 2 / 6 | 11111 / 5 complete 殼層填滿 |
| N | 2 | | |

current (n) electric charges (p.138) in motion form a current. A flow of electrons is the current of electric charge. Electric current is measured in amperes; it is said to flow from positive to negative.

conduct (v) to give direction to the flow of a fluid (p.11), electric current (↑) or heat, e.g. a pipe conducts water, a copper wire conducts an electric current. conductor (n), conducting (adj).

conductivity (n) a measure of the ability of a solution of given concentration to conduct electric current; its symbol is κ. The conductance of a solution is equal to 1/resistance; its symbol is G. $\kappa = \frac{G}{a}$ where a is a constant (p.106) for a particular electrolytic cell (↓).

electrolyte (n) a compound (p.8) which when dissolved in water will conduct (↑) electric current (↑). An electrolyte will also conduct electric current when molten (p.10). Acids, alkalis, inorganic salts (p.46) are generally electrolytes. The electrolyte is decomposed (p.65) by the current.

non-electrolyte (n) a compound (p.8) which when dissolved in a solvent (p.86) does not conduct electric current (↑). When molten, a non-electrolyte also does not conduct an electric current. Organic (p.55) compounds generally are non-electrolytes.

electrolytic cell a vessel in which electrolysis (↓) takes place. See voltameter, p.129.

electrolysis (n) the decomposition (p.65) of an electrolyte (↑) caused by passing an electric current through the solution. electrolytic (adj) describes any process (p.157) connected with electrolysis (↑).

electrode (n) a piece of material that conducts (↑) an electric current (↑) used in an electrolytic cell (↑) or voltameter (p.129) or simple cell (p.129). Electric current enters and leaves the solution of the electrolysis through the electrodes; the electrodes are connected (p.24) to a source (p.138) of electric current in an electrolytic cell or voltameter.

電流 (名) 流動成為電荷 (第 138 頁) 形成電流。電子流動成為電荷流。電流的量度單位為安培；電流由正流向負。

傳導 (動) 給予流體 (第 11 頁)、電流 (↑) 或熱量流動定向，例如，用管子引導水，電流 (↑) 或熱量流動定向，例如，用管子引導水，電銅線傳導電流。(名詞為 conductor，形容詞為 conducting)

導電率 (名) 特定濃度溶液的導電能力標準，符號為 k。溶液的導電率等於 1 / 電阻；其符號為 G。$\kappa = \frac{G}{a}$，其中 a 為特定電解池 (↓) 的常數 (第 106 頁)。

電解質 (名) 溶解於水或溶劑 (第 86 頁) 時能傳導 (↑) 電流 (↑) 的一種化合物 (第 8 頁)。酸、鹼和無機鹽 (第 46 頁) 類通常都是電解質。電解質可為電流所分解 (第 65 頁)。

非電解質 (名) 溶解在其中發生電解作用 (↓) 的一容器。參見 "伏特計" (第 129 頁)。溶解在溶劑 (第 86 頁) 中的一種化合物 (第 8 頁)。有機 (第 55 頁) 化合物通常是非電解質。

電解池 (名) 指在其中發生電解作用 (↓) 的一容器。參見 "伏特計" (第 129 頁)。

電解作用 (名) 指電流通過溶液引起電解質 (↑) 的分解作用 (第 65 頁)。(形容詞為 electrolytic)

電解的 (形) 描述與電解作用 (↑) 有關的任何過程 (第 157 頁)。

電極 (名) 電解池 (↑) 或伏特計 (第 129 頁) 或簡單電解池 (第 129 頁) 中一塊能傳導 (↑) 電流 (↑) 的材料。電流通過電極流入和流出電解質溶液；在電解池或伏特計中，電極 (第 24 頁) 與電源 (第 138 頁) 相連。

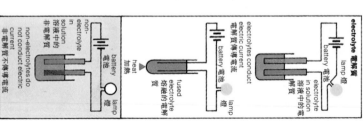

electrolyte 電解質

electric current 電解電流傳導電流
electrolytes conduct
electrolyte in solution 溶液中的電解質
battery 電池
lamp 燈

non-electrolyte 非電解質

non-electrolytes do not conduct electric current 非電解質不傳導電流
non-electrolyte in solution 溶液中的非電解質
battery 電池
lamp 燈

fused electrolyte 熔融的電解質
electrolyte in solution 溶液中的電解質
heat 加熱
lamp 燈

electrolytic cell 電解池

anode 陽極
electrode 電極
battery 電池
electrode 電極
cathode 陰極

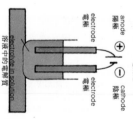

electrode 電極
electrolyte in solution 溶液中的電解質

**ammeter** (*n*) an instrument (p.23) for measuring the strength of an electric current (↑) in amperes.

安培計 (名) 以安培為單位的量度電流 (↑) 強度的一種儀器 (第 23 頁)。

**anode** (*n*) a positive electrode (↑); the electrode at which electric current is said to enter an electrolytic cell; the electrode from which electric current leaves a simple cell. **anodic** (*adj*).

陽極 (名) 即正電極 (↑)。為電流流入電解池的電極；電流流出簡單電池的電極。(形容詞為 anodic)

**cathode** (*n*) a negative electrode (↑); the electrode from which current is said to leave an electrolytic cell; the electrode at which electric current flows into a simple cell after conduction by conductors.

陰極 (名) 即負電極 (↑)。為電流流出電解池的電極；電流由導體傳導之後流入簡單電池的電極。

**electrolyze** (*v*) to pass an electric current (↑) through a solution of an electrolyte (↑), or a molten (p.10) electrolyte, in order to decompose it, e.g. copper metal is obtained by electrolyzing a solution of copper (II) sulphate. **electrolysis** (*n*), **electrolytic** (*adj*).

使電解 (動) 使電流 (↑) 流過電解質 (↑) 溶液或熔化的 (第 10 頁) 電解質使之分解。例如以使硫酸銅 (II) 溶液電解取得銅金屬。(名詞為 electrolysis，形容詞為 electrolytic)

**ion** (*n*) an ion is formed by an atom gaining or losing one or more valency (p.133) electrons (p.110). An atom, or a group of atoms in an ion, is neutral; the loss or gain of electrons results in an electric charge on the ion. Loss of electrons produces a positive charge; gain of electrons produces a negative charge. The magnitude of the charge depends upon the number of electrons concerned. The process of forming ions is ionization (↓). **ionize** (*v*).

離子 (名) 離子係由原子獲得或失去一個或多個價 (第 133 頁) 電子 (第 110 頁) 所形成。原子或離子中的原子團是電中性的，由於在離子上失去或獲得電子的結果而帶電。失去電子的帶正電荷，獲得電子的帶負電荷。電荷量取決於獲得或失去的電子數目。形成離子之過程稱為電離作用 (↓)。(動詞為 ionize)

**ionization** (*n*) the process of forming ions (↑). Ions are formed: (1) when energy is supplied to atoms; the energy can be heat or radiation; a high voltage (p.126) will also cause ions to be formed. (2) the attraction for electrons by atoms of some elements, e.g. a chlorine atom attracts one electron from a sodium atom. Crystals of electrolytes (↑) have a lattice consisting of ions; dissolving the crystals in water separates the ions, but this is not ionization. Hydrogen chloride is a covalent (p.136) compound, when it dissolves in water ionization takes place because the chlorine atom attracts an electron from the hydrogen atom. **ionize** (*v*), **ionized** (*adj*), **ionizing** (*adj*).

電離作用 (名) 形成離子 (↑) 的過程。離子是這樣形成的：(1) 給原子提供能量，提供的能量；輻射能或高電壓 (第 126 頁) 都可導致形成離子；(2) 某些元素之原子吸納，例如氯原子從鈉原子吸取一個電子。電解質 (↑) 之晶體含有由離子構成的晶格。晶體溶解在水中會分離出離子，但這並非電離作用。氯化氫是一種共價 (第 136 頁) 化合物，溶於水時發生電離作用，因氯離子從鈉原子吸取一個電子。(動詞為 ionize，形容詞為 ionized、ionizing)

**ammeter** 安培計

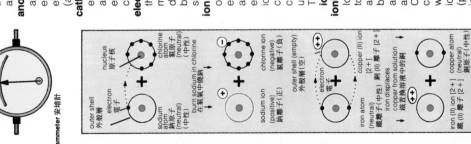

outer shell 外殼層
electron 電子
nucleus 原子核
sodium atom (neutral) 鈉原子 (中性)
chlorine atom (neutral) 氯原子 (中性)
burn sodium in chlorine 在氯氣中燒鈉
sodium ion (positive) 鈉離子 (正)
chlorine ion (negative) 氯離子 (負)
outer shell (empty) 外殼層 (空)
electron 電子
iron atom (neutral) 鐵原子 (中性)
copper from solution 從置換溶液中的銅
iron displaces 鐵置換
copper (II) ion [2+] 銅 (II) 離子 [2+]
iron (II) ion [2+] 鐵 (II) 離子 [2+]
copper atom (neutral) 銅原子 (中性)

**ions and ionization** 離子和電離作用

**attract** (v) to pull one object towards another object, e.g. a positive ion (p.123) attracts a negative ion towards itself; positive and negative electric charges attract each other. **attraction** (n), **attractive** (adj).

**repel** (v) to push one object away from another object, e.g. two positive ions (p.123) repel each other, e.g. two negative ions repel each other. **repulsion** (n), **repulsive** (adj).

**repulsion** (n) the force which causes objects to repel each other, e.g. a force of repulsion exists (p.213) between two positive electric charges and between two negative electric charges. **repulsive** (adj).

**ionic theory** the theory which explains the behaviour of electrolytes (p.122). It accounts for electrolysis (p.122) and the action of simple cells (p.129) by the existence (p.213) of ions (p.123) in all electrolytes.

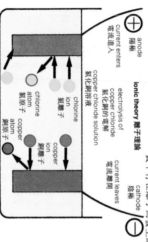

| | |
|---|---|
| anode | cathode |
| 陽極 | 陰極 |
| current enters | current leaves |
| 電流進入 | 電流離開 |

**electrolysis of copper chloride** 氯化銅的電解
copper chloride solution 氯化銅溶液

**ionic theory** 離子理論
electrolysis of copper chloride 氯化銅的電解

chlorine ion 氯離子
chlorine atom 氯原子
copper ion 銅離子
copper atom 銅原子

**discharge** (v) to give up an electric charge. When an ion (p.123) comes into contact (p.217) with an electrode (p.122), the ion either receives or gives electrons and becomes a neutral atom; the ion has given its charge to the electrode and has become discharged. In the case of metals, it may be deposited on the electrode, in the case of non-metals, it is generally liberated (p.69) as a gas.

**strength** (n) a measure of the degree (p.227) of ionization (p.123) of an electrolyte (p.122) in a solution; also a measure of the characteristic reactivity (p.62) of acids (p.45) and alkalis (p.45).

吸引：吸取：吸納 (動) 將一個物體拉向另一個物體。例如正、負電荷相互吸引，形容詞為 attractive)。

排斥 (動) 將一個物體推開一個物體。例如兩個正離子 (第 123 頁) 相互排斥；兩個負離子也相互排斥。(名詞為 repulsion，形容詞為 repulsive)

排斥力 (名) 使物體相互排斥而發生的力量。例如兩個正電荷之間和兩個負電荷之間都存在 (第 213 頁) 排斥作用的力。(形容詞為 repulsive)

離子理論 解釋電解質 (第 122 頁) 行為的理論。該理論論認為一切電解質中都存在 (第 122 頁) 和簡單電解池 (第 129 頁) 的作用都是由於電解質中存在著離子而發生的。

放電 (動) 放出電荷。當離子 (第 123 頁) 和電極 (第 122 頁) 接觸 (第 217 頁) 時，離子接收或給出電子而成為一個中性的原子；離子把其電荷給予電極並放電。如為金屬則沉積在電極上；如為非金屬通常成為一種氣體放出 (第 69 頁)。

強度 (名) 電解質 (第 122 頁) 在一種溶液中電離 (第 123 頁) 程度 (第 227 頁) 的一個量度；也是酸 (第 45 頁) 和鹼 (第 45 頁) 特徵反應性 (第 62 頁) 的一個量度。

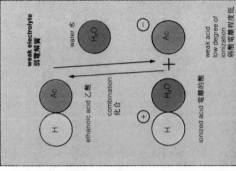

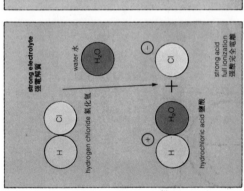

**strong electrolyte** an electrolyte (p.123) which is completely ionized (p.123) even in concentrated solutions. Strong acids (p.45) and strong alkalis (p.45) are strong electrolytes. Most inorganic salts are strong electrolytes, but *see hydrolysis[1] (p.66)*. Strong electrolytes are good conductors (p.122) of electric current.

**weak electrolyte** an electrolyte which is only partly ionized (p.123) in water, or other ionizing solvents (p.86). At great dilutions the ionization becomes almost complete. Solutions of weak electrolytes are poor conductors (p.122) of electric current (p.123). Weak acids (p.45) and weak alkalis (p.45) are weak electrolytes. Examples of weak electrolytes are: ethanoic acid (and other organic acids); ammonia solution (a weak alkali).

**anion** (*n*) an ion which carries a negative charge; it is attracted to the anode (p.123) in electrolysis (p.122). Examples of anions are: chloride ion $Cl^-$; sulphate ion $SO_4^{2-}$; zincate (II) ion $ZnO_2^{2-}$.

**cation** (*n*) an ion which carries a positive charge; it is attracted to the cathode (p.123) in electrolysis (p.122). Examples of cations are: copper (II) $Cu^{2+}$; sodium $Na^+$, iron (III) $Fe^{3+}$

**強電解質** 即使在濃溶液中也能完全電離（第 123 頁）的一種電解質（第 122 頁）。強酸（第 45 頁）和強鹼（第 45 頁）和大多數無機鹽是強電解質，參見 " 水解作用 "（第 66 頁）。強電解質是電解電流的良導體（第 122 頁）。

**弱電解質** 在水或其他離子化溶劑（第 86 頁）中只能部分電離的（第 123 頁）一種電解質。在高倍稀釋下電離作用近乎完全。弱電解質溶液是電流（第 122 頁）的不良導體（第 122 頁）。弱酸（第 45 頁）和弱鹼（第 45 頁）都是弱電解質。乙酸（和其他有機酸類）和氨水（一種弱鹼）是弱電解質的例子。

**陰離子（名）** 帶負電荷的離子。電解（第 122 頁）時，陰離子被吸向陽極（第 123 頁）。例如氯離子（$Cl^-$）、硫酸根離子（$SO_4^{2-}$）以及鋅（II）酸根離子（$ZnO_2^{2-}$）都是陰離子。

**陽離子（名）** 帶正電荷的離子。電解（第 123 頁）時，陽離子被吸向陰極（第 123 頁）。例如，銅（II）離子（$Cu^{2+}$）、鈉離子（$Na^+$）、鐵（III）離子（$Fe^{3+}$）都是陽離子。

**voltage** (n) a measurement in volts of the electrical force, or pressure, that drives an electric current (p.122) through an electrolyte (p.122) or conductor (p.122).

**voltmeter** (n) an instrument (p.23) that measures voltage (↑) in volts.

**decomposition voltage** the smallest voltage (↑) that will cause decomposition (p.65) of an electrolyte (p.122) by electrolysis (p.122) using platinum or other electrodes (↓). This voltage is necessary to overcome polarization (↓) and overvoltage (p.213) the effects of polarization (↓) and overvoltage (↓). With some electrodes and electrolytes it is 0 volts.

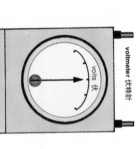

current 電流

copper (II) sulphate
copper electrodes
硫酸銅(II)
銅電極

decomposition voltage
分解電壓

voltage 電壓

dilute sulphuric acid
platinum electrodes
稀硫酸
鉑電極

decomposition voltage
分解電壓

**platinum electrode** an electrode consisting of platinum; such electrodes produce polarization (↓) and overvoltage (↓). *Platinized electrodes* consist of platinum electrodes with a layer of finely divided platinum called *platinum black*; such electrodes do not produce polarization.

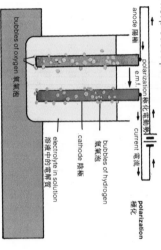

anode 陽極

bubbles of oxygen 氧氣泡

polarization 極化
e.m.f.

cathode 陰極

bubbles of hydrogen 氫氣泡

electrolyte in solution 溶液中的電解質

current 電流

polarization
極化

電壓（名） 電力或電壓力以伏特為單位的一個量度。電壓推動電流（第 122 頁）流經電解質（第 122 頁）或導體（第 122 頁）。

伏特計（名） 指測量電壓（↑）用的一種儀器（第 23 頁），測量單位為伏特。

分解電壓 用鉑作電極（↓）或用其他電極使電解質（第 122 頁）電解作用（第 122 頁）所需的最低電壓（↑）。這是克服極化（↓）和超電壓（第 213 頁）的電壓。使用某些電極和電解質（↓）效應所需電壓可為零伏特。

鉑電極 由鉑構成的一種電極；此種電極會產生極化作用（↓）和引致超電壓（↓）。"鍍鉑電極"由帶有一層稱為鉑黑的細碎鉑塊的鉑電極構成。此種電極不產生極化作用。

voltmeter 伏特計

volts 伏

**polarization** (*n*) the production of hydrogen and oxygen gases which collect on an electrode, changing the nature of the electrode. In electrolysis (p.122) a primary cell (p.129) is formed with hydrogen and oxygen electrodes; this produces an *electromotive force* (p.129) acting against the electrolyzing current and causes a decomposition voltage (↑) to be necessary. In primary cells, the electromotive force produced by polarization acts against the electromotive force of the cell and reduces its voltage (↑).

**overvoltage** (*n*) the further voltage necessary to release (p.69) a gas at an electrode above its *electrode potential* (p.128). The magnitude of the overvoltage depends on the nature of the electrode and the gas being discharged (p.124). The cause of overvoltage has not been explained.

**electrodeposit** (*n*) a coat (↓) of metal on an electrode (p.122) produced during electrolysis by the discharge (p.124) of ions of the metal at the electrode. **electrodeposition** (*n*).

**electroplating** (*n*) the process (p.157) of putting an electrodeposit (↑) on an article. In electroplating with silver, a metal article is made the cathode (p.123) of a *plating bath* and a solution of a silver salt is used as the electrolyte. A thin coat of silver is formed on the article.

**plate** (*v*) to carry out electroplating (↑), especially with silver.

**coat** (*n*) a thin outer layer on an object. **coat** (*v*).

**極化作用** (名) 生產氫氣和氧氣時，因氫氣和氧氣聚集在電極上而改變電極之性質。電解電池 (第 122 頁) 時，為電極和氧電極構成一個原電池 (第 129 頁)；原電池產生逆阻電解電流的 "電動勢" (第 129 頁) 並引起所產生之分解電壓 (↑)。在原電池中，極化作用產生的電動勢逆阻電池之電動勢逆降低其電壓 (↑)。

**超電壓** (名) 在一個高於其 "電極電位" (第 128 頁) 的電極釋放出 (第 69 頁) 一種氣體所需之額外電壓。超電壓之電壓值決定於電極和放電 (第 124 頁) 釋出的氣體之本質。超電壓的原因至今仍未能清楚解釋。

**電極沉積** (名) 電解過程中利用金屬離子在電極放電 (第 124 頁)，在電極 (第 122 頁) 上產生一層金屬鍍層 (↓)。(名詞為 electrodeposition)

**電鍍** (名) 在某物件件上鍍一層電解積層 (↑) 的過程 (第 157 頁)。用銀電鍍時，以金屬製品作為 "電鍍層" 之陰極 (第 123 頁)，銀鹽溶液作為電解質，在物件上形成一層薄的銀鍍。

**鍍** (動) 完成電鍍 (↑)。尤其是用銀電鍍。

**鍍層** (名) 在物件上的一層薄的外層。(動詞為 coat)

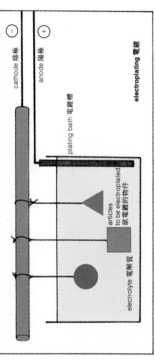

cathode 陰極 (−)

anode 陽極 (+)

plating bath 電鍍槽

articles to be electroplated 欲電鍍的物件

electrolyte 電解質

**electroplating 電鍍**

**anodic** (adj) describes anything happening at an anode (p.123) during electrolysis (p.122).

**cathodic** (adj) describes anything happening at a cathode (p.123) during electrolysis (p.122).

**cathodic reduction** a process (p.157) at a cathode (p.123) during electrolysis, in which a positive ion (p.123) gains one or more electrons from the cathode, e.g. hydrogen ions gain one electron to become hydrogen atoms, the ion is reduced (p.219) to the atom and discharged (p.123).

**anodic oxidation** a process (p.157) at an anode (p.123) during electrolysis (p.122), in which a negative ion (p.123) loses one or more electrons to the anode. A metal anode may go into solution, as its atoms can lose one or more electrons to become positive ions, e.g. negative ions:
$$2 Cl^- \rightarrow Cl_2 + 2e \text{ (chlorine discharged)}$$
atoms:
$$Cu \rightarrow Cu^{2+} + 2e \text{ (electrode goes into solution)},$$
metal by electrolysis (p.123). The metal is made the anode (p.123) of an electrolytic cell (p.122), a suitable electrolyte is used and oxygen is discharged (p.124) at the anode forming a coat of the oxide of the metal. With aluminium, dyes (p.162) can be absorbed in the oxide, producing coloured surfaces. **anodizing** (n).

**anodize** (v) to form a layer of oxide (p.48) on a

**electrochemical** (adj) describes any effect concerned with the electrical properties of solutions (p.86) and the ions (p.123) in solution.

**electrochemical equivalent** the mass in grams of an element liberated (p.69) during electrolysis (p.122) by 1 coulomb of electric charge. (1 coulomb of charge is the quantity of charge from a current of 1 ampere flowing for 1 second). An electrode potential is also formed between a metal and any electrolyte. Two metals will have two electrode potentials and the difference between these potentials produces the electromotive force (↓) of a primary cell.

**electrode potential** when a metal is put in a solution containing ions (p.123) of the metal, a voltage (p.126) difference forms between the metal and the solution; this is an electrode potential.

陽極的 (形) 描述電解 (第 122 頁) 過程中在陽極 (第 123 頁) 發生之任何事物。

陰極的 (形) 描述電解 (第 122 頁) 過程中在陰極 (第 123 頁) 發生之任何事物。

陰極還原 電解過程中，在陰極 (第 123 頁) 發生之過程 (第 157 頁)：在此過程中，一個正離子 (第 123 頁) 自陰極獲得一個或多個電子，例如氫離子獲得一個電子變爲氫原子，離子被還原 (第 219 頁) 成原子同時放電 (第 124 頁)。

陽極氧化 電解 (第 122 頁) 過程中，在陽極 (第 123 頁) 發生之過程 (第 157 頁)；在此過程中一個負離子 (第 123 頁) 失去一個或多個電子給陽極。金屬陽極在其原子可以失去一個或多個電子而成爲正離子時，便可以潛入溶液中。例如
負離子：
$$2Cl^- \longrightarrow Cl_2 + 2e \text{ (氯放電)}$$
原子：
$$Cu \longrightarrow Cu^{2+} + 2e \text{ (電極溶入溶液)}$$

陽極氧化處理 (動) 藉電解作用 (第 122 頁) 在一種金屬上形成一層氧化物 (第 48 頁)。金屬作爲電解池 (第 122 頁) 的陽極 (第 123 頁)，使用一種合適的電解質同時使氧氣在陽極放電 (第 124 頁) 釋出而形成金屬氧化物的表面。用鋁作爲陽極時，染料 (第 162 頁) 可被吸收在氧化物中，產生有色的表面。(名詞爲 anodizing)。

電化的 (形) 描述與溶液 (第 86 頁) 以及溶液中離子 (第 123 頁) 之電性質有關的任何效應。

電化當量 電解 (第 122 頁) 過程中，一庫侖電荷所能釋出某元素的質量克數。(一庫侖電量即一安培電流流過一秒鐘的電流量)。

電極電位 將一種金屬放入含有該種金屬之離子 (第 123 頁) 的溶液時，金屬與溶液之間產生電壓 (第 126 頁) 差；此電壓差即電極電位。金屬與任何電解質之間也可形成電極電位。兩種金屬就有兩種電極電位。兩電位間之差就成爲原電池的電動勢 (↓)。

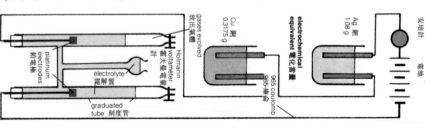

**electrochemical** 電化當量

ammeter 安培計
battery 電池
Ag 銀 1.08 g
Cu 銅 0.3175 g
965 coulomb 965 庫侖
gases evolved 放出氣體

H₂ 氫 0.01 g
O₂ 氧 0.16 g
**Hofmann voltameter** 霍夫曼電量計
Hofmann voltameter 霍夫曼電量計
platinum electrodes 鉑電極
electrolyte 電解質
graduated tube 刻度管

**series**[1] a way of connecting (p.24) parts of an electric circuit so that the current flows through each part in turn. The parts are said to be connected in series.

串聯 將電路的各部分相連接(第 24 頁)使電流依 次流過各部分的一種連接方式。將這些部分 稱爲以串聯相連接。

**parallel** (*n*) a way of connecting (p.24) parts of an electric circuit so that the current divides to go through each part at the same time. The parts are said to be connected in parallel.

並聯 (名) 將電路的各部分相連接(第 24 頁)使電 流同時分開流過各個部分的一種連接方式。 將這些部分連接以並聯相連接。

**voltameter** (*n*) an apparatus for determining the electrochemical equivalent (↑) of an element. The electrodes are large metal plates, weighed before and after passing a measured electric current for a measured time. A Hoffman voltameter is used to determine the electrochemical equivalent of gases. *See electrolytic cell (p.122).*

電量計 (名) 測量元素電化當量(↑)之儀器。儀 器所用的電極是一些大金屬板,讓被測量的 電流流過電極,並稱量電極在測量時間前後 之重量。霍夫曼電量計用於測量氣體之電化 當量。參見 "電解池"(第 122 頁)

**faraday** (*n*) the quantity of electric current that liberates (p.69) or forms 1 mole of monovalent (p.137) ions. Its value is 96 487 coulombs. *See electrochemical equivalent (↑).* One faraday of electric charge contains one mole of electrons.

法拉第 (名) 釋出(第 69 頁)或形成 1 摩爾一價 (第 137 頁)離子的電流量。1 法拉第的值等 於96487庫侖。參見 "電化當量"(↑)。1 法 拉第之電荷含有 1 摩爾電子。

**inactive electrode** an electrode which will not give ions (p.123) to the electrolyte (p.122) in an electrolytic cell (p.122), e.g. carbon and platinum are inactive electrodes.

不活潑電極 電解池(第 122 頁)中不將離子(第 123 頁)放給電解質(第 122 頁)的一種電極。 例如碳和鉑是不活潑電極。

**active electrode** an electrode which can give ions (p.123) to the electrolyte (p.122) in an electrolytic cell (p.122) if it is the anode. The cathode is not so important. For example, copper forms an active anode and may produce copper ions during electrolysis (p.122).

活潑電極 如爲陽極,則在電解池(第 122 頁)中可將離子(第 123 頁)放給電解質(第 122 頁)的一種電極。陰極則無關重要。例如, 電解第 122 頁過程中,銅形成活潑陽極並 產生銅離子。

**primary cell** a source (p.138) of electric current formed by putting two metals into an electrolyte. Also called a voltaic cell. The source of the electrical energy is the *electrode potentials* (↑) between each of the electrodes and the electrolyte in the cell, e.g. a simple voltaic cell has an electrode of zinc and one of copper in an electrolyte of dilute sulphuric acid; it has an electromotive force of 1.1 volt.

原電池 將兩種金屬放入一種電解質中形成一個 電流源(第 138 頁)。原電池又稱作打電池。 電池中各個電極與電解質之間的 "電極電 位"(↑)爲電能之源。例如,簡單伏打電池 有一個鋅電極和一個銅電極,於稀硫酸之電 解質中,此電池所具之電動勢爲 1.1 伏特。

**electromotive force** the electrical force of a primary cell (↑) which drives current through a connected circuit. The symbol for electromotive force is *E*, and the unit for it is the volt. The electromotive force is produced by the two *electrode potentials* (↑) of the cell. (*abbr.*) **e.m.f.**

電動勢 原電池(↑)中推動電流通過接電路的 電力。電動勢的符號爲 *E*,單位爲伏特。電 動勢是由電池的兩個 "電極電位"(↑)產生 的。英文縮寫成 **e.m.f.**

---

**copper voltameter**
銅電量計
ammeter 安培計
battery 電池
copper electrodes 銅電極
copper (II) sulphate solution 硫酸銅(II)溶液

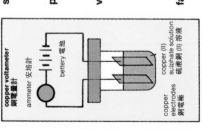

**primary cell** 原電池
voltameter 伏特計
1.1 V
negative 負
(electrolyte) 電解質
dilute sulphuric acid 稀硫酸
copper electrode 銅電極
zinc electrode 鋅電極
positive 正

**hydrogen electrode** a platinum electrode (p.126) is platinized with platinum black and hydrogen gas is put in contact with the electrode at standard pressure (p.102). This electrode is placed in a 1 M (p.88) solution of hydrogen ions (p.123) at 25°C to form a hydrogen electrode. The *standard electrode potential* (↓) of the hydrogen electrode is fixed at 0 volts.

**standard electrode potential** the electrode potential (p.128) of a metal, or a gas, in contact with 1 M (p.88) solution of its ions at 25°C, the magnitude is determined (p.222) against the standard hydrogen electrode (↑) of 0 volts.

**electrochemical series** a list of elements in order of their standard electrode potentials (↑) using a scale with the hydrogen electrode (↑) as zero. In this series, a metal such as zinc has a negative electrode potential as in a primary cell with a zinc and a hydrogen electrode, the zinc would be the cathode. *See below for the electrochemical series*. Similarly, in a primary cell with a chlorine and a hydrogen electrode, the chlorine would be the anode and so chlorine has a positive electrode potential.

| electrochemical series 電化序 | | |
| --- | --- | --- |
| element<br>元素 | electrode<br>potential<br>電極電位 | ionic reaction<br>離子反應 |
| potassium 鉀 | −2.92 v | K⁺ + e→K |
| calcium 鈣 | −2.87 v | Ca²⁺ + 2e→Ca |
| sodium 鈉 | −2.71 v | Na⁺ + e→Na |
| magnesium 鎂 | −2.38 v | Mg²⁺ + 2e→Mg |
| aluminium 鋁 | −1.66 v | Al³⁺ + 3e→Al |
| zinc 鋅 | −0.76 v | Zn²⁺ + 2e→Zn |
| iron 鐵 | −0.44 v | Fe²⁺ + 2e→Fe |
| tin 錫 | −0.14 v | Sn²⁺ + 2e→Sn |
| lead 鉛 | −0.1 v | Pb²⁺ + 2e→Pb |
| hydrogen 氫 | 0 v | H⁺ + e→H |
| copper 銅 | +0.34 v | Cu²⁺ + 2e→Cu |
| silver 銀 | +0.8 v | Ag⁺ + e→Ag |

氫電極 鉑電極（第 126 頁）鍍以鉑黑，氫氣在標準壓力（第 102 頁）下與電極接觸。將此電極置於 25℃、濃度 1 M（第 88 頁）的氫離子（第 123 頁）溶液中，即構成氫電極。氫電極之"標準電極電位"（↓）規定為 0 伏特。

標準電極電位 指一種金屬，或一種氣體，在 25℃ 時與含該金屬離子的 1 M（第 88 頁）溶液接觸時的電極電位（第 128 頁），是與標準氫電極的標準電極電位（↑）為 0 伏特相比而決定之大小。

電化序 以氫電極（↑）為零作為一種標準度，將各種元素按其標準電極電位（↑）順序排列的一張表。在電化序中，金屬，例如鋅電極的電位為負值，如同在由鋅電極和氫電極組成的原電池那樣，鋅是陰極。多見下表所列"電化序"。同樣，在一個由氫電極和氫電極構成的原電池中，氯是陽極，所以氯的電極電位為正值。

zinc electrode (negative)
鋅電極（負）

potential difference
of 0.76 v
電位差為 0.76 伏

1 M solution 1 M 溶液
Zn²⁺
(positive)（正）

temperature 25°C
溫度 25°C

**standard electrode potential** 標準電極電位

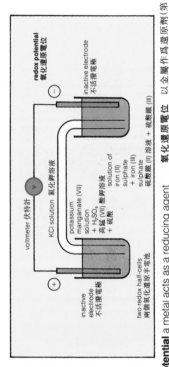

voltmeter 伏特計

redox potential
氧化還原電位

inactive electrode
不活潑電極

inactive
electrode
不活潑電極

two redox half-cells
兩個氧化還原半電池

KCl solution 氯化鉀溶液

potassium
manganate (VII)
solution
+ H₂SO₄
高錳酸(VII) 酸鉀溶液
+ 硫酸

solution of
iron (II)
sulphate
+ iron (III)
sulphate
硫酸鐵(II) 溶液
+ 硫酸鐵(III)

**redox potential** a metal acts as a reducing agent (p.71) and a non-metal, such as chlorine, acts as an oxidizing agent (p.71). The electrochemical series (↑) shows the relative (p.232) strengths of these agents. By using the apparatus of redox half-cells, other ions can be used to determine (p.222) their electrode potential and these potentials are called redox potentials. Redox potentials show whether an agent is able, or not, to reduce or to oxidize an ion. For example, the redox potential of chlorine to chlorine ions is + 1.36 V; the redox potential of $(Cr_2O_7)^{2-}$ is 1.33 V and of $(MnO_4)^-$ is + 1.52 V. Hence potassium manganate (VII) will oxidize chlorine ions to chlorine, but potassium dichromate (VI) will not usually oxidize chlorine ions.

**redox series** a list of redox reagents, which include the electrochemical series, in order of their redox potentials, measured at 25°C, in 1 M (p.88) solutions, forms the redox series. Hydrogen is given a redox potential of zero.

氧化還原電位 以金屬例如氯作爲還原劑(第 71 頁)，非金屬例如氯作爲化學氧化劑的相對(第 232 頁)強度。藉(使用這些氧化還原半電池裝置，可以測定(第 222 頁)其他離子的電極電位，因而這些電位稱爲氧化還原電位。氧化還原電位表示一種化學劑是否能還原或氧化一個離子。例如，氯對氯離子的氧化還原電位爲 + 1.36 伏；(Cr₂O₇)²⁻ 離子的氧化還原電位則爲 1.33 伏；而 (MnO₄)⁻ 離子的氧化還原電位爲 1.52 伏。因此，高錳(VII)酸鉀能將氯離子成氯，而重鉻(VI)酸鉀就不能氧化氯離子。

氧化還原序 指氧化還原試劑的一張列表，表中將每種試劑按其在 25°C 時於 1 M(第 88 頁)溶液中測得的氧化還原電位順序排列成氧化序而成氧化還原順序。氫的氧化還原電位規定爲零。

| redox series 氧化還原序 | |
|---|---|
| $Sn^{4+} + 2e \rightarrow Sn^{2+}$ | + 0.15 V |
| $Cu^{2+} + 2e \rightarrow Cu$ | + 0.34 V |
| $2H_2O + O_2 + 4e \rightarrow 4OH^-$ | + 0.44 V |
| $Fe^{3+} + e \rightarrow Fe^{2+}$ | + 0.77 V |
| $Cr_2O_7^{2-} + 14H_2^+ + 6e \rightarrow 2Cr^{3+} + 7H_2O$ | + 1.33 V |
| $Cl_2 + 2e \rightarrow 2Cl^-$ | + 1.36 V |
| $MnO_4 + 8H^+ + 5e \rightarrow Mn^{2+} + 4H_2O$ | + 1.52 V |

**complex ion** an ion (p.123) which has molecules (p.77) joined to it by coordinate bonds (p.136). A complex cation (p.125) has molecules or atoms such as water, a halogen, or cyanide (CN) joined by coordinate bonds to the metal ion or to hydrogen and carries a positive charge, e.g. tetrammine copper (II) ion [Cu(NH₃)₄]²⁺, tetrachlorodiaquo chromium (III) [CrCl₄(H₂O)₂]³⁺. A complex anion (p.125) has molecules or atoms such as water, a halogen or cyanide (CN) joined by coordinate bonds to a central metal atom and carries a negative charge, e.g. hexacyanoferrate (II) [Fe(CN)₆]⁴⁻ trichlorocuprate (I) [CuCl₃]²⁻. The central metal atom or ion in complex ions is usually a transitional element (p.121).

**aqua-ion** (n) a complex ion (↑) with water molecules joined to a metal ion by coordinate bonds, e.g. tetraquo copper (II) [Cu(H₂O)₄]²⁺.

**hydroxide ion** the ion formed from a water molecule by the removal of a hydrogen ion. It has a negative charge; it is discharged at an anode to produce oxygen:

$$4OH^- \rightarrow 2H_2O + O_2 + 4e$$

**transfer** (v) to move objects, materials, or energy from one place to another, without saying how they are moved. e.g. to transfer an electron from an ion to an electrode; to transfer a precipitate from a beaker. **transference** (n), **transferable** (adj).

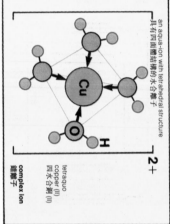

an aqua-ion with tetrahedral structure
具有四面體結構的水合離子

tetraquo
copper (II)
四水合銅(II)

**complex ion**
錯離子

**錯離子：複離子：絡離子** 含有分子(第77頁)靠配價鍵(第136頁)與離子相連的一種離子。錯陽離子(第125頁)含的分子或原子(如水、鹵素或氰(CN)靠配價鍵與金屬離子或原子相連並帶正電荷，例如四氨合銅(II)離子[Cu(NH₃)₄]²⁺、二水四氯合鉻(III)離子[CrCl₄(H₂O)₂]³⁺。錯陰離子(第125頁)含有的分子或原子(如水、鹵素或氰(CN)靠配價鍵連接到中心金屬原子並帶負電荷，例如六氰合鐵(II)離子[Fe(CN)₆]⁴⁻、三氯合銅(I)離子[CuCl₃]²⁻。錯離子中的中心金屬原子或離子通常都是一種過渡元素(第121頁)。

**水合離子(名)** 以水分子靠配價鍵與一個金屬離子相連的錯離子(↑)，例如四水合銅(II)離子[Cu(H₂O)₄]²⁺。

**氫氧離子(名)** 由一個水分子去除一個氫離子而形成的離子，帶負電荷，可在陽極釋放電生成氧：

$$4 OH^- \longrightarrow 2H_2O + O_2 + 4e$$

**轉移(動)** 使物體、材料或能量由一處移到另一處，但不說明其如何移動。例如，使沉澱物由燒杯的一處移到另一處，把電子轉移到電極，使電子轉移到一個電極或內。（名詞為 transference，形容詞為 transferable）

**bond** (n) in chemistry, a force which holds together two atoms, two ions, two molecules, or a combination of these. Bonds are broken and formed in chemical change. Energy is needed to break bonds; it is released or absorbed in forming bonds.

鍵 (名) 化學上指兩個原子、兩個離子、兩個分子或三者的組合結合在一起的力。化學變化時，鍵斷裂或形成。鍵斷裂需要能量，鍵形成時釋放或吸收能量。

**valency** (n) the ability of an atom, and hence an element, to form bonds (↑), e.g. carbon, valency 4, can form 4 bonds. *Valency* has now been replaced (p.68) by *electrovalency* (p.134) and *covalency* (p.136).

價 (名) 指一個原子亦即一個元素能形成鍵(↑)之能力。例如碳的價數為4，可形成四個鍵。"價"這個名詞已為"電價"(第134頁)和"共價"(第136頁)所取代(第68頁)。

**valency electron** an electron (p.110) which takes part in chemical bonds (↑).

價電子 指參加化學鍵(↑)的電子(第110頁)。

**valency shell** the outside shell (p.111) of an atom which contains valency electrons (↑).

價層 指含有價電子(↑)原子的外殼層(第111頁)。

**electron pair** an orbital (p.111) can hold two electrons (p.110) and the orbital is then full. An orbital in a valency shell (↑) can lose electrons or gain electrons to form an ion (p.123). An orbital can gain an electron by sharing an electron from another orbital; this forms a covalent bond. Full orbitals are stable (p.74).

電子對 一個軌道(第111頁)能容納2個電子(第110頁)，則此軌道即填滿。殼層(↑)中的價層電子可以失去或獲得電子而形成離子(第123頁)。軌道可由另一個軌道共享電子而得到電子，形成共價鍵。填滿的軌道是穩定的(第74頁)。

**lone pair** an orbital (p.111) in a valency shell (↑) containing two electrons not involved in bonding in a molecule. A lone pair can take part in a coordinate bond (p.136), e.g. nitrogen has 5 electrons in its valency shell, 2 in one orbital forming a lone pair and 3, each in one orbital, available (p.85) to form bonds (↑).

不共用電子對；孤對電子 指一個電子的價層(↑)的p軌道(第111頁)中含有兩個在分子中不參與成鍵的電子。不共用電子對可參加一個共價鍵(第136頁)。例如氮有5個電子，其中2個處於一個軌道形成一對孤對電子；另3個則分別處於一個軌道，可利用於(第85頁)形成鍵(↑)。

**octet** (n) the eight electrons which fill the s- and p-orbitals (p.112) of the valency shell (↑) of an atom. Except for hydrogen, helium and the transition elements (p.121), the valency shell of an atom has one s- and three p-orbitals. When these orbitals are full a very stable structure is formed as in the noble gases. Atoms with a valency shell of s- and p-orbitals tend to form ions with no electrons or an octet in the outer shell, giving a stable atomic structure.

八隅體 (名) 指填滿一個原子價層(↑)的s軌道和p軌道(第112頁)的8個電子。除了氫、氦和過渡元素(第121頁)之外，原子的價層有一個s軌道和3個p軌道。當這些軌道都填滿時猶如一種如惰性氣體那樣穩定的結構。具有s軌道和p軌道價層的原子傾向於形成在外層不含有電子的離子或在外殼層含有八隅體，產生一個穩定的原子結構。

**shared electron** two orbitals (p.111), each with one electron (p.110), overlap (p.218). Each orbital shares its electron with the other orbital. In this way each orbital has its own electron and a shared electron, so each orbital can be considered to have two electrons.

共享電子；共用電子 各有一個電子(第110頁)的兩個軌道(第111頁)重疊(第218頁)。每一軌道與另一軌道共用其電子。因此各有一個共有的一個電子和一個共用的電子，因此每個軌道都可看作各容納有兩個電子。

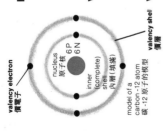

valency electron
價電子

valency shell
價層

nucleus 6P 6N
原子核

inner (complete) shell
內層（填滿）

model of a carbon-12 atom
碳-12 原子的模型

orbital
軌道

nucleus
原子核

model of an electron pair in an orbital
一個軌道中電子對的模型

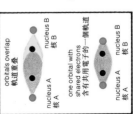

shared electron
共用電子

orbitals overlap
軌道重疊

nucleus A
核 A

nucleus B
核 B

one orbital with shared electrons
含有共用電子的一個軌道

nucleus A
核 A

nucleus B
核 B

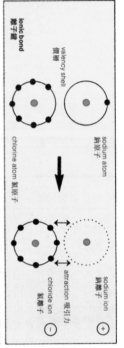

**ionic bond** 離子鍵

valency shell 價層

sodium atom 鈉原子

chlorine atom 氯原子

attraction 吸引力

sodium ion 鈉離子

chloride ion 氯離子

(+) (−)

**ionic bond** a kind of chemical bond (p.133) formed between ions (p.123). It is usually formed between the ions of a metal and a non-metal (p.116). The metal atom loses one or more electrons (p.110) to form a positive ion and the non-metal atom gains one or more electrons (p.110) to form a negative ion, e.g. sodium burns in chlorine to form sodium chloride; a sodium atom loses 1 electron to form a sodium ion and a chlorine atom gains 1 electron to form a chloride ion. In solution the ions are (solvated (p.90) and) free to move; in a crystal (p.91) the ions are held together by the attraction (p.124) of a positive charge for a negative charge and the ions form a giant structure (p.94). Attraction between opposite charges forms an ionic bond.

**electrovalent bond** = ionic bond.

**electrovalency** (*n*) the number of ionic bonds (↑) an atom can form. This is equal to the number of electrons (p.110) which it can lose or gain in its valency shell (p.133) to form an ion. The electro-valency of an atom is also the electrovalency (p.133) of the element. For example, a magnesium atom has two electrons in its valency shell, both can be removed (p.215) to form an ion with 2 positive charges, so magnesium has an electrovalency of + 2 (this shows the ion is positive). A chlorine atom has 7 electrons in its valency shell and it can gain one electron to form an ion with 1 negative charge, so chlorine has an electrovalency of − 1 (this shows the ion is negative). In both cases (magnesium and chlorine) the atoms form ions with a stable (p.74) octet (p.133) in the outside shell.

**離子鍵** (第 133 頁) 離子鍵通常是在金屬和非金屬 (第 116 頁) 離子之間形成的一種化學鍵 (第 133 頁)，在離子 (第 123 頁) 之間形成的一種化學屬原子失去一個或多個電子 (第 110 頁) 形成正離子，而非金屬原子則得到一個或多個電子而形成負離子。例如鈉在氯氣中燃燒生成氯化鈉，其中鈉原子失去一個電子形成鈉離子，而氯子子則得到一個電子形成氯離子。離子在溶液中 (被溶劑化 (第 90 頁) 且自由運動；在晶體中 (第 91 頁) 離子靠正及負電荷的相互吸引力 (第 124 頁) 結合在一起，同時這些離子形成巨大結構 (第 94 頁)。相反的電荷間的吸引力形成離子鍵。

**電價鍵** (名) 同離子鍵 (↑)

**電價** (名) 指一個原子可以形成的離子鍵 (↑) 的數目。電價等於在其價層 (第 133 頁) 中可以失或得電子而成一個離子的電子 (第 110 頁) 數目。原子的電價也是元素的電價。例如，鎂原子的價層中有兩個電子，兩個電子都可除去 (第 215 頁) 以形成一個帶 2 個正電荷的離子，所以鎂的電價為 + 2 (表示該離子帶正電)；氯原子的價層有 7 個電子，可得到一個電子形成一個帶負電荷的離子，所以氯的電價為 − 1 (表示該離子帶負電)。在此兩種情況下 (鎂和氯) 中，原子都在其外層中形成具有穩定 (第 74 頁) 八隅體 (第 133 頁) 的離子。

**氧化態** 由元素（第 116 頁）之離子（第 123 頁）的電荷決定。銅的電價（↑）為 +2，可形成 Cu²⁺ 離子，因此銅的氧化態為 +2。電價為 +1 的銅則形成 Cu⁺ 離子，其氧化態為 +1。同樣地，Cl⁻、S²⁻ 的氧化態分別為 −1、−2。氧化態指示（第 38 頁）已除去（第 215 頁）或加入一個原子中形成離子的電子數目。例如，氧化態 +3，則表示可從一個中性的原子除去 3 個電子。參見"氧化值"（第 78 頁）

**oxidation state** this is determined by the charge on an ion (p.123) of an element (p.116). Copper, with an electrovalency (↑) of + 2, forms an ion Cu²⁺; the oxidation state of copper is + 2. Copper with an electrovalency of + 1 forms an ion Cu⁺, the oxidation state is + 1. Similarly, the oxidation state of Cl⁻ is − 1 and of S²⁻ is − 2. The oxidation state indicates (p.38) the number of electrons that have been removed (p.215) from, or added to, an atom to form an ion, e.g. if the oxidation state is + 3, then three electrons have been removed from a neutral atom. *See oxidation number (p.78).*

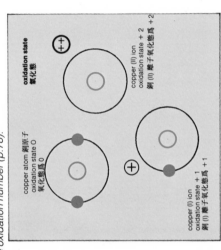

oxidation state
氧化態

copper atom 銅原子
oxidation state 0
氧化態爲 0

copper (I) ion
oxidation state + 1
銅(I) 離子氧化態爲 +1

copper (II) ion
oxidation state + 2
銅(II) 離子氧化態爲 +2

**能量（名）** 能量可用於做功。材料或物質（第 8 頁）的化學能可從儲存於其化學鍵（第 133 頁）的能量獲得（第 85 頁）。當鍵斷裂或形成新鍵時釋放出化學能量。在化學反應中，通常都將化學能轉變（第 144 頁）成熱能。例如，燃燒（第 58 頁）時，熱能可用於汽車引擎做功。在原電池（第 129 頁）中，化學能被轉變成電能；電能可用於做功。（動詞同爲 energize）

**energy** (n) energy can be used to do work. Chemical energy is available (p.85) in a material, or substance (p.8), from the energy stored in its chemical bonds (p.133). This energy is set free when the bonds are broken or new bonds made. Chemical energy is often transformed (p.144) into heat energy in a chemical reaction, e.g. in combustion (p.58); the heat energy can be used to do work as in a motor car engine. Chemical energy is transformed into electrical energy in a primary cell (p.129); the electrical energy can be used to do work. **energize** (v).

**covalent bond** a kind of chemical bond (p.133) formed between atoms (p.110) by shared electrons (p.133). In simple molecules, covalent bonds are usually formed by p-electrons (p.112) in p-orbitals (p.112). These orbitals are directed in space, so covalent bonds are directed in space and account for the shape of molecules. After an atom has formed covalent bonds in simple molecules, it has a complete octet (p.133) of electrons. Compounds with covalent bonds form molecular crystals (p.94) or very hard giant structures (p.94). **covalent** (adj).

**covalency** (n) the number of covalent bonds (↑) an atom (p.110) can form. This is equal to the number of electrons (p.110) which are single electrons in orbitals (p.111); such electrons are available (p.85) to become shared electrons (p.133), e.g. oxygen has a covalency of 2 because it has two unpaired electrons (p.110) able to form two covalent bonds. **covalent** (adj).

**bond energy** the energy given out, or taken in, when a covalent bond is formed between two free atoms. It is also the energy needed to break the bond; and is thus a measure of the strength of the bond, e.g. the bond energy of the C–H bond is 415 kJ mol⁻¹

**coordinate bond** a kind of covalent bond (↑) formed between two atoms when one atom gives both electrons to form the shared electrons (p.133) of the bond. The electrons that are given are a lone pair (p.133). The ammonium ion is formed by a coordinate bond between the nitrogen atom of ammonia and a hydrogen ion, see diagram.

**dative bond** another name for coordinate bond (↑), not used now.

**dative covalent bond** another name for coordinate bond (↑), not used now.

**semipolar bond** another name for coordinate bond (↑), not used now.

**donor** (n) the atom which gives a lone pair in a coordinate bond (↑).

**acceptor** (n) the atom which receives a lone pair in a coordinate bond (↑).

共價鍵（名） 原子（第 110 頁）之間通過共用電子（第 133 頁）形成的一種化學鍵（第 133 頁）。簡單分子中的共價鍵通常是由 p 軌道（第 112 頁）的 p 電子（第 112 頁）形成。這些 p 軌道在空間是有方向的，所以共價鍵也在空間有方向性，這造成分子有形狀而有助於構成分子中，原子形成共價鍵之後就有了完整的電子八隅體（第 133 頁）。具有共價鍵的化合物可形成分子晶體（第 94 頁）或形成極硬的巨大結構（第 94 頁）。

共價（名） 一個原子（第 110 頁）能形成的共價鍵（↑）數目。其數值等於軌道（第 111 頁）中單獨電子的電子（第 110 頁）數目；這些單獨的電子可用（第 85 頁）作共用電子（第 133 頁）。例如，氧有兩個不成對的電子（第 110 頁）能夠形成兩個共價鍵，故其共價為 2。

鍵能 兩個原子自由地之間形成的共價鍵所釋出或吸收的能量；也指將鍵破壞所需要的能量。因此也是鍵強度的量度。例如 C–H 鍵的鍵能為 415 kJ mol⁻¹。

配位鍵 一個原子提供兩個電子作為形成鍵的共用電子（第 133 頁），在兩個原子之間形成的一種共價鍵（↑）。所給出的電子是一對孤對電子（第 133 頁）。銨離子是氨分子中的氫原子與一個氫離子之間的配價鍵形成的（見圖）。

配價鍵 配位鍵（↑）之別稱，現已不用此名稱。

配位鍵 配位鍵（↑）之別稱，現已不用此名稱。

半極性鍵 配價鍵（↑）之別稱，現已不用此名稱。

給予體（名） 在一個配鍵（↑）中給出孤對電子的原子。

接受體（名） 在一個配價鍵（↑）中接受孤對電子的原子。

model of CH₄ 的模型

bond directed in space 在空間有方向的鍵

carbon has 8 electrons 碳有 8 個電子（共用）

each hydrogen has 2 electrons (shared) 各個氫原子有 2 個電子（共用）

carbon nucleus 碳的原子核

hydrogen nucleus 氫的原子核

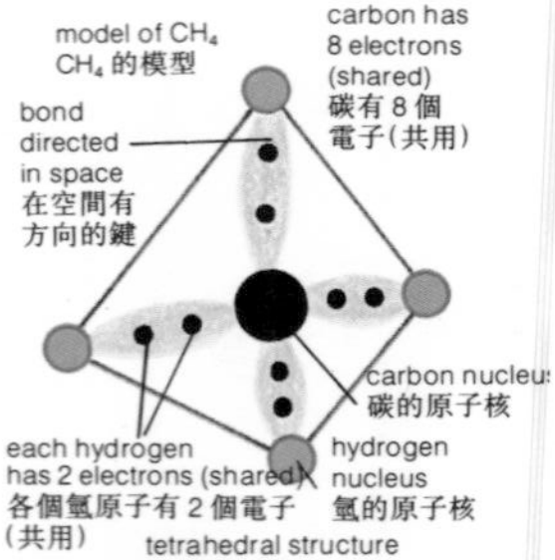

tetrahedral structure 四面體結構

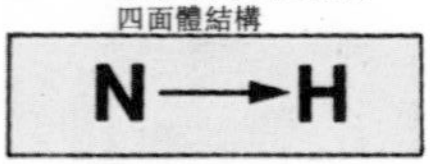

co-ordinate bond 配價鍵

N → H

covalent bond 共價鍵

diagram of CH₄ CH₄ 的圖示

H C H H

donor 給予體

H N H H

lone pair 孤對

acceptor 接受體

H ⁺ hydrogen ion 氫離子

donor 給予體

H N H H + H ⁺

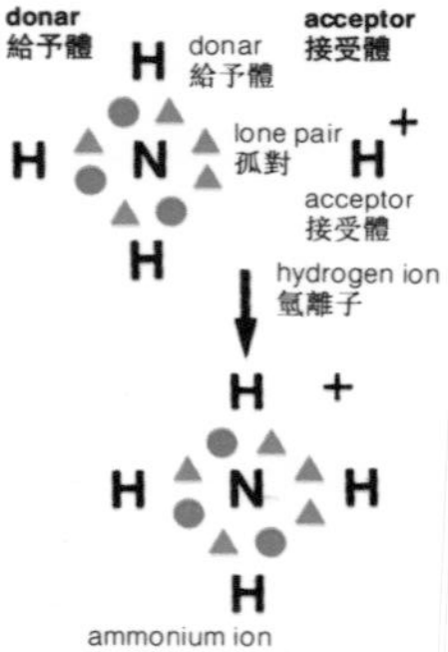

ammonium ion 銨根離子

H H H–N–H H ⁺

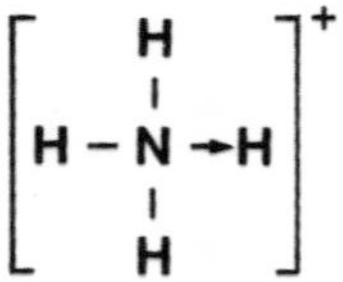

**metallic bond** a kind of chemical bond (p.133) found in metal crystals (p.95). The crystal consists (p.55) of positive ions of the metal in a lattice (p.92) with the valency electrons (p.133) free to move between the metal ions. The moving electrons form the metallic bond.

**van der Waals' bond** a weak chemical bond (p.133) which holds molecules together. It is an attractive (p.124) force arising from the movement of electrons in atoms and is ten to twenty times weaker than the attractive force between atoms. See *molecular crystals (p.94)*. Graphite is an example of molecules held by van der Waals' bonds. The carbon atoms form hexagons in layers, held by strong covalent bonds (↑). The layers are held together by van der Waals' bonds so that the layers slip over each other, explaining the softness of graphite.

**polar** (*adj.*) describes a molecule (p.77) in which the covalent bonds (↑) have the pair of electrons forming the bond nearer to one atom than another. This makes one atom slightly negative and the other atom slightly positive, each with a fraction of the charge on an electron (shown by $\delta+$ or $\delta-$), e.g. the water molecule is polar, *see diagram*. If the substance is liquid, a polar solvent (p.86) is formed. Polar solvents dissolve ionic (p.123) compounds.

**non-polar** (*adj.*) describes a molecule (p.77) which is not polar (↑). If the substance is liquid, a non-polar solvent (p.86) is formed. Non-polar solvents dissolve organic (p.55) compounds, which are generally non-polar themselves.

**monovalent** (*adj.*) describes an element with a valency of 1; it is better described as monoelectrovalent or monocovalent.

**divalent** (*adj.*) describes an element with a valency of 2; it is better described by the electrovalency or covalency.

**tervalent** (*adj.*) describes an element with a valency of 3; it is better described by the electrovalency or covalency.

**金屬鍵** 金屬晶體（第 95 頁）中所見的一種化學鍵（第 133 頁）。金屬晶體由排在晶格（第 92 頁）中的金屬正離子構成（第 55 頁），價電子（第 133 頁）則在金屬離子之間自由移動。這些移動的價電子形成金屬鍵。

**范德華耳斯鍵** 使分子結合在一起的一種弱化學鍵（第 133 頁），它是電子在原子中運動產生的一種引力（第 124 頁）力，離子之間吸引力較之強 10 至 12 倍，參見 " 分子晶體 "（第 94 頁）。石墨是分子靠范德華耳斯鍵結合在一起的例子。石墨分子中的碳原子形成六角形，靠強的共價鍵（↑）結合在一起。各層之間則靠范德華耳斯鍵結合在一起，各層之間可以互相滑過，這可解釋石墨之軟性。

**極性的（形）** 描述一個分子（第 77 頁）的共價鍵（↑）中，電子對形成的鍵偏向某一原子，使該原子稍靠近另一個原子則稍帶正電荷，每個原子只有一個電子上的一小部分電荷（用 $\delta+$ 或 $\delta-$ 表示），如水分子是極性的（見圖）。如成液體物質，則形成一種極性的溶劑（第 86 頁）。極性溶劑可溶解離子（第 123 頁）化合物。

**非極性的（形）** 描述一種沒有極性的分子（第 77 頁）。如成液體物質，則形成非極性的溶劑（第 86 頁）。非極性溶劑可溶解有機（第 55 頁）化合物，而有機化合物本身通常是非極性的。

**一價的（形）** 指一種元素的價數為 1；以一電價的或一共價的描述之爲佳。

**二價的（形）** 指一種元素的價數為 2；以其電價或共價描述之爲佳。

**三價的（形）** 指一種元素的價數為 3；以其電價或共價描述之爲佳。

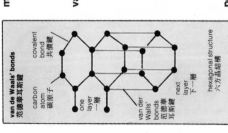

**van der Waals' bonds 范德華耳斯鍵**

covalent bond 共價鍵
carbon atom 碳原子
one layer 一層
van der Waals' bonds 范德華耳斯鍵
next layer 下一層
hexagonal structure 六方晶結構

**polar 極性的**

polar molecule of water (H₂O) 水（H₂O）的極性分子

$\delta+$　$\delta-$　$\delta-$

**radioactivity** (n) the property of spontaneous (p.75) nuclear (p.110) changes in which energy is released as radiation (↓) and a new nucleus is formed with a different number of protons and, in some cases, a different number of neutrons. **radioactive** (adj).

**radiation** (n) (1) a process in which energy is passed on in the form of a wave motion, with electromagnetic waves. Examples of radiation are light, ultra-violet light, X-rays, radio waves. (2) a form of energy, travelling in straight lines and causing ionization (p.123) of any material through which it passes. **radiant** (adj).

**radioactive** (adj) describes an element, or one of its compounds, which has the property of radioactivity (↑).

**natural radioactivity** the radioactivity (↑) of elements with atomic numbers (p.113) greater than 83. Such elements have radioactive compounds which occur (p.63) naturally (p.19).

**artificial radioactivity** the radioactivity of elements which have been made radioactive by bombardment (p.143) with particles (p.13) of high energy.

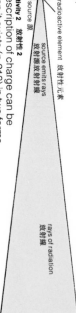

radioactive element 放射性元素

source 源

source emits rays 放射源放射射線

rays of radiation 放射線

radioactivity 2 放射性 2

**ray** (n) (1) a line which shows the direction of radiation (↑). (2) a line of radiant (↑) energy. (3) a stream of charged (↑) particles (p.13).

**source** (n) (1) a place from which something, such as radiation (↑), or an ore (p.154), has come. e.g. a radioactive element is a source of radiation; a primary cell is a source of electric current.

**emit** (v) to give out radiation, gas, odour, sound, from a source (↑), e.g. a bell emits sound when hit; a radioactive element emits radiation. **emission** (n), **emissive** (adj), **emitter** (n).

**charge** (n) no description of charge can be made other than it exists (p.213) in two forms, positive and negative. It is a property of sub-atomic particles (p.13); an electric current is a flow of charge. **charged** (adj).

---

放射性（名）　原子核（第 110 頁）自發（第 75 頁）蛻變的性質，在蛻變時以輻射（↓）形式釋放出能量，產生有不同質子數目的新的原子核。新核的質子數目也不同。（形容詞為 radioactive）

輻射：輻射線（名）　（1）能量以電磁波傳遞的過程。光、紫外光、X 射線、無線電波皆為輻射的例子。（2）能量的一種形式，以直線傳播並在通過的任何材料引致發生電離（第 123 頁）。（形容詞為 radiant）

放射性的（形）　描述一種元素或其中的一種化合物具有放射性（↑）性質。

天然放射　指原子序數（第 113 頁）大於 83 的元素所具有的放射性（↑）。這些元素有天然（第 19 頁）出現（第 63 頁）的各種放射性化合物。

人工放射　用高能粒子（第 13 頁）轟擊（第 143 頁）一種元素，使之具有放射性元素的放射性。

radioactivity + 輻射 radioactivity 1 放射性 1

radon-220 氡 -220

134 n 86 p

radium-224 鐳 -224

136 n 88 p 136 個中子 88 個質子

134 n 86 p 134 個中子 86 個質子

射線（名）　（1）表示輻射（↑）方向的一條線（↑）或出來源（↑）的地方。例如放射性元素是源頭。（3）帶電的（↑）粒子（第 13 頁）流。

源（名）　（1）某些事物如輻射（↑）或礦石（第 154 頁）的地方。例如放射性元素是一種來源，一種電池是一種電源。

電荷（名）　電荷存在（第 213 頁）正電荷和負電荷這兩種形式。電荷是次原子粒子（第 13 頁）的屬性；電流即電荷流。（形容詞為 charged）

發射出（動）　從一個源（↑）放出輻射、氣體、氣味、聲音，例如敲鐘被敲擊時發出聲音；放射性元素發射出輻射。（名詞為 emission，形容詞為 emissive，名詞 emitter 意為發射體）

**Becquerel rays** the rays (↑) emitted by a radioactive source. The kind of ray is not described.

**alpha ray** a ray (↑) consisting (p.55) of alpha particles (↓).

**beta ray** a ray (↑) consisting (p.55) of beta particles (↓).

**gamma ray** a ray (↑) of gamma radiation (↓).

**alpha particle** the nucleus of a helium atom consisting (p.55) of two protons and two neutrons; it has a relative atomic mass of 4.0029 and a charge of +2. An alpha particle has a range (p.140) in air of 7 cm and ionizes (p.123) through which it passes. Alpha particles cause scintillation (p.140) when they hit fluorescent (p.140) surfaces.

**α-particle** another way of writing alpha particle (↑).

**beta particle** an electron (p.110) or positron (p.110) emitted (↑) from the nucleus of a radioactive atom; it has a very high speed, up to 99% of the speed of light. Loss of an electron results in the nucleus increasing its number of protons by 1 and decreasing its neutrons by 1. A beta particle has a range (p.140) in air of 750 cm and can penetrate (p.144) thin pieces of metal foil. It ionizes (p.123) gases through which it passes but the effect is weaker than that of alpha particles (↑).

**β-particle** another way of writing beta particle.

**gamma radiation** radiation (↑) emitted by radioactive (↑) elements; it has wavelengths shorter than those of X-rays. Gamma radiation is emitted with either alpha or beta radiation. The radiation is very penetrating (p.144) with a range of 150mm in lead; it carries no charge, so is not deflected by electric or magnetic fields. The wavelength of a radiation is a characteristic (p.9) of the nucleus emitting it. Gamma radiation has a very weak ionizing (p.123) effect.

**γ-radiation** a way of writing gamma radiation.

**貝克勒爾射線** 自放射性源放射出的射線（↑）。類射線是不記述的。

**α 射線** 由 α 粒子（↓）組成（第 55 頁）的一種射線（↑）。

**β 射線** 由 β 粒子（↓）組成（第 55 頁）的一種射線（↑）。

**γ 射線** γ 輻射（↓）的射線（↑）。

**阿爾法粒子** 由兩個質子和兩個中子組成（第 55 頁）的一個氦原子的核；相對原子質量為 4.0029、電荷為 +2。α 粒子在空氣中的能量範圍（第 140 頁）為 7 cm，可使它所通過的氣體離電（第 123 頁）。α 粒子衝擊螢光的第 140 頁表面時產生閃爍（第 140 頁）。

**α粒子** 阿爾法粒子（↑）的另一寫法。

**貝塔粒子** 放射性原子放射出（↑）的電子（第 110 頁）或正子（第 110 頁）；其速度極高，達光速之 99%。原子失去一個電子導致核的質子數增加 1、中子數則減少 1。β 粒子在空氣中的能量範圍（第 140 頁）為 750 cm，可穿透金屬薄片。β 粒子可使其通過的氣體離電（第 123 頁），但效應比 α 粒子（↑）效應很弱。

**β粒子** 貝塔粒子的另一寫法。

**伽瑪輻射** 放射性（↑）元素放射出的輻射線（↑），波長短於 X 射線。γ 輻射伴隨 α 輻射或 β 輻射一起放射出。γ 輻射線的穿透性（第 144 頁）極強，可穿透 150 mm 範圍的鉛；γ 輻射不帶電荷，因此不為電場或磁場所偏轉。γ 射線的波長是放射或該原子核的特徵（第 9 頁）。γ 射線放射 β 射線或 α 射線和 γ 射線。

**γ輻射** 伽瑪輻射的另一寫法。

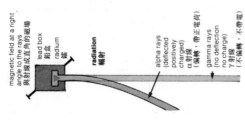

magnetic field at a right angle to the rays
與射線成直角的磁場

lead box
鉛盒

radium
鐳

alpha rays (deflected positively charged)
α射線（偏轉、帶正電荷）

gamma rays (no deflection no charge)
γ射線（不偏轉、不帶電）

radiation
輻射

a radioactive source emits alpha rays or alpha rays and gamma rays
放射性源放射 α 射線或 α 射線和 γ 射線

lead box
鉛盒

protactinium
鏷

magnetic field at a right angle to the rays
與射線成直角的磁場

beta rays (deflected, negatively charged)
β 射線（偏轉、帶負電荷）

gamma rays (no deflection, no charged)
γ 射線（不偏轉、不帶電）

radiation
輻射

a radioactive source emits beta rays or beta rays and gamma rays
放射性源放射 β 射線或 β 射線和 γ 射線

**alpha emission** the emission of an alpha particle (p.139) from the nucleus of a radioactive (p.138) atom. The resulting nucleus has an atomic number (p.113) which is 4 less and a mass number (p.113) which is 4 less, e.g.

$$^{220}_{86}Rn \rightarrow\ ^{216}_{84}Po + ^{4}_{2}He \text{ (α-particle)}$$

Alpha emission may be accompanied by gamma radiation; it is never accompanied by beta emission (↓).

**beta emission** the emission of a beta particle (p.139) from the nucleus of a radioactive (p.138) atom. The resulting nucleus has an atomic number (p.113) which is 1 more, but the mass number (p.113) remains the same, e.g

$$^{228}_{88}Ra \rightarrow\ ^{228}_{89}Ac + ^{0}_{-1}e$$

In the nucleus, a neutron change produces a proton and an electron; the proton remains and the electron is emitted. Beta emission may be accompanied (p.213) by gamma radiation; it is never accompanied by alpha emission (↑).

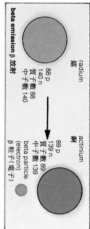

beta emission β 放射

radium 鐳
88 p
140 n
質子數 88
中子數 140

actinium 錒
89 p
139 n
質子數 89
beta particle (electron)
β 粒子 (電子)

**range** (n) (1) the greatest distance an object can travel when it has been given energy, e.g. the range of an aeroplane on a tank full of fuel (p.160). (2) a set of values between an upper and lower limit, e.g. the temperature range between 0°C and 100°C. **range** (v).

**scintillation** (n) very short, bright, quick flashes of light. **scintillate** (v).

**fluorescent** (adj) a material which emits light when radiation, of a shorter wavelength than light, falls on the material. **fluorescence** (n).

**spinthariscope** (n) a device consisting of a surface coated (p.127) with zinc sulphide. When alpha particles hit the surface, it scintillates (↑). The number of scintillations can be counted, giving the number of alpha particles emitted by a radioactive source (p.138).

α 發射　由放射性（第 138 頁）原子的核放射出 α 粒子（第 139 頁）。產生的新核，原子序數（第 113 頁）減少 2，質量數（第 113 頁）則減少 4。例如：

$$^{220}_{86}Rn \longrightarrow\ ^{216}_{84}Po + ^{4}_{2}He \text{ (α粒子)}$$

α 放射可能伴隨有 β 放射（↓）。

β 發射　由放射性（第 138 頁）原子的核放射出 β 電子（第 139 頁）。產生的新核，原子序數（第 113 頁）增加 1，質量數（第 113 頁）保持相同。例如：

$$^{228}_{86}Ra \longrightarrow\ ^{228}_{89}Ac + ^{0}_{-1}e$$

在核中，一個中子變化產生一個質子和一個電子；中子保留而電子則被放射出。β 放射可能伴隨（第 213 頁）有 γ 輻射，但絕不會伴隨有 α 放射（↑）。

程：範圍（名）　(1) 給予一個物體能量而能使該物體移動的最大距離。例如：飛機油箱裝滿燃油（第 160 頁）時的航程；(2) 上限和下限之間的一組數值。例如0°C 至 100°C 之間的溫度範圍。（動詞曰 range）

閃爍（名）　指時間極短、明亮而快速的閃光。（動詞曰 scintillate）

螢光的（形）　當波長比光波短的輻射放射在其上時會發光的材料。（名詞曰 fluorescence）

閃爍鏡（名）　表面塗（第 127 頁）有硫化鋅的一種裝置。當 α 粒子撞擊此表面時，它便會閃爍（↑）。由閃爍的次數可以計算出放射性源（第 138 頁）所放射的 α 粒子數目。

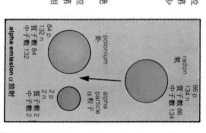

radon 氡
86 p
134 n
質子數 86
中子數 134

polonium 釙
84 p
132 n
質子數 84
中子數 132

alpha particle α粒子
2 p
2 n
質子數 2
中子數 2

alpha emission α放射

spinthariscope 閃爍鏡
glass lens 玻璃透鏡
radioactive source 放射性源
zinc sulphide screen 硫化鋅
brass tube 銅管

**Geiger counter** a device for the detection (p.225) of ionizing (p.123) radiations. It consists (p.55) of a metal tube, with a window, and a wire electrode (p.122) in the centre of the tube. The tube is filled with argon (an inert gas) at a low pressure. There is a high electric potential between the electrode and the tube, with the electrode made positive. When radiation enters through the window, the argon atoms are ionized and the electrode is discharged, giving a pulse of current, which can be detected.

蓋格計數器 檢測（第225頁）電離（第123頁）輻射用的一種儀器。由金屬管組成（第55頁），管中心有一窗口和金屬絲電極（第122頁）。管內充滿低壓的氫氣（一種惰性氣體）。電極和管之間存在着高電位。電極爲正極。當輻射經窗口進入計數器時，氫原子被電離同時電極放電，產生可檢測的電流脈衝。

**radioactive disintegration** the disintegration (p.65) of the nucleus of a radioactive element into two parts. The disintegration usually results in the formation of another nucleus and an alpha or beta particle.

放射性蛻變 放射性元素的原子核分成兩部分的蛻變（第65頁）。蛻變結果通常形成另一種原子核和放射出 α 粒子或 β 粒子。

**radioactive decay** the decrease, with time, of radioactivity (p.138) from a specimen of a radioactive element.

放射性衰變 放射性元素的樣本之放射性（第138頁）隨時間而減少。

**half-life** (n) the time taken for one half of the atoms in a specimen of a radioactive element to disintegrate (↑). It is an important characteristic (p.9) of a radioactive element, and is a constant for that element, e.g. examples of half-lives are: uranium-238, 4.5 × 10⁹ years; radium-226, 1620 years; radium-221, 30 seconds. The half-life period is independent of temperature, pressure, concentration and nature of the material.

半衰期（名）放射性元素的樣本有數一半衰變之時間（↑）減少到原有數一半所需之時間。是放射性元素的一個重要特徵（第9頁）。是放射性元素的一個常數。例如下列元素的半衰期：鈾-238爲4.5×10⁹年；鐳-226爲1620年；鐳-221爲30秒。半衰期與溫度、壓力以及材料的濃度和性質無關。

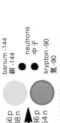

barium-144 鋇-144
56 p / 88 n
neutrons 中子
krypton-90 氪-90
36 p / 54 n

neutron 中子
92 p / 143 n
uranium-235 鈾-235

**nuclear fission** 核裂變

**nuclear fission** the disintegration of a nucleus into smaller nuclei, brought about by bombardment (p.143) with neutrons, e.g. uranium-238 bombarded by low energy neutrons splits into barium-144 and krypton-90, releasing two neutrons:
$$^{235}_{92}U + ^{1}_{0}n \longrightarrow\ ^{144}_{56}Ba + ^{90}_{36}Kr + 2^{1}_{0}n$$
In this process, energy is released; 1 kg of uranium by nuclear fission produces approximately 10¹⁰ kJ.

核裂變：核分裂 用中子轟擊（第143頁）使原子核蛻變成較輕的原子核。例如用低能量的中子轟擊使鈾-235分裂成鋇-144和氪-90並釋出兩個中子：
$$^{235}_{92}U + ^{1}_{0}n \longrightarrow\ ^{144}_{56}Ba + ^{90}_{36}Kr + 2^{1}_{0}n$$
此過程中釋放出能量。1 kg 鈾核裂變可產生接近於 10¹⁰ kJ 的能量。

**fissile** (adj) describes any element, or substance, which can undergo fission, particularly nuclear fission (↑).

可裂變的（形）描述可進行裂變或發生裂變，尤其是核裂變（↑）的任何元素或物質。

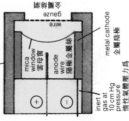

protective cover 保護蓋
mica window 雲母窗
anode wire 陽極金屬絲
metal cathode 金屬陰極
inert gas at 10 cm Hg pressure 惰性氣體壓力爲 10 cm Hg
wire gauze 金屬絲網

**Geiger counter** 蓋格計數器

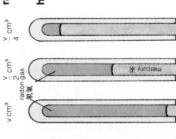

v cm³  V/2 cm³  V/4 cm³
mercury 汞
radon gas 氡氣
3.8 days 天
radioactive decay of radon half-life period of 3.8 days
氡的放射性衰變半衰期爲3.8天

**half-life** 半衰期

**nuclear fusion** the combination of two nuclei of small mass number to produce one nucleus of a higher mass number; the nuclear reaction takes place only at very high temperatures (about $10^8$ K). A large quantity of energy is released, e.g. $^2_1H + {^3_1}H \rightarrow {^4_2}He + 5 \times 10^5$ MJ. Deuterium bombarded with protons forms helium-3, releasing large quantities of energy.

**curie** (n) the unit for the rate of radioactive disintegration (p.141). 1 curie $= 3.7 \times 10^{10}$ atoms disintegrating per second. 2996 kg uranium-238 produce $3.7 \times 10^{10}$ atomic disintegrations per second, hence this mass of uranium has an activity of 1 curie. The safe limit of radioactivity for human beings is estimated to be 10 micro-curies. The symbol for curie is Ci.

**radioactive series** one of three groups of naturally occurring radioactive (p.138) elements, each series named after the element with which the series starts. The uranium series begins with uranium-238 and passes, by 15 nuclear changes, to lead-206, which is stable (p.74). The thorium series begins with thorium-232 and passes by 11 nuclear changes to lead-208, which is stable. The actinium series starts with protoactinium-231 and passes through 9 nuclear changes to lead-207, which is stable.

**disintegration series** alternative name for radioactive series (↑).

**Fajans and Soddy law** the emission of an alpha particle during radioactive change produces an element two places to the left in the periodic table (p.119). The emission of a beta particle, however, produces an element one place to the right in the periodic table.

**nuclide** (n) an atomic species which is defined by the number of protons and neutrons in the nucleus, and by its kind of radioactive decay and the half-life (p.141) of the rate of decay, e.g. radium-221 has 88 protons and 133 neutrons in its nucleus, it decays by alpha-emission with a half-life of 30 seconds. Radium-223 has 88 protons and 135 neutrons in its nucleus, it decays by alpha-emission accompanied (p.213) by gamma-radiation and has a half-life of 11.7 days.

核聚變　使質量數小的兩個原子核合併產生一個質量數較高的原子核。核反應在極高溫度(約 $10^8$ K)下才會發生。反應放出大量能量。例如：$^2_1H + {^3_1}H \longrightarrow {^4_2}He + 5 \times 10^5$ MJ。氘受質子轟擊形成氦-3 並放出大量能量。

居里(名)　放射性蛻變(第 141 頁)速率的單位。1 居里 = 每秒鐘蛻變 $3.7 \times 10^{10}$ 個原子。2996 kg 鈾-238 每秒鐘可產生 $3.7 \times 10^{10}$ 個原子蛻變，因此此質量的鈾有 1 居里。人類的放射性安全限值估計前為 10 微居里。居里的符號為 Ci。

放射系列(名)　天然存在的三組放射性(第 138 頁)元素中的一組，每一系列都以該系列中排頭之元素命名。鈾系列以鈾-238 開始，經 15 步核變化成為穩定的鉛-206。釷系列以釷-232 開始，經 11 步核變化成為穩定的鉛-208。錒系以鏷-231 開始，經 9 步核變化成為穩定的鉛 -207。

蛻變系列(名)　放射系列(↑)之別稱。

法揚斯-索迪定律　在放射性變化過程中，放射一個 α 粒子產生的新元素向週期表(第 119 頁)中的左方移兩位，而放射 β 粒子產生的新元素向週期表中的右方移一位。

核素(名)　由原子核中質子和中子的數目以及原子放射性衰變的種類和衰變速率的半衰期(第 141 頁)確定的原子種類，例如，錒-221 的原子核有 88 個原子和 133 個中子，其以 α 發射衰變的半衰期為 30 秒。錒-223 的原子核中有 88 個原子和 135 個中子，其 α 發變伴隨着(第 213 頁)γ 輻射，半衰期為 11.7 天。

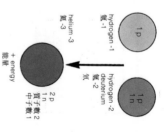

actinium series
錒系

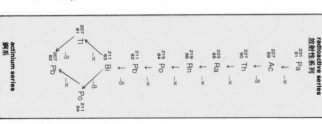

radioactive series
放射性系列

atomic fusion 原子核聚變

hydrogen-1 氫-1

helium-3 氦-3

hydrogen-2 deuterium 氫-2 氘

+ energy 能量

**chamber** (*n*) a hollow space surrounded by walls, usually with a small entrance.

**track** (*n*) visible (p.42) signs that an object has travelled along a path. The signs are continuous.

**室** (名) 四周有壁包圍着的一個空間，一般都有一個小入口。

**徑迹** (名) 指物體沿一路徑移行的可見 (第 42 頁) 痕迹。

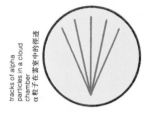

tracks of alpha particles in a cloud chamber
α 粒子在雲室中的徑迹

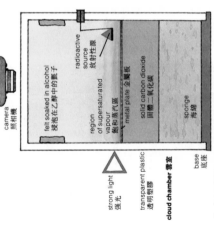

camera 照相機

felt soaked in alcohol 浸泡在乙醇中的氈子

radioactive source 放射性源

region of supersaturated vapour 飽和蒸汽區

metal plate 金屬板

solid carbon dioxide 固體二氧化碳

sponge 海綿

strong light 強光

transparent plastic 透明塑膠

**cloud chamber 雲室**

base 底座

**cloud chamber** a round chamber (↑), made of transparent (p.16) plastic which has a firm base. Above the base, solid carbon dioxide is supported by a sponge. A metal plate rests on the solid carbon dioxide. Round the top of the chamber is a strip of felt, soaked in alcohol (ethanol). The ethanol vaporizes (p.11) and saturates dust-free air in the chamber. The metal plate is very cold and forms a region of supersaturated vapour, just above the plate. A radioactive source produces radiation which ionizes (p.123) the molecules in the air. Drops of liquid condense on the ions, forming a track (↑) which is photographed by a camera, using strong light, as shown in the diagram. Condensation takes place only in the region of supersaturated vapour. Cloud chambers are used to study radiation from radioactive sources and nuclear reactions (p.144).

**bombard** (*v*) to hit a large object repeatedly with many small objects, each small object possessing considerable energy. **bombardment** (*n*).

**雲室** 由牢固的底座和透明 (第 16 頁) 塑膠構成的一個圓形室 (↑)。固體二氧化碳置於底座的海綿之上，固體二氧化碳上面擱置一塊金屬板。室頂部周圍有一條浸泡在酒精 (乙醇) 中的毛氈。乙醇蒸發室中蒸發 (第 11 頁) 並飽和室中的無塵空氣。金屬板極冷，因而恰好在板上方形成一個過飽和的蒸汽區域。放射性源產生輻射使空氣中的分子電離 (第 123 頁)。液滴在離子上冷凝形成一條迹徑 (↑)，用一架照相機以強光拍攝下此迹徑。如圖所示。冷凝作用僅在過飽和的蒸汽區域中發生。雲室用於研究放射性源和核反應 (第 144 頁) 的輻射。

**轟擊** (動) 用多個小物體 (每個小物體都含有很大的能量) 重複撞擊一塊大的物體。(名詞為 bombardment)

**penetrate** (v) to go through an outer cover, using force or energy, or similarly, to reach inside a solid material, e.g. gamma rays penetrate steel; a neutron penetrates an atomic nucleus. **penetration** (n), **penetrating** (adj).

**radio opaque** (adj) describes a material which does not allow radiation (p.138) to pass through it because the radiation is absorbed. Radio opaqueness depends on the density of the material and on the wavelength of the radiation. The shorter the wavelength, the greater the penetration (↑). **radio opaqueness** (n).

**transform** (v) to change one form of energy into another form. **transformation** (n).

**radiology** (n) the study of X-rays and radioactivity (p.138). In particular, it is the use of radiations (p.138) in medical science for curing diseases. **radiologist** (n), **radiological** (adj).

**nuclear reaction** a reaction in which changes in the nucleus of an atom take place. Nuclear changes are caused by the bombardment (p.143) of a nucleus with subatomic (p.110) and other particles, e.g.

$$^{63}_{29}Cu + ^{1}_{1}H \rightarrow ^{63}_{30}Zn + ^{1}_{0}n$$

in which copper-63 is bombarded with protons and forms zinc-63 and neutrons. Aluminium-27 bombarded with alpha particles forms phosphorus-30 and neutrons:

$$^{27}_{13}Al + ^{4}_{2}He \rightarrow ^{30}_{15}p + ^{1}_{0}n$$

The products of nuclear reactions are generally radioactive isotopes (p.114) of the element.

proton 質子

2 p
14 n
質子數 2
中子數 2

alpha particle α粒子

13 p
14 n
質子數 13
中子數 14

aluminium nucleus
鋁原子核

29 p
34 n
質子數 29
中子數 34

copper nucleus 銅原子核

nuclear reactions
核反應

30 p
33 n
質子數 30
中子數 33

zinc nucleus 鋅原子核

15 p
15 n
質子數 15
中子數 15

phosphorus nucleus
磷原子核

neutron
中子

**穿透** (動) 利用力或能量或類似方法穿過外圍達固體材料之內部。例如 γ 射線能吸收輻射並穿透原子核。（名詞爲 penetration，形容詞爲 penetrating）

**不透射線的** (形) 描述一種因能吸收輻射而不讓射線通過的材料。不透射線性決定於材料的密度和波長。波長越短，則穿透性 (↑) 越強。（名詞爲 radio opaqueness）

**轉變** (動) 使一種能量形式改變爲另一種能量形式。（名詞爲 transformation）

**放射學** (名) 研究 X 射線和放射性 (第 138 頁) 的一門學科，特別是指醫科上使用放射線 (第 138 頁) 治病。（名詞爲 radiologist，形容詞爲 radiological）

**核反應** 在原子核中發生變化的一種反應。核變化是用次原子 (第 110 頁) 粒子和其他粒子轟擊 (第 143 頁) 原子核而產生的。例如其中銅-63 被質子轟擊而生成鋅-63 和中子。

$$^{63}_{29}Cu + ^{1}_{1}H \rightarrow ^{63}_{30}Zn + ^{1}_{0}n$$

用 α 粒子轟擊鋁-27 生成磷-30 和中子：

$$^{27}_{13}Al + ^{4}_{2}He \rightarrow ^{30}_{15}p + ^{1}_{0}n$$

核反應之產物一般都是該元素的放射性同位素 (第 114 頁)。

block of wood
木塊

3cm

**penetrate 穿透**

a nail penetrates a
block of wood to a
depth of 3 cm
釘子穿入木塊至深度 3 cm

**mass spectrograph**
質譜儀

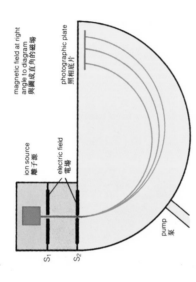

magnetic field at right angle to diagram
與圖成直角的磁場

ion source
離子源

electric field
電場

photographic plate
照相底片

$S_1$

$S_2$

pump
泵

mass spectrograph
質譜儀

22　23　24　25　26　27

mass spectrograph of magnesium
鎂的質譜圖

**mass spectrograph** a piece of apparatus for determining (p.222) the mass number of isotopes (p.114). An ion (p.123) source, *see diagram,* produces positive ions of an element. The ions pass through two slits (p.211), $S_1$ and $S_2$, in electrodes. The electrodes are given a negative potential (voltage) to accelerate (p.219) the ions. The ions then enter a semi-circular chamber (p.143) which has a strong magnetic field acting at a right angle to the path of the ions. The ions follow a circular path, due to the magnetic field; the lower the mass number of the ion, the smaller is the radius of the path. The ions hit a photographic plate and produce a mark. The width of the mark corresponds to the abundance (p.231) of an isotope. The spectrograph is calibrated (p.26) using carbon-12, so mass numbers (p.113) can be read directly. The diagram shows a photograph of the isotopes of magnesium. The apparatus operates (p.157) with all the air pumped out, forming a high vacuum. **mass spectrography** (*n*).

**mass spectrometer** a piece of apparatus similar to a mass spectrograph (↑); it measures accurately the relative proportion (p.76) of each isotope in a naturally-occurring element.

**質譜儀** 測定（第 222 頁）同位素（第 114 頁）質量數用之儀器。由一個離子（第 123 頁）源（見圖）產生元素之正離子。離子流過兩電極中的兩個狹縫（第 211 頁）$S_1$ 和 $S_2$。給電極提供負電位（電壓）使離子加速（第 219 頁）。然後離子進入一個半圓形室（第 143 頁）。在此用強磁場以直角作用於離子的路徑。離子因電場作用而沿一條圓形路徑前行；離子之質量數越低，則路徑之半徑越小。離子碰撞照相底片並產生一個標記。標點的寬度和同位素的豐度（第 231 頁）一致。利用碳-12 校準（第 26 頁）質譜儀，因而可直接讀出質量數（第 113 頁）。圖示鎂的同位素照片。儀器在抽除去全部空氣形成高真空下樂作（第 157 頁）。各詞 mass spectrography 意爲 質譜法。

**質譜計** 與質譜儀（↑）相類之儀器設備。質譜儀可準確量度天然存在的元素中各種同位素的相對比例（第 76 頁）。

**heat of reaction** the heat energy given out, or taken in, when a chemical reaction takes place between the masses of reactants (p.62) shown by the equation (p.78) for the reaction. The heat energy is measured in joules (p.153).

**heat of combustion** the heat energy given out when one mole of a substance is completely burned in oxygen, e.g. ethyne (acetylene) is burned completely to carbon dioxide and water, 1 mole of ethyne produces 1558 kJ of heat energy. This is written as:
$$C_2H_2 + 5/2 O_2 = 2 CO_2 + H_2O, \Delta H = -1558 \text{ kJ}$$
The symbol (p.77) $\Delta H$ indicates (p.38) the heat change during the reaction. The negative sign shows heat is given out, i.e. lost by the reactants.

**heat of neutralization** the heat energy given out when one mole of hydrogen ions in an acid solution is neutralized by a base, with the reaction taking place in a dilute aqueous solution. In reactions between strong acids and strong bases, both reactants are completely ionized (p.123) and the reaction is between one mole of hydrogen ions and one mole of hydroxyl (p.132) ions. The heat of neutralization for strong acids with strong bases is 57.3 kJ; for weak acids or weak bases, the heat energy given out is less than 57.3 kJ, as energy is used to complete ionization of the acid or base.

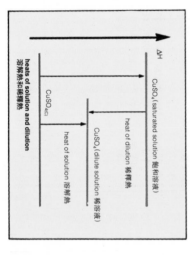

heats of solution and dilution
溶解熱和稀釋熱

CuSO₄ (saturated solution 飽和溶液)

heat of dilution 稀釋熱

CuSO₄ (dilute solution 稀溶液)

heat of solution 溶解熱

ΔH

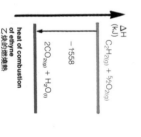

heat of combustion
of ethyne
乙炔的燃燒熱

C₂H₂₍g₎
ΔH
(kJ)

2CO₂₍g₎ + H₂O₍l₎

−1558

**反應熱** 用反應方程式 (第 78 頁) 表示大量反應物 (第 62 頁) 之間進行化學反應時所放出或吸入的熱能。熱能的量度單位爲焦耳 (第 153 頁)。

**燃燒熱** 1 摩爾物質在氧氣中完全燃燒時所放出的熱能。例如，乙炔完全燃燒生成二氧化碳和水時，1 摩爾乙炔產生 1558 千焦耳熱能。熱化學方程式爲：
$$C_2H_2 + 5/2 O_2 = 2 CO_2 + H_2O,$$
$$\Delta H = -1558 \text{ kJ}。$$
符號 (第 77 頁) ΔH 表示 (第 38 頁) 反應過程中的熱量變化。負 (−) 號表示放出熱量；即反應物失去熱量。

**中和熱** 稀酸水溶液中 1 摩爾氫離子與鹼中和時放出的熱能。強酸和強鹼進行中和反應時，兩種反應物都完全電離 (第 123 頁)，反應是在 1 摩爾氫離子和 1 摩爾氫氧根 (第 132 頁) 離子之間進行的。強酸及強鹼之中和熱局 57.3 kJ。弱酸或弱鹼進行中和反應時，因能量用於酸或鹼之完全電離，所以放出的熱能低於 57.3 kJ。

**heat of dilution** the change in heat energy when a solution is diluted (p.81). *See diagram.*

**heat of solution** the heat energy given out, or taken in, when one mole of a substance is dissolved in such a large volume of water that further dilution (p.81) produces no heat change.

**heat of formation** the heat change when one mole of a compound is formed from its elements, under stated conditions of temperature and pressure.

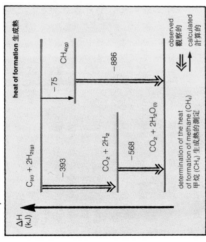

heat of formation 生成熱

$\Delta H$ (kJ)

$-75$  $CH_{4(g)}$

$C_{(c)} + 2H_{2(g)}$

$-393$

$-886$

$CO_2 + 2H_2$

$-568$

$CO_2 + 2H_2O_{(l)}$

observed 觀察的

calculated 計算的

determination of the heat of formation of methane (CH₄) 甲烷 (CH₄) 生成熱的測定

$CH_{4(g)} + 2O_{2(g)} = CO_{2(g)} + 2H_2O_{(l)}$
$\Delta H_{298} = -889$ kJ

The reactants are gaseous methane and oxygen; the products are gaseous carbon dioxide and liquid water. The reaction between 1 mole of methane and two moles of oxygen gives out 889 kJ of heat energy at 298 K (25°C). For crystalline solids, the symbol (c) (p.77) is written, e.g. S_(c) for crystalline sulphur.

**heat of ionization** the heat change necessary to produce complete ionization (p.123) of a substance in an aqueous (p.88) solution.

**thermochemical equation** an equation showing the chemical reaction between whole numbers of moles of the reactants, giving information about the states of the reactants, the temperature of measurement, and the heat change, e.g.

**稀釋熱** 溶液被稀釋 (第 81 頁) 時的熱能變化。(見圖)。

**溶解熱** 1 摩爾某一種物質溶解於大量水中所放出或吸入的熱能，此溶液進一步稀釋 (第 81 頁) 亦不產生熱量變化。

**生成熱** 在規定溫度和壓力條件下，1 摩爾化合物由其組成的元素形成時的熱量變化。

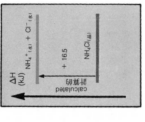

heat of ionization 電離熱

$\Delta H$ (kJ)

$NH_4^+{}_{(水)} + Cl^-{}_{(水)}$

$+ 16.5$

calculated 計算的

$NH_4Cl_{(固)}$

**電離熱** 使一種物質在水 (第 88 頁) 溶液中產生完全電離作用 (第 123 頁) 所需的熱量變化。

**熱化學方程式** 表示整數摩爾數的反應物之間化學反應的方程式，式中指明有關反應物狀態，計量的溫度和熱量變化的數據。例如：

$CH_{4(氣)} + 2O_{2(水)} = CO_{2(氣)} + 2H_2O_{(液)}$
$\Delta H_{298} = -889$ kJ

反應物為氣體甲烷和氧氣，產物為氣體二氧化碳和液體的水。1 摩爾甲烷和 2 摩爾氧在 298K (25°C) 下反應放出 889 kJ 熱能。對於結晶的固體，寫上符號 (c) (第 77 頁)，如 S_(c) 表示結晶的硫。

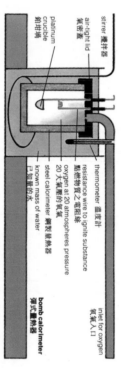

stirrer 攪拌器
air-tight lid 氣密蓋
platinum crucible 鉑坩堝

thermometer 溫度計
resistance wire to ignite substance 點燃物質之電阻絲
oxygen at 20 atmospheres pressure 20 大氣壓的氧氣
steel calorimeter 鋼製量熱器
known mass of water 已知量的水
inlet for oxygen 氧氣入口

**bomb calorimeter**
彈式量熱器

**bomb calorimeter** a device for measuring heats of combustion (p. 146). It consists (p.55) of a thick-walled steel calorimeter, with an air-tight screw top. A known amount of substance is put in a platinum crucible and oxygen is pumped in to 20 atmospheres pressure. The substance is ignited (p.32) by a heated resistance wire. The heat evolved (p.40) heats the calorimeter, and thus the water. The temperature rise of the water is measured by a sensitive thermometer and the heat evolved is calculated, and hence the heat of combustion of the substance is found.

**exothermic** (*adj*) describes a reaction in which heat energy is given out.

**endothermic** (*adj*) describes a reaction in which heat energy is taken in.

**Hess's law** the heat energy given out, or taken in, in a chemical change is the same, no matter how the change takes place. For example, ammonium chloride solution can be made in two ways both starting with gaseous ammonia and gaseous hydrogen chloride:

(1) $NH_{3(aq)}$ + aq → $NH_{3(aq)}$; $HCl_{(g)}$ + aq → $HCl_{(aq)}$
$NH_{3(aq)}$ + $HCl_{(aq)}$ → $NH_4Cl_{(aq)}$
where (aq) is an aqueous solution.

(2) $HCl_{(g)}$ + $NH_{3(g)}$ → $NH_4Cl_{(c)}$
$NH_4Cl_{(c)}$ + aq → $NH_4Cl_{(aq)}$

The changes of heat energy are shown in the diagram opposite. Both ways of preparation produce the same result.

**thermochemistry** (*n*) the study of the changes in heat energy that take place during a chemical reaction. Heat energy given out heats the products, while heat energy taken in cools the products.

彈式量熱器 測量燃燒熱(第 146 頁)之裝置，由一個厚壁的鋼製量熱計配備一個氣密的螺旋蓋所組成(第 55 頁)。淋已知量之物質放入一個鉑坩堝中，同時打入氧氣至達到 20 大氣壓。用一根熱電阻絲將物質點燃(第 32 頁)，所放出的(第 40 頁)熱量用來加熱量熱器及水。用一支靈敏的溫度計測量水的溫升，並計算出所放出的熱量，從而求出該物質的燃燒熱。

放熱的(形) 描述一種放出熱能的反應。

吸熱的(形) 描述一種吸收熱能的反應。

赫斯定律 在一個化學反應中，無論該反應是一步還是分幾步完成，所釋放出或吸收入的熱量總是相等。例如，氯化銨溶液可以從氨氣或氫化氯氣體兩種方式開始反應製造：

(1) $NH_{3(氣)}$ + 水 —→ $NH_{3(水)}$；
$HCl_{(氣)}$ + 水 —→ $HCl_{(水)}$；
$NH_{3(水)}$ + $HCl_{(水)}$ —→ $NH_4Cl_{(水)}$
式中(水)指一種水溶液。

(2) $HCl_{(氣)}$ + $NH_{3(氣)}$ —→ $NH_4Cl_{(晶)}$
$NH_4Cl_{(晶)}$ + 水 —→ $NH_4Cl_{(水)}$

熱能的變化如右圖所示。兩種製備方式產生同樣的結果。

熱化學(名) 研究化學反應過程中發生熱能變化之學科。放出的熱可使產物加熱，而吸收熱則使產物冷卻。

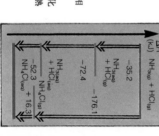

ΔH (kJ)

$NH_{3(g)}$ + $HCl_{(g)}$

−35.2

$NH_{3(aq)}$ + $HCl_{(g)}$

−72.4

$NH_{3(aq)}$ + $HCl_{(aq)}$

−176.1

$NH_4Cl_{(c)}$

−523

$NH_4Cl_{(aq)}$

+16.31

**rate of reaction** the rate of reaction is measured by the rate at which the reactants are used up or by the rate at which the products are formed.

**反應速率** 反應速率可用反應物消耗的速率來量度，也可以用產物生成的速率來量度。

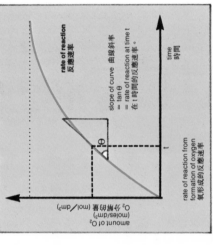

amount of O₂ 氧分解的質量 (moles/dm³)

slope of curve 曲線斜率
= tan θ
= rate of reaction at time t
在 t 時間的反應速率。

rate of reaction
反應速率

time
時間

rate of reaction from
formation of oxygen
氧形成的反應速率

**rate constant** if an amount x of a substance is used up in a reaction and the initial (p.85) amount was a, then the rate of reaction (量) at a given time is:$\frac{dx}{dt} \propto (a - x)$ i.e. the rate is proportional to (a – x). This can be written as: $\frac{dx}{dt} = k(a - x)$ where k is the *rate constant*.

The rate constant is independent of pressure in reactions in solution, dependent on pressure in gaseous reactions, but is changed by catalysis and temperature changes.

**velocity constant** = rate constant (↑).
**law of mass action** the rate of a reaction, at a constant temperature, is proportional to the product of the active masses of the reacting substances. In a reaction: A + B → products (A and B are two substances) the rate of reaction ∝ [A] × [B], where [A] represents the active mass of A. For gases, the active mass is measured by the partial pressure (p.108) of the gas; for solutions, it is measured by the concentration in moles per dm³. With solids, the active mass is difficult to measure.

**反應速率常數** 如在一反應中消耗的物質量爲 x，初始（第 85 頁）量爲 a，則在某一指定時刻的反應速率（量）爲（$\frac{dx}{dt} \propto (a - x)$，即速率與 (a – x) 成正比。可寫成 $\frac{dx}{dt} = k(a - x)$

式中 k 爲反應速率常數。

在溶液中反應時，反應速率常數與壓力無關；在氣體中反應時則和壓力有關，並受催化和溫度變化所改變。

**反應速度常數** 同反應速率常數 (↑)。
**質量作用定律** 在恆定溫度條件下，化學反應速率與反應物各物的有效質量之乘積成正比。在 A + B → 產物（反應中 A、B 分別代表兩種物質）反應的反應率 ∝ [A] × [B]，式中 [A] 表示 A 物質的有效質量。對氣體而言，有效質量用氣體的分壓（第 108 頁）量度；而對液體而言，則用濃度的量度，單位爲摩爾（分米）⁻³。固體的有效質量則難以量度。

**equilibrium** (*n*) in a reversible reaction (p.64), the products start to react the moment they are formed. The rate of reaction of the reactants (p.62) is high as their concentration is high, while the rate of reaction of the products (p.62) is low because their concentration is low; this is because the law of mass action (p.149) is obeyed (p.107). Eventually, the two rates of reaction will be the same, provided the products are not removed (p.215), and the reactants and products will both be present. This is a state of *equilibrium* and thereafter there is no change in concentration of reactants and products.

**dynamic equilibrium** a state of equilibrium (↑) in which the rate of change of two processes (p.157) is equal and opposite, e.g. (a) in a closed vessel (p.25) the rate of evaporation (p.11) of a liquid, and the rate of condensation (p.11) of its vapour (p.11), are equal at a constant temperature; the vapour exerts (p.106) its vapour pressure (p.103) for that temperature. (b) phosphorus pentachloride (PCl₅) decomposes (p.65) on heating to form phosphorus trichloride (PCl₃) and chlorine (Cl₂). In a closed vessel, PCl₃ and Cl₂ combine to form PCl₅. At equilibrium (↑) all three substances are present, with the two opposite chemical processes being in dynamic equilibrium.

$$PCl_5 \rightleftharpoons PCl_3 + Cl_2$$

**equilibrium mixture** the relative (p.232) concentrations of reactants and products when equilibrium (↑) is reached in a reversible reaction (p.64).

**equilibrium constant** in a reversible reaction (p.64):

$$pA + qB \rightleftharpoons rC + sD$$

where A, B, C, D are substances and p, q, r, s are mole fractions, the equilibrium constant is given by:

$$K = \frac{[C]^r [D]^s}{[A]^p [B]^q}$$

[A], [B], [C], [D] represent the concentrations of the substances. The value of *K* changes with temperature.

---

平衡 (名) 在可逆反應 (第 64 頁) 中，產物在其生成的瞬間即開始發生反應。由於反應物的濃度高，故其反應速率也快；而產物 (第 62 頁) 的濃度低，故其反應速率也低，也慢，而且反應時將遵從 (第 107 頁) 質量作用定律 (第 149 頁)。只要產物不被除去 (第 215 頁)，而且反應物又同時存在，兩種狀態。此後反應速率都會達到相等而處於 "平衡" 狀態。此後反應物和產物的濃度都不再變化。

動態平衡 兩個過程 (第 157 頁) 的變化速率相等而方向相反的一種平衡 (↑) 狀態。例如，(a) 在恆溫條件下的密閉容器 (第 25 頁) 中，液體蒸發 (第 11 頁) 的速率等於其蒸汽 (第 11 頁) 冷凝的速率 (第 103 頁) 此溫度下冷凝的蒸汽壓 (第 106 頁) 在此溫度下冷凝的蒸汽壓 (第 103 頁)。(b) 五氯化磷 (PCl₅) 加熱分解 (第 65 頁) 成三氯化磷 (PCl₃) 和氯 (Cl₂)；在平衡 (↑) 時，由於所有三種物質都同時存在。

$$PCl_5 \rightleftharpoons PCl_3 + Cl_2$$

平衡混合物 可逆反應 (第 64 頁) 達到平衡 (↑) 時，反應物和產物的相對 (第 232 頁) 濃度。

平衡常數 在以下可逆反應 (第 64 頁) 中：

$$pA + qB \rightleftharpoons rC + sD$$

式中 A、B、C、D 分別是四種物質，p、q、r、s 為摩爾分數，平衡常數用下式表示：

$$K = \frac{[C]^r [D]^s}{[A]^p [B]^q}$$

[A]、[B]、[C]、[D] 分別表示四種物質的濃度。K 值隨溫度而變化。

---

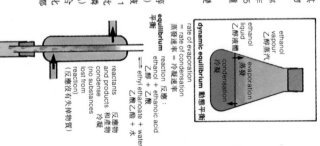

**equilibrium** 平衡
ethanol liquid 乙醇液體
ethanol vapour 乙醇蒸汽
evaporation 蒸發
condensation 冷凝

**dynamic equilibrium** 動態平衡
rate of evaporation 蒸發速率 = rate of condensation 冷凝速率

reaction 反應：
ethanol + ethanoic acid 乙醇 + 乙酸 = ethyl ethanoate + water 乙酸乙酯 + 水

reactants and products 反應物和產物
condense 冷凝
(no substances lost from reaction) (反應沒有失掉物質)

ethanol 乙醇
ethanoic acid 乙酸
ethyl ethanoate 乙酸乙酯
water 水

heat 加熱

equilibrium 平衡
ethanol 乙醇
+
ethanoic acid 乙酸

equilibrium mixture 平衡混合物

**Le Chatelier's principle** whenever changes take place in the external conditions of a chemical reaction *in equilibrium*, then changes occur (p.63), if possible, in the reaction which tend (p.216) to counteract (p.216) the effect of those external changes. The principle can be illustrated from the reversible reaction:

$$N_{2(g)} + 3H_{2(g)} \rightleftharpoons 2NH_{3(g)} \quad \Delta H = -50 \text{ kJ}$$
1 vol.　3 vols.　2 vols.

The reaction between nitrogen and hydrogen is exothermic (p.148) with 50kJ of heat energy given out. The decomposition of ammonia is endothermic (p.148) with 50 kJ heat absorbed. One volume of nitrogen reacts with 3 volumes of hydrogen to produce 2 volumes of ammonia, hence there are 4 volumes of the reactants and 2 volumes of the products. If an equilibrium mixture has its temperature raised the mixture absorbs heat to counteract the rise in temperature; this favours (p.214) the endothermic reaction, i.e. the decomposition of ammonia. Raising of the temperature quickens both rates of reaction, so equilibrium is reached more quickly. Increasing the pressure favours a smaller volume of gases, so the equilibrium moves towards forming more ammonia. The effect of external conditions on equilibrium is summarized in the table below.

勒沙特利原理 一個處於 "平衡狀態" 的化學反應系統，每當外界條件發生變化時，反應中可能出現 (第 63 頁) 的種種變化傾向於 (第 216 頁) 抵消 (第 216 頁) 那些外界變化的影響。這個原理可以用以下可逆反應來說明：

$$N_{2(g)} + 3H_{2(g)} \rightleftharpoons 2NH_{3(g)} \quad \Delta H = -50 \text{ kJ}$$
1 體積　3 體積　2 體積

氮和氫之間的反應是放熱反應 (第 148 頁)，放出之熱能為 50 kJ。氨分解為氮和氫的反應是吸熱的 (第 148 頁) 反應，吸收的熱能為 50 kJ。一個體積的氮和 3 個體積的氫反應生成 2 個體積的氨，即有 4 個體積的反應物和 2 個體積的產物。如果將平衡混合物的溫度升高，那麼該混合物將吸收熱量以抵消溫度的升高，這有利於 (第 214 頁) 吸熱反應，即有利於氨的分解。溫度上升加快這兩種反應的速率，使反應更快速到達平衡。升高壓力有利於體積較小的氣體，使平衡向生成氨的方向移動。外界條件對平衡的影響概述如下表：

| Reaction 反應： | A + B (larger volume) (體積較大) | C + D, (smaller volume) (體積較小) | + heat given out 放出熱量 |
|---|---|---|---|
| EXTERNAL CHANGE 外部變化 | RATE OF REACTION 反應速率 | EQUILIBRIUM MIXTURE 平衡混合物 | EQUILIBRIUM CONSTANT 平衡常數 |
| increase of temperature 溫度升高 | increased 加快 | more A + B 更多的 A + B | changed 變 |
| decrease of temperature 溫度降低 | decreased 減慢 | more C + D 更多的 C + D | changed 變 |
| increase of pressure 壓力升高 | increased for gaseous reactions 加快對氣體的反應 | more C + D 更多的 C + D | unchanged 不變 |
| decrease of pressure 壓力降低 | decrease for gaseous reactions 減慢對氣體的反應 | more A + B 更多的 A + B | unchanged 不變 |
| addition of catalyst 加入催化劑 | increased 加快 | unchanged 不改變 | unchanged 不變 |

**energy level** an electron possesses energy according to its distance from the nucleus (p.110) of an atom; the nearer it is to the nucleus, the lower is the energy it possesses. This energy is the energy level of an electron.

**ground state** the condition of an atom (p.110) when all its extranuclear (p.113) electrons are in their positions of lowest energy, i.e. at their lowest energy levels (↑).

**excitation** (*n*) the changing of an atom from its ground state (↑) by supplying energy in the form of heat, radiation, or bombardment (p.143) by sub-atomic (p.110) particles. This causes one or more electrons to move to orbitals further from the nucleus. Such electrons may then enter into covalent bonds (p.136), or with sufficient energy, be removed from the atom, forming an ion (p.123). **excitatory** (*adj*).

**energy barrier** a measure of the quantity of energy that must be supplied in order that a chemical reaction can take place. If a lesser quantity of energy is supplied, there is no reaction.

**activation energy** the additional energy that must be supplied to reactants before a reaction can take place. The atoms are in a state of excitation (↑). The activation energy is the energy barrier to a reaction; until the activation energy is supplied, no reaction can take place.

**activated state** the state of molecules, atoms, or ions in which they are able to react.

**reaction profile** a diagram which shows the relation between the ground states (↑) of reactions and products, and the energy barrier to a reaction.

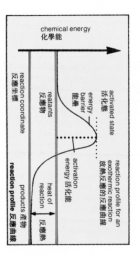

chemical energy
化學能

activated state
活化態

energy barrier
能障

reactants
反應物

reaction coordinate
反應坐標

activation energy
活化能

reaction profile for an exothermic reaction
放熱反應的反應曲線

heat of reaction
反應熱

products
產物

**reaction profile 反應曲線**

能級：能階 電子隨其與原子核(第 110 頁)距離而具有的能量；此能量稱爲電子能級。

基態 指一個原子(第 110 頁)的全部核外(第 113 頁)電子都處於其最低能量位置，即處於其最低能級(↑)的狀態。

激發(名) 以熱、輻射形式提供能量給原子(第 110 頁)粒子轟擊(第 143 頁)，結果使一個原子的基態(↑)改變，這些使一個或多個電子移到距離核較遠的軌道，這些電子可以參加共價鍵(第 136 頁)，或者具有足夠的能量逸出原子，使原子成爲一個離子(第 123 頁)。(形容詞 excitatory)

能障 能障 使化學反應得以進行所必須提供的能量之量度，提供之能量少於此能量，則不發生反應。

活化能 使化學反應進行前所必須提供給反應物之額外能量。原子都是處於激發(↑)的狀態。活化能是反應的能障，直至給予化學反應提供了活化能才能進行反應。

活化態 分子、原子或離子能起反應之狀態。

反應曲線圖 顯示反應物與產物的基態(↑)間的關係及顯示反應能量的曲線圖。

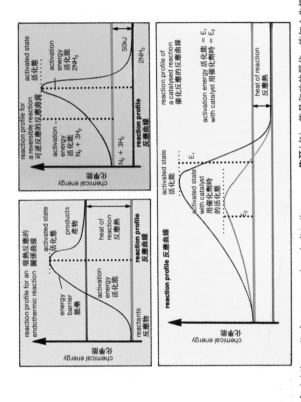

**joule** (*n*) the unit of energy and work. It is the work done when the point of application of a force of one newton is displaced through a distance of one metre in the direction of the force. It is also the work done when a current of 1 ampere flows between a voltage difference of 1 volt for 1 second. (A force of one newton gives a mass of 1 kg an accelera-tion of $1\,ms^{-2}$.) The symbol for joule is J.

**calorie** (*n*) a measurement of heat energy; it is that quantity of heat that will raise the temperature of 1 g of water through 1°C. The unit is no longer used. 1 calorie $\simeq$ 4.18 joules.

**electron-volt** a unit of energy that is not included in S.I. units, but is useful in many chemical cal-culations on radiation (p.138) energy. It is the quantity of energy gained by an electron when it falls through a voltage difference of 1 volt.

$$1\ \text{electron-volt} = 1\,eV$$
$$= 1.6 \times 10^{-19}\ \text{coulomb} \times 1\ \text{volt}$$
$$= 1.6 \times 10^{-19}\ \text{joule.}$$

焦耳 (名) 能量和功的單位。施加 1 牛頓作用力於一質點，使此質點在力的方向上移動 1 米距離所作的功。也指 1 安培電流在 1 伏特電壓差間流動 1 秒所作的功。(1 牛頓的力給予 1 kg 質量物體的加速度爲 $1\,ms^{-2}$)。焦耳的符號爲 J。

卡 (名) 熱能的量度；使 1 克水之溫度上升 1°C 所需之熱量。現已不用此單位。1 卡 $\simeq$ 4.18 焦耳。

電子伏特 國際制單位中不包括這個能量單位，但在有關輻射 (第 138 頁) 能的許多化學計算上常用此單位。係指使一個電子通過 1 伏特電壓差位降所需的能量。

$$1\ \text{電子伏特} = 1\ eV$$
$$= 1.6 \times 10^{-19}\ \text{庫侖} \times 1\ \text{伏特}$$
$$= 1.6 \times 10^{-19}\ \text{焦耳。}$$

**raw materials** the materials which are used as the starting point of chemical processes (p.157). They include such materials as ores (↓), minerals (↓), limestone, common salt, coal, mineral oil.

**mineral** (n) any material which occurs (↓) naturally but does not come from animals or plants. Examples of minerals are: metal ores (↓), limestone (↓), rock salt (↓), coal (p.156), mineral oil (p.156). A mineral has a known chemical constitution (p.82) and definite physical and chemical properties.

**ore** (n) a mineral (↑) which is a compound (p.8) of a metal and from which the metal can be extracted (p.164).

**deposit** (n) a mass of a mineral found in the earth, in a large enough quantity to be worth mining (↓).

**seam** (n) a thick, fairly level, layer of a deposit (↑) of a mineral (↑), e.g. a seam of coal.

**lode** (n) a large, thin deposit (↑) of an ore (↑) which is between the walls of a deep crack in the Earth's structure.

**mine** (v) to dig a hole in the ground to reach a deposit, seam or lode (↑) and to get minerals (↑) out. **mine** (n), **miner** (↑), **mining** (n).

**supply** (n) the raw materials (↑), electric power, fuel (p.160), necessary for the working of a chemical process (p.157). **supply** (v).

**reserve** (n) a quantity of a material or mineral which is not needed at present, but can be used in the future, e.g. a reserve deposit of mineral oil which can be used when present supplies (↑) are finished.

**occur²** (v) to be in a particular place, e.g. petroleum (p.160) occurs in Saudi Arabia in great quantities. To contrast occur and exist (p.213): sulphur occurs in large deposits in North America; sulphur exists in two crystalline (p.15) allotropes (p.118).

mine 礦

seam of coal 煤層

原料　用作化學過程（第 157 頁）起始的材料。包括礦石（↓）、礦物（↓）、石灰石、食鹽、礦物油之類材料。

礦物（名）　天然存在（↓）而非來自動物或植物的任何材料。金屬礦石（↓）、煤（第 156 頁）、礦物油（第 156 頁）都是礦物。礦物有已知的化學構造（第 82 頁）和確定的物理和化學性質。

礦石（名）　為金屬化合物（第 8 頁）的一種礦物（↑）。從礦石可提取（第 164 頁）金屬。

礦床（名）　在地球中發現的一大片礦物，其數量在地球上值得開採（↓）。

礦層（名）　一層厚而相當平的礦物（↑）的礦床（↑）。例如煤層。

礦脈（名）　在地球結構深裂隙壁間的大而薄的礦石（↑）礦床（↑）。

開採（動）　從地面掘洞深至礦床、礦層或礦脈（↑）並取出礦物（↑）。名詞為 mine 礦、miner 礦工、mining 採礦。

供應（名）　進行化工過程（第 157 頁）所需之原料（↑）、電力、燃料（第 160 頁）。（動詞為 supply）

儲藏量（名）　當前並不需但可供將來用的一定量的材料或礦物。例如，當現有供應（↑）用盡之後可開採的礦物油儲油層。

埋藏（發現有）（動）　在一個特定地方存在有，例如在沙烏地阿拉伯埋藏（發現有）大量石油（第 160 頁）。occur（發現有）與 exist（存在）（第 213 頁）之比較：北美洲埋藏（發現有）（occur）硫礦的大礦床；硫存在（exist）兩種晶體（第 15 頁）同素異形體（第 118 頁）。

**assay** (*v.t.*) to make chemical tests (p.42) to find the mineral content (p.85) of a material; in particular to determine (p.222) the relative amount of a metal in an ore (↑).

**native** (*adj*) describes an element found in the Earth and not combined in a mineral, e.g. gold, copper, and sulphur are found uncombined in deposits (↑); they occur (↑) native, and are native elements.

**blende** (*n*) a mineral which is a sulphide (p.51) of a metal, e.g. zinc blende which is zinc sulphide.

**pyrites** (*n*) a mineral which is a sulphide of a metal, e.g. iron pyrites is iron sulphide.

**chalk** (*n*) a soft white rock consisting (p.55) of calcium carbonate, formed from the shells of shellfish. The shells were deposited (↑) on the sea-bed over a long time and pressure formed the rock from the shells.

**limestone** (*n*) a rock consisting (p.55) mainly of calcium carbonate, but sometimes containing magnesium carbonate. It is harder than chalk and is used as a building material. On heating, it is converted (p.73) to lime (p.169).

**marble** (*n*) a hard, crystalline (p.15) rock, formed from limestone (↑) by heat and pressure within the Earth. It consists (p.55) of calcium carbonate. Marble is used as a building material.

**dolomite** (*n*) a white crystalline (p.15) mineral (↑) consisting of calcium and magnesium carbonates. The mineral is found in great abundance (p.231).

**common salt** the substance sodium chloride; it occurs (↑) in sea water which has a salt content (p.85) of about 3%. Common salt is obtained by evaporating large quantities of sea water.

**rock salt** a mineral consisting of sodium chloride. Large deposits (↑) are found in many countries. It is the raw material (↑) for the manufacture (p.157) of chlorine and many sodium salts (p.46).

**sulphur** (*n*) a yellow solid element, occurring (↑) native (↑) in the earth, also found in many ores (↑) such as blendes (↑) and pyrites (↑). It is the raw material (↑) for the manufacture (p.157) of sulphuric acid, the most important industrial (p.157) chemical.

化驗 (動、及) 進行化學試驗 (第 42 頁) 以找出材料中的礦物含量 (第 85 頁)；尤其是測定 (第 222 頁) 礦石 (↑) 中金屬的相對含量。

天然的 (形) 描述在地球中發現且未結合在礦物中的一種元素。例如，在礦床 (↑) 中發現未結合的金、銅和硫；人們發現 (↑) 這些元素天然存在，所以是天然元素。

閃鋅礦 (名) 含金屬硫化物 (第 51 頁) 之一種礦物。例如閃鋅礦係含有硫化鋅。

黃鐵礦 (名) 含金屬硫化物之一種礦物。例如黃鐵礦係含硫化鐵。

白堊 (名) 由碳酸鈣組成 (第 55 頁) 的一種鬆軟的白色岩石，是由貝類的殼形成的。這些貝殼沉積 (↑) 在海床上，經過很長時間及壓力使之由貝殼形成岩石。

石灰石 (名) 主要是由碳酸鈣組成 (第 55 頁) 的一種岩石，有時還含有碳酸鎂。石灰石比白堊硬，常用作建築材料。加熱時轉化 (第 73 頁) 成石灰 (第 169 頁)。

大理石 (名) 石灰石 (↑) 在地球內部受熱和壓力作用形成一種堅硬的晶體 (第 15 頁) 岩石，係由碳酸鈣組成 (第 55 頁)，可用作建築材料。

白雲石 (名) 由碳酸鈣和碳酸鎂構成的一種白色晶體 (第 15 頁) 礦物 (↑)。此種礦物的蘊藏量極豐。

食鹽 海水中藏 (↑) 的含氯化鈉物質，海水的鹽含量 (第 85 頁) 約為 3%。蒸發大量海水便可獲得食鹽。

岩鹽 由氯化鈉組成的一種礦物。許多國家都發現有大的岩鹽礦床 (↑)。岩鹽是製造 (第 157 頁) 氯和許多種鈉鹽 (第 46 頁) 的原料 (↑)。

硫磺 (名) 在地球中發現 (↑) 天然 (↑) 存在的一種黃色固體元素，多種礦石 (↑) 例如閃鋅礦 (↑) 和黃鐵礦 (↑) 中也存在有硫。硫是製造 (第 157 頁) 硫酸的原料 (↑)，硫酸是最重要的工業 (第 157 頁) 化學品。

chalk 白堊
chalk cliffs 白堊懸崖

limestone 建築物中的石灰石
limestone in building
limestone 石灰石

marble 大理石

**carboniferous** (adj) describes any material having carbon in it, or being a form of coal (↓). Also describes any material producing carbon when heated, or any plant life which formed coal (↓).

**carbonaceous** (adj) describes any mineral (p.154) containing carbon, but not carbonates, e.g. coal (↓) is a carbonaceous mineral.

**coal** (n) a black, fairly hard mineral which occurs in large deposits (p.154) and seams (p.154). Coal has been formed from plant life which existed (p.213) many millions of years ago. Chemical action, heat and pressure of the earth formed the mineral. It consists (p.55) mainly of carbon, but contains many other useful chemical compounds (p.8).

**mineral oil** a dark liquid mixture of various compounds (p.8) occurring (p.154) in deposits in many parts of the world; it varies widely in composition (p.82) depending on the place from where it comes. It is thought that mineral oil has been formed from decayed plants and animals. Deposits of mineral oil usually occur with natural gas.

**bitumen** (n) a naturally occurring black material which contains hydrocarbons (p.172). It is a solid or a very thick liquid. Bitumen is also obtained as a product of distillation from coal (↑).

**charcoal** (n) a form of carbon obtained by heating wood, or other plant materials or animal materials, when no air is present (p.217). Charcoal is black in colour and generally very porous (p.15). Charcoal absorbs (p.35) gases and decolorizes coloured liquids if the colour is from organic (p.55) materials.

**coke** (n) a solid material obtained by heating coal with no air present (p.217). It is mainly carbon.

**liquid air** air is cooled below its critical temperature (p.104) and then liquefied (p.11). Liquid air is the source (p.138) of nitrogen and oxygen. Liquid air absorbs (p.35) silicon dioxide. It occurs in many minerals, e.g. quartz, sand. It is the source of silicon, an element used in electronic circuits.

**silica** (n) a very hard white substance; silicon

含碳的 (形) 描述含有碳或是屬煤 (↓) 形式的任何材料。也描述在加熱時產生碳的任何植物或已成爲煤 (↓) 的任何植物。

礦質的 (形) 描述含有碳、但並非碳化物的任何礦物。例如煤 (↓) 爲一種礦質礦物。

煤 (名) 一種相當堅硬的黑色礦物，此種礦物埋藏於大片礦床 (第154頁) 和礦層 (第154頁) 中。煤是由幾百萬年之前就已存在 (第213頁) 的植物所形成。理在地球中的熱力和壓力形成煤礦物。煤主要是由碳所組成的 (第55頁)，但也含有許多其他有用的化合物 (第8頁)。

礦物油 世界許多地域的沉積層中都發現存在的一種黑色的由各種不同化合物 (第8頁) 組成的液體混合物。其成分在 (第82頁) 因產地而差別很大。礦物油是動物和植物的腐爛而形成的，其沉積層通常都發現含有天然氣。

(地) 瀝青 (名) 天然存在的含烃 (第172頁) 黑色物質，是一種固體或極黏稠的液體。瀝青也作爲煤的 (↑) 蒸餾的產物而獲得。

炭 (名) 木材或其他動物、植物材料、在不提供 (第217頁) 空氣下加熱得的一種固體的形式。炭是黑色的，通常都極其多孔 (第15頁)。炭可吸收 (第35頁) 氣體、含有機 (第55頁) 物質的有色液體可用炭脫色。

焦煤：焦炭 (名) 煤在不提供 (第217頁) 空氣下加熱所獲得的一種固體材料。其成分主要是碳。

液態空氣 使空氣冷卻到其臨界溫度 (第104頁) 以下、然後液化 (第11頁) 而成。液態空氣是氮和氧的來源 (第138頁)。

矽石 (名) 係一種極堅硬的白色物質二氧化矽：硅石。它存在於諸如英和矽的許多種礦物中。爲矽：硅石。它存在於諸如英和矽的許多種礦物中。爲電子電路所用之一種元素──矽 (或稱硅) 的來源。

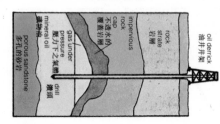

oil derrick 油井井架
impervious rock 不透水的岩層
strata 岩層
rock 岩石
gas under pressure 壓力下之氣體
mineral oil 礦物油
porous sandstone 多孔的矽岩
drill 鑽頭

**process** (n) (1) in chemistry, a method of making a substance in large quantities; the method includes a description of the chemical reaction, the conditions (p.103) under which the reaction takes place and the plant (↓) needed for the process. (2) the different methods of preparing (p.43) and isolating (p.43) a substance, e.g. distillation, sublimation. (3) generally, a process is a number of events taking part in a connected and continuous change. **process** (v.t.).

**plant** (n) all the building, pipes, furnaces (p.164), machinery, special apparatus (p.23) and devices (p.23) used to make substances in large quantities for industry (↓).

**industrial** (adj) describes anything connected with the making of chemical substances, machinery, or materials in large quantities. Industrial chemicals are divided into heavy chemicals (used in other industrial processes), pharmaceutical chemicals (p.20) (used in curing diseases), fine chemicals (used in chemical analysis (p.82), fertilizers and plastics. **industry** (n).

**operate** (v.t.) to make an industrial (↑) process work, e.g. to operate a process for the manufacture (↓) of ammonia is to provide the reactants and obtain the products using the necessary plant (↑). **operation** (n).

**carry out** to work on a process, an experiment or an investigation and to complete it.

**manufacture** (v.t.) to make chemical substances, materials, machinery by an industrial (↑) process (↑). **manufacture** (n).

**main product** the product for which a particular process (↑) is planned. In some processes it is the only product, e.g. sulphuric acid in the Contact process (p.166) is the only product.

**by-product** (n) a substance produced in the manufacture (↑) of a main product (↑); it has an industrial (↑) use, e.g. in the Castner-Kellner process (p.169) chlorine is a by-product.

**waste product** a substance that is produced in a chemical process and which has no industrial (↑) use, e.g. calcium chloride, produced in the Solvay process (p.169), has no industrial use.

過程：方法 (名) (1)化學上指大量製造某種物質的方法，包括化學反應之描述、進行反應之條件 (第103頁) 以及過程所需之廠房設備 (↓) 某種物質的不同方法 (第43頁) 和分離 (第43頁) 某植物質的不同方法，例如蒸餾、昇華；(3)一般而言，過程是指參與一個連貫的和連續變化的一系列事件。(及物動詞為 process)

廠房及設備 (名) 工業 (↑) 上大量製造各種物質所用之全部建築物、管道 (第23頁)、爐、特殊儀器 (第23頁) 和裝置 (第23頁)。

工業的 (形) 描述與大量製造化學物質、機械或材料有關的任何事物。工業化學品分成重化學品 (第20頁) (用於冶病)、精細化學品 (用於化學分析 (第82頁)、化學肥料和塑膠製品等。(名詞為 industry)

操作 (動、及) 做一個工業 (↑) 過程的工作。例如，為操作一個製造 (↓) 氨的過程必須提供反應物並使用必要的廠房設備 (↑) 以獲得產品。(名詞 operation)

實現：進行 (動、及) 完成一個過程、一項實驗或研究的工作。

製造 (動、及) 利用工業 (↑) 過程 (↑) 生產化學物質、材料和機械。(名詞為 manufacture)

主要產物 (名) 一個特定過程 (↑) 所計劃好的產物。在某些過程中是唯一的產物。例如在接觸法 (第166頁) 中，硫酸是唯一的產物。

副產物 (名) 在製造 (↑) 一種主要產物 (↑) 的同時所產生的一種物質。副產物有工業 (↑) 用途，例如，在卡斯特納-凱耳納法 (第169頁) 中，氯為副產物。

廢產物 指化工過程中所生產而又無工業 (↑) 用途的物質。例如無工業用處。是一種廢產物。

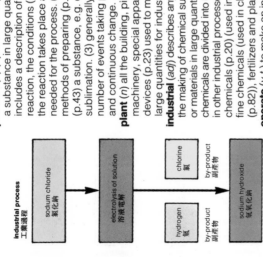

industrial process 工業過程

sodium chloride 氯化鈉

electrolysis of solution 溶液電解

hydrogen 氫 by-product 副產物

chlorine 氯 by-product 副產物

sodium hydroxide 氫氧化鈉 main product 主要產物

industrial process for the manufacture of sodium hydroxide 製造氫氧化鈉的工業過程

**end product** a product formed after an original (p.220) substance has taken part in several reactions or processes, e.g. petrol is one end product from the refining (↓) of petroleum.

**mill** (n) (1) a machine for breaking large lumps (p.13) of solid materials and turning the material into a powder (p.13), e.g. a mill for making flour. (2) a work place where cloth is made, e.g. a cotton mill.

**pulverize** (v.t.) to turn pieces of solid into a powder (p.13) by hitting them repeatedly. A mill (↑) can pulverize a solid. **pulverization** (n).

**spraying** (n) the action of sending many small drops of liquid from small holes in a pipe so as to cover a large space. **spray** (n), **spray** (v.t.).

**lixiviation** (n) the process of removing one substance from a mixture of substances by using a suitable solvent (p.86), e.g. salts containing iodine are obtained from heated seaweed by lixiviation with water. **lixiviate** (v.t.).

**leaching** (n) the removal (p.215) of soluble (p.17) substances by water from a mixture of solids when the water washes out the substance. To contrast lixiviation and leaching: if a mixture is treated (p.38) with water, the process is lixiviation; if water flows through the mixture to wash away a substance the process is leaching. Solvents other than water can be used in leaching. **leach** (v).

**slaking** (n) adding water to quicklime (calcium oxide) to form slaked lime (calcium hydroxide). **slake** (v).

**scum** (n) any solid substance, especially dirt or waste (p.170), which floats on the surface of a liquid, e.g. impurities floating on the surface of a molten (p.10) metal.

**sludge** (n) soft wet solid material, usually waste or unwanted material.

**froth flotation** a process for the separation (p.34) of ores (p.154) from earthy material. Oil and water form a froth (p.100) with particles of ore trapped in the liquid-air interfaces (p.18) of the bubbles (p.40). The foam is stabilized by frothing agents (p.63). The ore is removed with the froth.

最終產物　—種原始（第 220 頁）物質參與幾次反應或過程之後生成的產物。例如汽油為石油精煉（↓）所得之最終產物。

碾磨機：布廠 （名）　(1)將大塊（第 13 頁）固體材料破碎成粉末（第 13 頁）的機器；例如碾磨粉的的碾磨機。(2)織布的工作場所，例如紗廠。

粉碎（動）（及）　將—塊固體反复碰擊使之變為粉末（第 13 頁）。碾磨機(1)可將碎固體。（名詞為 pulverization）。

噴霧 （名）噴射（1）使許多微細微滴自管中小孔出並覆蓋—個大空間的作用。（名詞為 spray，動詞為 spray）。

浸濾作用 （名）溶濾作用 使用適當溶劑（第 86 頁）去除物質混合之一種物質之過程。例如藉水的浸濾作用自加熱的海藻中取得含碘鹽。（動詞為 lixiviate）

濾濾作用 （名）用水沖洗物質時，水移去（第 215 頁）固體混合物中的可溶解（第 17 頁）物質。浸濾作用與溜濾作用間的差別：若混合物經水處理用，此過程為浸濾作用；若水流過混合物以洗掉某種物質，此過程為濾濾作用。除用水濾濾外，還可用溶劑濾濾。（動詞為 leach）

消化 ： 熟化（名）　生石灰（即氧化鈣）中加入水使之成為消石灰（即氫氧化鈣）。（動詞為 slake）

浮渣 （名）浮在液體表面的任何固體物質，尤指污物或廢物（第 170 頁）—種混合物（第 10 頁）金屬表面上的雜質。

污泥 ：污泥（名）指軟而濕的固體材料，通常都是廢料或無用的材料。

泡沫浮選　自泡狀材料分離（第 34 頁）礦石（第 154 頁）的—種方法。油和水形成的泡沫（第 100 頁）中帶有捕集在氣泡（第 40 頁）液氣界面（第 18 頁）的礦石顆粒。用起泡劑（第 63 頁）使泡沫穩定。礦石隨泡沫除去。

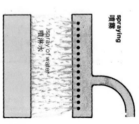

spraying 噴霧

spray of water 噴淋水

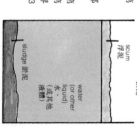

scum/sludge 浮渣 / 淤泥

scum floating and sludge setting in a liquid 浮渣在液體中漂浮和沉降的淤泥

scum 浮渣

sludge 淤泥

water (or other liquid) 水、（或其他液體）

**stage** (*n*) a length of space or time in a process, e.g. (a) the primary, intermediate and secondary stages of an education system (p.212); (b) the stages in the Bosch process (p.168) for manufacturing hydrogen are (i) mixing water gas and steam, (ii) passing the gases over a catalyst, (iii) removing carbon dioxide.

**refine** (*v*) (1) to remove impurities from a substance, e.g. to refine copper by electrolysis (p.122). (2) to separate the constituents (p.54) of a mixture to get pure specimens of some, or all, of the constituents, e.g. to refine petroleum (p.160) to get petrol, kerosene, etc. (3) to make a method, apparatus, or a technique more efficient or more suitable for its purpose, e.g. (a) a refined method of determining the equilibrium constant of a reaction; (b) a more refined apparatus for carrying out (p.157) chromatography. **refinement** (*n*).

**liquation** (*n*) a method of refining (↑) a metal in which a mixture of metals is heated until one metal melts (p.10) and flows away from the mixture.

**dross** (*n*) the impurities, or other waste material, that floats on top of molten (p.10) metal as a scum (↑). Dross is removed (p.215) in refining (↑) metals.

**optimum** (*adj*) describes a condition which is the most favourable (p.214) for a reaction, e.g. the optimum temperature for the Contact process (p.166) is 500°C. At this temperature the best yield (↓) is obtained.

**yield** (*n*) the quantity of a product obtained from an industrial (p.157) process or preparation (p.43). The actual yield is often compared, as a percentage, with the theoretical stoichiometrical (p.82) yield, e.g. the yield in the Haber process is only about 6%, i.e. only 6% of nitrogen and hydrogen combine to form ammonia. **yield** (*v*).

**replenish** (*v*) if a material is being used in a process, further quantities have to be added to replace (p.68) the losses. To replenish is to make these additions of the material. **replenishment** (*n*).

階段 (名) 指一個過程的空間或時間長度。例如 (a) 教育制度 (第 212 頁) 的小學、中學及中學階段:(b) 波希法第 168 頁) 製氫包括以下階段:(i) 水煤氣與水蒸汽混合 (ii) 混合氣體從催化劑上面流過 (iii) 除去二氧化碳。

精煉 (使精確) (動) (1) 由一種物質中將雜質除去,例如,用電解法精煉銅;(2)分離混合物之成分 (第 54 頁),取得某部分成分或全部成分的純標本,例如精煉石油 (第 160 頁) 取得汽油、煤油等;(3) 使一種方法、儀器或技術更有效或更適合於其用途。例如 (a) 一種測定反應平衡常數的精確方法;(b) 實現 (第 157 頁) 色譜分離法的一種更精確的儀器。(名詞爲 refinement)

熔析 (名) 一種精煉 (↑) 金屬的方法,將金屬的混合物加熱至其中一種金屬熔化 (第 10 頁) 並從混合物中流走。

廢渣 (名) 浮在雜質熔融 (第 10 頁) 金屬上層成爲浮渣 (↑) 的雜質或其他廢物料。精煉 (↑) 金屬時都要除去 (第 215 頁) 廢渣。

最佳的 (最適宜的) 描述最有利於 (第 214 頁) 一種反應的條件。例如接觸法 (第 166 頁) 的最佳溫度爲 500°C。此溫度時可達到 (↓) 最高的產率。

產率 (名) 自工業 (第 157 頁) 過程或製備 (第 43 頁) 中取得的產品數量。實際產率以百分率表示,往往與理論化學計量 (第 82 頁) 產率作比較。例如哈伯法的產率約爲 6%,即只有 6% 的氮和氫化合生成氨。(動詞爲 yield)

補足 (動) 在一個工業過程中,必須再加一定量的某種材料以取代 (第 68 頁) 損耗的量。所謂補足就是加入這些材料。(名詞爲 replenishment)

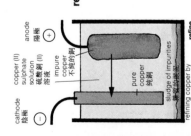

cathode 陰極 (—)
anode 陽極 (+)
copper (II) sulphate solution 硫酸銅 (II) 溶液
impure copper 不純的銅
pure copper 純銅
sludge of impurities 雜質的淤渣
refining copper by electrolysis 用電解法精煉銅

**refine 精煉**

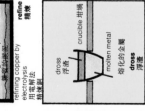

crucible 坩堝
dross 浮渣
molten metal 熔化的金屬

**dross 浮渣**

**fuel** (n) any material or substance that is burned to give heat, and through heat, to give power. Common fuels in industry are: coal, coke, kerosene, fuel oil, coal gas, natural gas.

**petroleum** (n) mineral oil (p.172) and some compounds of sulphur and nitrogen. Petroleum is fractionally distilled (p.201) to obtain fractions (p.202), each of which has a particular use.

**petrol** (n) a volatile (p.18) liquid of low boiling point (p.12) obtained from petroleum (↑) by distillation (p.33). Petrol distils between 20°C and 150°C. It is a mixture of hydrocarbons (p.172) from hexane to decane. Petrol is highly flammable (p.21); it is used in internal combustion engines.

**gasoline** (n) name for petrol in the U.S.A.

**kerosene** (n) a volatile liquid similar to petrol (↑) but with a higher boiling point. It is obtained from petroleum (↑) by fractional distillation; kerosene distils between 150°C and 250°C. It is used for heating, lighting and in jet engines.

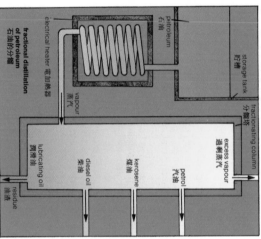

petroleum 石油
electrical heater 電加熱器
storage tank 貯油

**fractional distillation of petroleum 石油的分餾**

vapour 蒸汽
fractionating column 分餾塔
excess vapour 過剩蒸汽
petrol 汽油
kerosene 煤油
diesel oil 柴油
lubricating oil 潤滑油
residue 油渣

燃料(名) 用於燃燒產生熱量並利用此熱量取得動力的任何材料或物質。煤、焦炭、煤油、煤氣和天然氣都是工業上常用的燃料。

石油(名) 由礦物油(第 172 頁)和一些硫的化合物及氮的化合物組成(第 55 頁)的礦物油(第 156 頁)。石油分餾(第 201 頁),可獲得各種餾出物(第 202 頁),各個餾出物都有其特定用途。

汽油(名) 藉蒸餾作用(第 33 頁)自石油(↑)取得之低沸點(第 12 頁)揮發性(第 18 頁)液體。汽油在溫度 20°C 至 150°C 之間蒸餾。是己烷至癸烷的烴類(第 172 頁)混合物。可燃性(第 21 頁)高,常用作內燃機的燃料。

汽油 在美國稱汽油及 gasoline。

煤油:火油(名) 類似汽油(↑)但沸點較高之揮發性液體。煤油係由石油(↑)在溫度 150°C 至 250°C 之間分餾而得。煤油用於加熱、照明和噴氣式發動機。

| | |
|---|---|
| **petrol 汽油** | fuel in motor car engines 汽車引擎燃料 |
| **kerosene 煤油** | for heating and lighting 供加熱和照明用<br>fuel in jet engines 噴氣發動機燃料 |
| **gas oil 粗柴油** | fuel in diesel engines 柴油機燃料 |
| **lubricants 潤滑油** | lubricating moving parts 潤滑各種機械之運<br>of machinery 動機件 |
| **pitch 瀝青** | surfacing roads 鋪路面，<br>sealing roofs 密封屋頂 |

**asphalt** (*n*) a black, sticky, solid material consisting (p.55) mainly of hydrocarbons (p.172). It occurs naturally in various places and is also found in some kinds of petroleum (↑). Asphalt is used for road surfaces and for making roofs waterproof.

**naphtha** (*n*) an organic (p.55) liquid obtained by dry distillation (p.203) of coal or wood. It can contain (p.55) a number of different compounds, but has no particular composition (p.82). Naphtha is a flammable (p.21), volatile (p. 18) liquid.

**paraffin¹** (*n*) another name for kerosene (↑); paraffin oil is also used as a name.

**gas oil** a liquid obtained from petroleum (↑) by fractional distillation. It is less volatile than kerosene and distils between 250°C and 300°C. Gas oil is used in diesel engines.

**lubricants** (*n.pl.*) various kinds of lubricating oils. Lubricants are used to reduce friction. They are distilled from petroleum at between 300°C and 400°C. Lubricants with a high boiling point are soft solids such as petroleum jelly. Lubricants with lower boiling points are thick, sticky liquids.

**doctor solution** a solution which is used to remove bad-smelling compounds from petroleum (↑).

**pitch** (*n*) a black, sticky liquid left after fractional distillation of petroleum (↑). It is the same material as asphalt (↑).

**石油瀝青：柏油**（名） 主要是由烴類（第 172 頁）組成的（第 55 頁）一種黑色、黏稠性固體材料。許多地方都含天然存在石油瀝青，一些種類的石油（↑）中也存在石油瀝青。可用於鋪路面和屋頂防水層。

**石油腦：粗揮發油：粗汽油**（名） 煤或木材乾餾（第 203 頁）獲得的一種有機（第 55 頁）液體。石油腦含（第 55 頁）多種不同的化合物，但無特定之組成（第 82 頁）。石油腦是一種可燃的（第 21 頁）揮發性（第 18 頁）液體。

**煤油**（↑）的另一英文名稱爲 **parafin**，也稱爲石蠟油。

**瓦斯油：粗柴油：氣油** 石油（↑）分餾獲得的一種液體。揮發性比煤油低，在溫度 250°C 至 300°C 之間蒸餾而得。瓦斯油供柴油機使用。

**潤滑劑**（名・複） 指不同種類的潤滑油。潤滑劑用於減少摩擦。由石油於溫度 300°C 至 400°C 之間蒸餾而得。高沸點潤滑劑爲如同石油脂的軟質固體。低沸點潤滑劑爲稠而黏液的。

**脫硫液**（名） 用於去除石油（↑）中不良味化合物的一種溶液。

**瀝青**（名） 石油（↑）分餾後留下的一種黑色黏性液體。係與石油瀝青（↑）相同之材料。

**tar** (*n*) a thick, black, sticky liquid formed during the destructive distillation (p.203) of coal. It contains many different organic (p.55) substances from which many useful compounds are manufactured (p.157).

**scrubber** (*n*) a device for removing (p.215) ammonia and benzene from impure coal gas. The impure gas is passed up the scrubber, and a spray of water flows down the scrubber, washing out the ammonia and benzene. Ammoniacal liquor, containing these two substances, collects in the scrubber. **scrub** (*v*).

**gasometer** (*n*) a very large iron vessel (p.25) for the storing of coal gas.

**vat** (*n*) a large, open vessel (p.25) used for making wine or soap, dyeing cloth, etc.

**dye** (*n*) a substance used to give colour to cloth, plastics or paper. Some dyes are made from plants, but most are synthetic (p.200). **dye** (*v*).

**pigment** (*n*) a solid substance used to give colour to paint or varnish. A pigment is not soluble in water. Pigments can be either organic (p.55) or inorganic (p.55) substances.

**mordant** (*n*) a substance which is used with dyes (↑) that do not dye a material directly. The material is treated (p.38) first with the mordant; the treated material is then put in a vat (↑) with a dye. With acid dyes, the mordant is aluminium or tin hydroxide. With basic dyes, vinegar (ethanoic acid) or tannic acid is used as a mordant. The mordant causes the dye to dye the material.

**soap** (*n*) any salt of a metal and a fatty acid. The water-soluble soaps are the sodium and potassium salts; sodium forming hard soaps and potassium forming soft soaps. These soaps have a cleansing action, removing dirt from surfaces. Other soaps have different properties. The most common soaps have water-soluble soaps are stearic acid (C₁₇H₃₅COOH) and palmitic acid (C₁₅H₃₁COOH), **soapy** (*adj*).

**salting out** the addition of a concentrated solution of sodium chloride to a solution of an organic (p.55) compound in water or ethanol, in order to throw the organic compound out of solution.

焦油：（名）煤分解蒸餾（第 203 頁）過程中生成的一種稠的黑色黏性液體。其中含有多種不同的有機（第 55 頁）物質，這些物質可用於製造（第 157 頁）多種有用的化合物。

滌氣器（名）用於（第 215 頁）不純煤氣中所含之氨和苯的一種裝置。不純的煤氣向滌氣器向上輸入，噴淋水則向下噴，從而除去氨和苯。含有此兩種物質的氨液收集在滌氣器內。（動詞爲 scrub）

貯氣器：氣體（名）儲存煤氣用之極大型鐵製容器（第 25 頁）。

大槽（名）釀酒或製皂、染布等用之大型無蓋的容器（第 25 頁）。

染料（名）使布、塑膠或紙染上顏色的一種物質。有些染料是由植物製成的，但大多數是合成的（第 200 頁）。（動詞爲 dye）

顏料（名）使油漆或清漆具有顏色的所用的一種固體物質。顏料不溶於水。它或爲有機（第 55 頁）物質或爲無機（第 55 頁）物質。

媒染劑（名）和不能直接染上材料上的染料（↑）一起使用的一種物質。首先將材料用媒染劑處理（第 38 頁）然後將處理過的材料放入一個有染料的大染槽（↑）內。酸性染料用氫氧化鋁或氫氧化錫作爲媒染劑；鹼性染料則用醋（即乙酸）或鞣酸酸作爲媒染劑。媒染劑使染料染到材料上。

肥皂（名）爲金屬和脂肪酸的任何鹽。水溶性皂爲鈉鹽和鉀鹽；鈉皂爲硬肥皂、鉀皂爲軟肥皂。這些肥皂都有清潔作用，可除去表面的污垢。其他肥皂具有不同的性質。水溶性肥皂最常用的脂肪酸是硬脂酸（C₁₇H₃₅COOH）以及棕櫚酸（C₁₅H₃₁COOH）。（形容詞爲 soapy）

鹽析 在有機（第 55 頁）化合物的水或乙醇溶液中加入濃氯化鈉溶液，使有機化合物由溶液中析出。

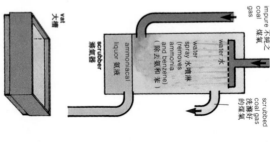

impure coal gas 不純之煤氣

water spray 水噴淋

water (removes ammonia and benzene) 水（除去氨和苯）

ammoniacal liquor 氨液

scrubber 滌氣器

scrubbed coal gas 洗滌好的煤氣

vat 大槽

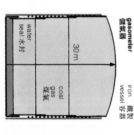

gasometer 儲氣器

water seal 水封

30m

coal gas 煤氣

iron vessel 鐵製容器

**blast furnace** a vertical furnace (p.164) for obtaining iron from its ores. A strong current (a blast) of air is blown up the furnace. Coke (p.156) is mixed with limestone (p.155) and the iron ore. The furnace is kept at a temperature of about 1800°C in the hottest part. The coke is converted to carbon monoxide in two stages:

$$C + O_2 \rightarrow CO_2 \qquad CO_2 + C \rightarrow 2CO$$

The carbon monoxide reduces the iron ore to iron. Molten (p.10) iron runs out of the bottom of the furnace.

**tuyere** (n) a pipe, usually made of copper, through which air is blown into a blast furnace.

**pig iron** (n) iron, straight from a blast furnace (↑), which has cooled in moulds (p.210). It contains (p.55) 2–5% carbon and smaller quantities of other impurities making it hard and brittle (p.14).

**cast iron** pig iron which has been melted, mixed with steel scrap (p.171), and cooled in moulds (p.210) to give it a shape. Cast iron is impure iron and is hard and brittle.

**wrought iron** wrought iron is obtained from pig iron by heating it in a reverberatory furnace (p.164) with limestone and stirring the molten (p.10) mass with long iron rods. Wrought iron is almost pure, containing (p.55) less than 0.2% carbon. It is soft, malleable and easily welded.

**puddling process** cast iron is fused in a reverberatory furnace (p.164) with haematite (an ore of iron oxide) lining the hearth (p.164). The oxygen in haematite oxidises the carbon in cast iron and forms nearly pure wrought iron (↑).

**slag** (n) a waste product (p.157) from a blast furnace (↑). It consists (p.55) of calcium silicate formed by the reaction between limestone and the earthy parts of iron ore.

**clinker** (n) a hard mass of material which is not combustible (p.58), formed in furnaces and boiler fires. It consists of silicates formed by fusion (p.32) of earthy materials in fuels and ores.

**roast** (v) to heat metals or ores in air at a temperature too low for fusion. The roasting removes impurities (p.20) by oxidation (p.70) with atmospheric oxygen. Roasting removes (p.215) sulphur and sulphur dioxide from sulphide ores.

鼓風爐：高爐 熔煉鐵礦石以取得鐵用的直立式熔爐（第164頁）。一股強烈氣流（鼓風）向高爐頂部吹入，爐內焦碳（第156頁）與石灰（第155頁）及鐵礦石混合一起。高爐的最熱部分，溫度保持約1800°C。焦碳分兩步轉化成一氧化碳：

$$C + O_2 \rightarrow CO_2 \qquad CO_2 + C \rightarrow 2CO$$

一氧化碳將鐵礦石還原成鐵。熔化的鐵（第10頁）鐵由爐底流出。

風嘴：吹風管嘴（名） 通常是銅製管。空氣由此管吹入鼓風爐中。

生鐵：銑鐵（名） 直接從高爐（↑）熔煉得，並在模型（第210頁）中冷卻的鐵。生鐵含（第55頁）碳2至5%以及少量其他雜質，這些雜質使鐵堅硬而脆（第14頁）。

鑄鐵：鑄鐵是將生鐵放入反射爐（第164頁）碎鋼屑混合的（第171頁）而熔的生鐵，倒在模型（第210頁）中冷卻成一種形狀。鑄鐵是不純的鐵，它硬而脆。

熟鐵：鍛鐵 熟鐵是將生鐵放入反射爐（第164頁）中和石灰一起加熱並用長鐵棒攪拌熔化的（第10頁）鐵塊取得的。熟鐵幾乎是純的鐵，含（第55頁）碳低於0.2%。它柔軟、可展、易焊。

煉鐵法：攪煉法 鑄鐵在反射爐（第164頁）中和赤鐵礦（一種氧化鐵礦石）一起熔煉，赤鐵礦中的氧氣使鑄鐵中的碳氧化並形成近接於純的熟鐵（↑）。

熔渣（名） 自鼓風爐（↑）出來的廢產物（第157頁）。熔渣中含有（第55頁）矽酸鈣（也稱硅酸鈣）。它是由石灰和鐵礦石中的土質部分反應生成的。

熔塊（名） 在爐堆和鍋爐火管中生成的不可燃（第58頁）材料硬塊。熔塊中含有由石灰料及礦石中的土質材料熔融（第32頁）而成的矽酸鹽。

煅燒（動） 將金屬或礦石在空氣溫度低至不足於熔化的情況下加熱。煅燒的結果可藉大氣中氧的氧化作用（第70頁）除去雜質（第20頁）。煅燒可除去（第215頁）硫化物礦石中的硫和二氧化硫。

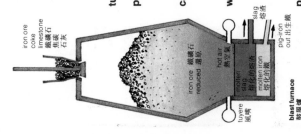

iron ore 鐵礦石
coke 焦碳
limestone 石灰

slag 熔渣
pig-iron out 出生鐵

iron ore reduced 鐵礦石 還原
hot air 熱空氣
molten slag 熔化的熔渣
molten iron 熔化的鐵

tuyere 風嘴

**blast furnace** 鼓風爐

**smelt** (v) to separate a metal from its ore by heating the ore with a suitable reducing agent (p.71). The metal becomes molten (p.10) and impurities separate from the metal, e.g. iron ore is smelted in a blast furnace (p.163). **smelting** (n).

**extract** (v) to obtain an element from the Earth, either as a native element or by mining the ore and obtaining the element by chemical action, e.g. extracting aluminium by mining bauxite (aluminium oxide) and electrolyzing it to obtain aluminium. **extraction** (n).

**furnace** (n) a brick construction inside which great heat is produced by burning fuels (p.160) or by electricity; mainly used in the extraction of metals.

**flue** (n) a pipe leading hot air and smoke away from a fire.

**reverberatory furnace** a furnace (↑) with a low roof above the hearth (↓) so that the heat from the fire is directed onto the reactants on the hearth. Reverberatory furnaces are used for smelting ores.

**hearth** (n) the floor on which a fire burns or a reaction takes place in a furnace.

**open-hearth furnace** a shallow, rectangular furnace (↑) which is heated by burning a gaseous (p.11) fuel. The hot gases from the burning fuel pass over the hearth (↑) on which an ore is placed. The hearth is covered with limestone or dolomite (p.155).

**ash** (n) a residue (p.31) which is a powder, left after a material has been burned completely, e.g. the ash left after wood or plants have been burned. **ashen** (adj).

**converter** (n) a large iron vessel in which pig iron (p.163) is converted (p.73) to steel in the Bessemer process (↓).

**metallurgy** (n) the science of the extraction of metals from their ores (p.154), refining (p.159) the metals, and forming alloys (p.55). **metallurgist** (n).

**weld** (v) to join together two pieces of metal by (a) heating them so that they melt (p.10) and join or (b) by beating them with a hammer to make them soft enough to join together. **welding** (n).

熔煉（動）將礦石和適當的還原劑（第 71 頁）一起加熱，自礦石中分離出金屬。金屬熔化（第 10 頁），雜質則從金屬中分離出。（名詞為 smelting）

提取（動）自地球中取得一種元素，它或者是以天然元素的形式取得，或者是開採取得礦石後再利用化學作用取得元素。例如開採鋁土礦（即氧化鋁）提煉鋁，並將之電解而獲得鋁。（名詞為 extraction）。

熔爐（名）一種內有磚砌的結構，可在其內產生巨大熱量的一種爐。主要用於提煉金屬。

烟道（名）引導熱空氣和烟離開火之源的管道。

反射爐（名）爐內有低矮爐頂，使火之熱量能直接作用於爐膛（↓）上之反應物的一種爐（↑）。反射爐用於熔煉礦石。

爐膛（名）爐中燃火或進行反應的底板。

平爐：開爐靠燃燒氣體（第 11 頁）燃料來加熱的一種淺的矩形爐。燃料產生的熱氣體從放置礦石的爐膛（↑）上方流過。爐膛用石灰石或白雲石（第 155 頁）覆蓋。

爐灰（名）材料充分燃燒後留下的粉末殘渣（第 31 頁）。例如，木材或植物燃燒後留下的灰。（形容詞為 ashen）

轉爐（名）以稻塞麥法（使第 163 頁）在其中（↓）轉化（第 73 頁）成鋼的一種巨大的鐵製容器。

冶金術（名）自金屬礦石（第 154 頁）中提取金屬、精煉（第 159 頁）金屬和形成合金（第 55 頁）的科學。（名詞 metallurgist 意為冶金學家）

熔接（動）藉以下方法將兩件金屬連接在一起：(a) 將兩件金屬加熱使之熔化（第 10 頁）和連接或 (b) 用鎚敲打兩件金屬使之軟化而連接在一起。（名詞形式為 welding）

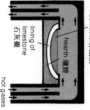

open-hearth furnace
平爐

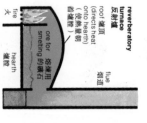

reverberatory furnace
反射爐
roof (that
directs heat
onto hearth)
（使熱量朝
着爐膛）
fire
火
flue
烟道
hearth
爐膛
ore for
smelting 煉鋼
的礦石
pig-iron 生鐵
+ scrap steel 廢鋼
+ haematite 赤鐵礦
lining of
limestone
石灰襯
hearth 爐膛
hot gases
熱氣體

**steel** (*n*) steel is obtained from pig iron by first oxidizing all the impurities away and then adding a known amount of carbon to the molten (p.10) iron. Other metals, such as manganese or chromium, can be added to make different kinds of steel. Steel contains between 0.15% to 1.5% carbon, depending on the kind of steel that is wanted. Steel is hard and elastic.

**quench** (*v*) to harden steel by heating it until red-hot, then quickly putting it into cold water or cold oil. **quenching** (*n*).

**temper** (*v*) quenched (↑) steel is heated to a temperature of 200°C–600°C, depending upon the kind of tempering needed. The steel is kept at that temperature for 30 minutes and then allowed to cool. Tempering makes a steel elastic as well as hard. **tempering** (*n*).

**pickling** (*n*) steel is put in vats with concentrated sulphuric acid. This removes rust (p.61) and any other surface impurities. The steel can then be galvanized (p.166), tinned (p.166), or painted. **pickle** (*v*).

**Bessemer process** a process to make steel (↑) from pig iron (p.163) or cast iron (p.163). A converter (↑) is lined with calcium and magnesium oxides; pig iron and scrap (p.171) steel are added as molten (p.10) metal. Air is blown through the metal oxidizing all the impurities. Carbon, as needed, is added to make the steel.

**Siemens-Martin process** pig iron (p.163), scrap (↑) steel and haematite (iron (III) oxide) are heated in an open-hearth furnace (↑). The hearth (↑) is lined with limestone. The proportion of iron haematite and steel are calculated to give the correct proportion of carbon in the product.

**open-hearth process** another name for the Siemens-Martin process.

**Linz-Donawitz process** molten, impure pig iron (p.163) from the blast furnace is used to make steel. Oxygen is blown over the molten (p.10) iron causing the oxidation of carbon and other impurities. Lime powder is added to form a slag (p.163). This is a rapid process (it takes 10–20 minutes) and steel as pure as the steel from the open-hearth process is formed.

鋼 (名) 鋼係由生鐵熔煉而成。首先，將生鐵裏面去除全部雜質，然後於熔化的（第 10 頁）鐵中加入已知量的碳而製得。也可加入其他金屬如錳或鉻製成不同種類的鋼。鋼含碳 0.15% 至 1.5%，視所需之鋼種類而定。鋼既硬又富有彈性。

使…淬火 (動) 將鋼件加熱至赤熱之後迅速放入冷水或冷油中使之硬化。（名詞爲 quenching）

回火 (動) 按所要求的回火種類，將已淬火的鋼加熱至 200–600℃。回火處理使硬鋼下保持 30 分鐘再使之冷卻。回火處理使鋼件富有彈性而又硬。（名詞爲 tempering）

浸酸 (名) 鋼件放入裝有濃硫酸的缸裡，除去鐵鏽（第 61 頁）和表面的任何其他雜質。浸酸處理過的鋼件可以鍍鋅（第 166 頁）、鍍錫（第 166 頁）或上油漆。（動詞爲 pickle）

栢塞麥煉鋼法 用生鐵（第 163 頁）或鑄鐵（第 163 頁）煉鋼的一種方法。轉爐（第 163 頁）內襯以氧化鈣和氧化鎂；將生鐵和廢（第 171 頁）鋼格化（第 10 頁）後加入，空氣吹過使所有雜質氧化。按需求加入碳以製造鋼。

西門子 - 馬丁爐煉鋼法 將生鐵（即氧化鐵(III)）（第 171 頁）鋼和赤鐵礦（即氧化鐵(III)）置於平爐（↑）中加熱。爐膛以石灰石（↑）要準確計算生鐵赤鐵礦中鐵與鋼的比例，使產品中碳之比例正確。

平爐煉鋼法 西門子 - 馬丁爐煉鋼法的另一名稱。

氧氣頂吹轉爐煉鋼法（即 LD 煉鋼法） 將未純化的不純生鐵（第 163 頁）用於煉鋼。爐取得之熔化的不純生鐵（第 10 頁）鐵，使碳和其他雜質氧化。加入石灰粉以形成爐渣（第 163 頁）。這是一種快速過程（只需 10 至 20 分鐘）。煉得的鋼和平爐法煉得者一樣純。

cold steel 冷的鋼件
water or oil 水或油
red-hot steel 赤熱的鋼件
tank 槽

**quenching steel**
淬火鋼

lining of calcium and magnesium oxides 氧化鈣和氧化鎂的
axie 軸
molten pig-iron 熔融的生鐵

air 空氣

**Bessemer converter**
栢塞麥煉鋼法

**spelter** (n) zinc that has not been refined
(p.159); it contains other metals (e.g. lead) and
some impurities.

**galvanize** (v) to cover an iron surface with a coat
(p.127) of zinc, usually spelter (↑); a metal object
is dipped into molten (p.10) zinc. **galvanized** (adj).

**sherardize** (v) to cover an iron surface with a coat
(p.127) of zinc. A metal object is heated in a
closed vessel containing zinc dust at a
temperature just below the melting point of zinc.

**tin** (v) to coat (p.127) an iron article with a thin
layer of tin. The iron is first pickled (p.165) and
then dipped in molten (p.10) tin. **tinning** (n).

**contact process** a process for making concentrated sulphuric acid by converting (p.73)
sulphur dioxide to sulphur trioxide and then
converting the trioxide to sulphuric acid. The
process is shown in the diagram below. Sulphur
dioxide and oxygen (from air) are passed over
a catalyst (p.72) of vanadium pentoxide at
500°C, forming sulphur trioxide. The sulphur
trioxide is absorbed in concentrated sulphuric
acid forming oleum (↓), which is diluted to
form sulphuric acid of 98% concentration.

**contact process**
接觸法

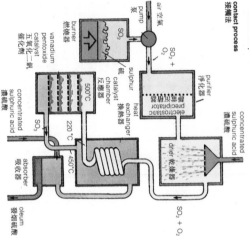

air 空氣
pump 泵
burner 燃燒器
SO₂
sulphur 硫
catalyst chamber 催化器換熱器
heat exchanger 換熱器
vanadium pentoxide catalyst 五氧化二釩催化劑
500°C
220°C
450°C
SO₃
absorber 吸收器
purifier 淨化器
electrostatic precipitator 靜電除塵器
drier 乾燥器
SO₂ + O₂
concentrated sulphuric acid 濃硫酸
oleum 發煙硫酸
concentrated sulphuric acid 濃硫酸
SO₂ + O₂

生鋅(名) 未精煉的(第 159 頁)鋅,含有其他金屬(例如鉛)和某些雜質。

鍍鋅:浸鍍鋅(動) 鐵之表面覆蓋一層鋅,一般為生鋅(↑)的鍍層(第 127 頁)。金屬物件浸入熔化的(第 10 頁)鋅中。(形容詞為
galvanized)

粉鍍鋅(動) 鐵之表面覆蓋有鋅粉的鍍層(第 127
頁)。將金屬物件置於一個裝有鋅粉的密閉
容器內,以略低於鋅熔點的溫度加熱。

鍍錫:錫浸鍍(動) 將鐵物件(件)鍍上一層
薄的錫鍍層。將鐵物件(件)鍍(第 127 頁)之後再浸
入熔化的(第 10 頁)錫中。(名詞為 tinning)

接觸法 製濃硫酸的一種方法:先將二氧化硫轉
化(第 73 頁)成三氧化硫後再轉化成硫酸。
其過程如下圖。二氧化硫和(空氣中的)氧氣
在 500°C 下流過五氧化二釩催化劑(第 72
頁)上面,生成三氧化硫。三氧化硫被吸收
入濃硫酸中形成發煙硫酸(↓),發煙硫酸經
稀釋即成為濃度為 98% 的濃硫酸。

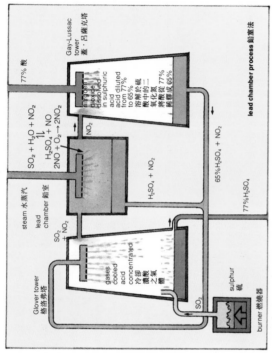

**lead chamber process 鉛室法**

Glover tower
格洛弗塔
gases cooled
acid concentrated
冷部酸濃縮之氣體

steam 水蒸汽
lead chamber 鉛室

SO₂ + H₂O + NO₂
H₂SO₄ + NO
2NO + O₂ → 2NO₂

$SO_2 + NO_2 \rightarrow SO_3 + NO$

sulphur 硫
burner 燃燒器

SO₂ + NO₂

SO₂

$77\% H_2SO_4$

$65\% H_2SO_4$ + NO₂

$H_2SO_4$ + NO₂

$H_2SO_4$ + NO₂

Gay-Lussac tower
蓋·呂薩克塔

nitrogen dioxide dissolved in sulphuric acid diluted from 77% to 65%
溶解於硫酸中的二氧化氮 酸從77%稀釋成65%

77% 酸

**lead chamber process an** older process for the manufacture of concentrated sulphuric acid. Sulphur dioxide is oxidized to sulphur trioxide by nitrogen dioxide: $SO_2 + NO_2 \rightarrow SO_3 + NO$ The nitrogen oxide (NO) is oxidized by oxygen to nitrogen dioxide; the nitrogen compounds act like a catalyst. The process is shown in the diagram. This process produces a 77% concentration of acid, not as concentrated and not as pure as the acid from the Contact process, but still useful for manufacturing processes.

**oleum (n)** concentrated sulphuric acid containing dissolved sulphur trioxide; also known as fuming sulphuric acid.

**spent oxide** iron (III) oxide is used to remove sulphur compounds from coal gas; the oxide eventually contains a high percentage of sulphur and is no longer of use for purifying (p.43) the coal gas; it then becomes spent oxide, i. e. it has been used up. Spent oxide is burned to form sulphur dioxide for use in the manufacture of sulphuric acid.

**鉛室法** 是一種古老的濃硫酸製法。用二氧化氮將二氧化硫氧化成三氧化硫：$SO_2 + NO_2$ —→ $SO_3 + NO$。再用氧將一氧化氮 (NO) 氧化成二氧化氮；氮的化合物起催化劑作用。其過程如上圖所示。此法可生產濃度爲77%的濃硫酸，濃度和純度雖不如接觸法，但仍用於生產過程。

**礬油 (名)** 含有溶解的三氧化硫的濃硫酸，也稱爲發煙濃硫酸。

**廢氧化物** 用氧化鐵 (III) 除去煤氣中的含硫化合物；氧化物中含硫的百分率高到不能再用於淨化 (第 43 頁) 煤氣，就成爲廢氧化物，即氧化物已用盡。可將廢氧化物燃燒生成二氧化硫供生產硫酸用。

**water gas** a mixture of hydrogen and carbon monoxide formed by blowing steam over red-hot coke (p.156). The reaction cools the coke.

**producer gas** a mixture of nitrogen and carbon monoxide formed by blowing air through coke (p.156). The reaction heats the coke.

**semi-water gas** a mixture of hydrogen, carbon monoxide and nitrogen, formed by blowing steam and air alternately over coke (p.156). One reaction heats, and the other reaction cools, the coke, so its temperature is maintained.

**Bosch process** a process for making hydrogen from water gas (↑). Water gas and steam are passed over a catalyst of iron with traces (p.20) of chromium (III) oxide as a promoter (p.72). At 450°C, the reaction is:

$$H_2 + CO + H_2O \rightleftharpoons CO_2 + 2H_2$$

The carbon dioxide is removed by washing with hot potassium carbonate solution. Natural gas, containing methane, is now used in this process instead of water gas.

**mercury cathode cell** a method manufacturing sodium hydroxide and chlorine from sodium chloride. A solution of sodium chloride is electrolyzed (p.123) using carbon anodes (p.123) and a mercury cathode. Chlorine is taken away by pipes from the cell. The cathode forms a sodium amalgam (p.55), which is removed and taken to another tank where sodium hydroxide is formed and hydrogen set free.

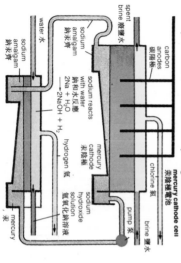

spent brine 廢鹽水
water 水
sodium amalgam 鈉汞齊
sodium amalgam 鈉汞齊
carbon anodes 碳陽極
sodium reacts with water
鈉和水反應
2Na + H₂O ──→ 2NaOH + H₂
mercury cathode 汞陰極
hydrogen 氫
sodium hydroxide solution 氫氧化鈉溶液
chlorine 氯
brine 鹽水
pump 泵
mercury 汞
**mercury cathode cell** 汞陰極電池

水煤氣　將水蒸汽吹過赤熱焦碳（第156頁）上面所形成的氫和一氧化碳混合物。反應使焦碳冷卻。

發生爐煤氣　空氣吹過焦碳（第156頁）所形成的氮和一氧化碳混合物。反應使焦碳加熱。

半水煤氣　水蒸汽和空氣交替地吹過焦碳（第156頁）上面所形成的氫、一氧化碳和氮的混合物。其中一個反應使焦碳加熱，而另一反應使之冷卻，因而保持其溫度。

波希法　從水煤氣（↑）製取氫的一種方法。水煤氣和蒸汽從鐵催化劑上面流過，催化劑中含有微量（第20頁）三氧化二鉻（III）作為促化劑（第72頁）。反應在450°C溫度進行，反應式如下：

$$H_2 + CO + H_2O \rightleftharpoons CO_2 + 2H_2$$

用熱的碳酸鉀溶液洗滌去二氧化碳。現今是在這一過程中使用含甲烷的天然氣，而不用水煤氣。

汞陰極電解池　從氯化鈉製造氫氧化鈉和氯的一種方法。用碳作陽極（第123頁）、汞作陰極的氯化鈉溶液電解（第123頁）。電解池產生的氯氣由管道導出，在陰極上生成的鈉汞齊（第55頁）則去除並移送到另一個槽中，在此產生氫氧化鈉，同時放出氫氣。

**lime** (*n*) lime is either **quicklime**, calcium oxide, or **slaked lime**, calcium hydroxide.

**soda** (*n*) soda is either **caustic soda**, sodium hydroxide, or **washing soda**, sodium carbonate.

**brine** (*n*) a solution of sodium chloride.

**Castner-Kellner process** another name for the mercury cathode cell (↑).

**Kellner-Solvay process** another name for the mercury cathode cell (↑).

**Solvay process** a process for the manufacture of sodium carbonate ($Na_2CO_3.1OH_2O$) using sodium chloride and calcium carbonate as raw materials. Brine (↑) is saturated with ammonia and passed down a column, up which is passed carbon dioxide gas. The reaction which takes place is:

$NaCl + NH_3 + H_2O + CO_2 \longrightarrow NaHCO_3 + NH_4Cl$

The sodium hydrogen carbonate ($NaHCO_3$) is heated to form sodium carbonate and carbon dioxide. The sodium carbonate is recrystallized to form washing soda ($Na_2CO_3.1OH_2O$). The carbon dioxide and ammonia from the ammonium chloride are put back into the cycle (p.64) of operations (p.157) as shown in the diagram below. Calcium chloride is produced as a waste product.

石灰 (名) 石灰或者是生石灰 (即氧化鈣),或者是消石灰 (即氫氧化鈣)。

蘇打 (名) 蘇打或者是苛性鈉 (即氫氧化鈉),或者是洗滌鹼 (即碳酸鈉)。

鹽水 (名) 為氯化鈉的溶液。

卡斯納·克爾納法 汞陰極電池 (↑) 之別稱。

克爾納·索爾維法 汞陰極電池 (↑) 之別稱。

索爾維法 以氯化鈉和碳酸鈣為原料製造碳酸鈉 ($Na_2CO_3 · 1OH_2O$) 的一種方法。用氨將鹽水 (↑) 飽和並使之由反應塔頂流下,二氧化碳氣則自下而上通入。反應式為:

$NaCl + NH_3 + H_2O + CO_2$
$\longrightarrow NaHCO_3 + NH_4Cl$

將碳酸氫鈉 ($NaHCO_3$) 加熱,使之生成碳酸鈉和二氧化碳。碳酸鈉則重結晶生成洗滌鹼鈉 ($Na_2CO_3 · 1OH_2O$)。將二氧化碳和由氯化銨產生的氨返回操作 (第 157 頁) 循環 (第 64 頁) 中,如下圖所示。產生的氯化鈣成為一種廢產物。

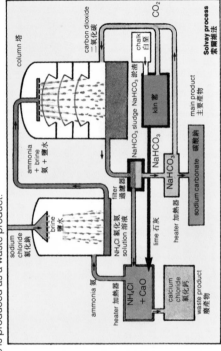

**Solvay process**
索爾維法

**Haber process** a process for the manufacture of ammonia by synthesis (p.200) from nitrogen and hydrogen. The two elements combine in the presence of a catalyst (iron with potassium oxide as a promoter (p.72)) under high pressure at 450°C – 500°C. The reaction is reversible (p.64) and only 6% of the elements combine. The ammonia is liquefied by cooling and then removed (p.215); the uncombined gases are passed back to the catalyst chamber.

**kiln** (*n*) a furnace (p.164) for making bricks, ceramics (↓) and heating chalk to form lime and carbon dioxide.

**spent** (*adj*) describes something which has been used and is finished; also something which has had an important constituent used, so it can no longer carry out its purpose, e.g. spent brine which is too dilute to be of use.

**waste** (*adj*) describes something that has been made but is not wanted, e.g. paper that is no longer wanted is waste paper, the paper is usable but not wanted for its original purpose.

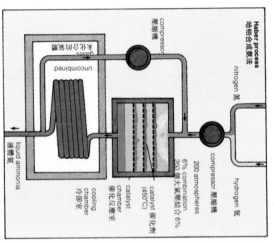

Haber process
哈格合成氨法

compressor 壓縮機
uncombined gases 未化合的氣體

nitrogen 氮

compressor 壓縮機

hydrogen 氫

liquid ammonia 液體氨

200 atmospheres 200 個大氣壓結合
6% combination 6%
and 450°C 450°C
catalyst chamber 催化反應室
catalyst 催化劑
cooling chamber 冷卻室

哈格合成氨法 利用氮和氫合成（第200頁）法製造氨氣的一種方法。氮和氫兩種元素在高壓力和450°C至500°C以及催化劑存在〔用鐵和氧化鉀作為助催化劑（第72頁）〕下使和氮結合。反應是可逆的（第64頁）且只有6%的元素化合。氨經冷卻液化之後除去（第215頁），未化合的氣體則返回催化反應室。

窰（名）製磚、陶瓷器（↓）用的一種爐（第164頁）。此種窰可用於將白堊加熱以生產石灰和二氧化碳。

用過的（形）描述經已用過及已完成的東西，也描述某些東西中的一種重要成分已被用過，因而不再用於其目的，例如已用過的鹽水，已太稀不能再用了。

廢的（形）描述某種物件雖已製成而卻不需要，例如不再要的紙是廢紙，此種紙是可用的但已不想作原先的用途。

**obsolete** (*adj*) describes something which is usable but has been replaced (p.68) by something which is better, and so is no longer used, e.g. retorts (p.28) are obsolete, as distillation flasks (p.28) have replaced them.

已廢棄的 (形) 描述某種物件仍可使用但已爲更好的某物所取代 (第 68 頁)。因而不再使用了。例如曲頸瓶 (第 28 頁) 已廢棄不用，因爲已爲蒸餾瓶 (第 28 頁) 所取代。

**scrap** (*n*) describes something which is no longer used, e.g. when a car is no longer used it becomes scrap. The steel in machines which are no longer used, is scrap steel. Metals are most commonly described as scrap.

廢料 (名) 已不能再使用的某物。已不能再使用的汽車就成了廢物。已不使用之機器中的鋼是廢鋼。金屬最常被描述爲廢物。

**Thermit process** a process for the extraction (p.34) of chromium, manganese or tungsten from their oxides or for welding (p.164) iron objects. Aluminium powder is mixed with the metal oxide and a piece of magnesium is set alight to start the reaction. The aluminium reduces the oxide and, in the case of iron oxide, produces molten (p.10) iron to weld two iron objects together.

鋁熱法 從鉻錳鎢的氧化物提取 (第 34 頁) 鉻、錳或鎢的方法或銲接 (第164 頁) 鐵製物件的方法。將鋁粉和金屬氧化物混合並把一塊鎂燒着使反應開始。鋁將氧化物還原，如用氧化鐵則鐵溶化 (第 10 頁) 使兩件鐵器銲在一起。

**detergent** (*n*) any substance used for removing (p.215) dirt. Soaps are detergents. Other substances used as detergents are made from sulphonic acids. Their molecules consist of hydrocarbon chains attached to acidic groups. See *sulphonate* (*p.193*).

去垢劑；清潔劑(名) 用於除去 (第 215 頁) 污垢的任何物質。肥皂是去垢劑。作去垢劑的其他物質都是用磺酸類組成的。其分子是由烴鏈和酸基連接構成的。參見"磺酸鹽"(第 193 頁)。

**vulcanization** (*n*) the process for changing rubber from a weak material into a hard, strong material, e.g. the rubber used in car tyres has been hardened by vulcanization. Vulcanization is usually carried out by heating rubber with sulphur. **vulcanize** (*v*).

硫化 (名) 使橡膠由強度低之材料變成硬變韌而強固之材料的過程。例如汽車輪胎的橡膠已藉硫化作用變硬。一般是將橡膠和硫磺一起加熱完成硫化。(動詞爲 vulcanize)

**ceramics** (*n*) the manufacture of earthenware and porcelain objects. **ceramic** (*adj*).

陶瓷工藝 (名) 製造陶器和瓷器。(形容詞爲 ceramic)

**tamp** (*v*) to push soft, powdery solids into a hole to fill the hole completely, e.g. in the Thermit process, the aluminium powder and metal oxide is tamped into the crucible. **tamping** (*n*).

填塞 (動) 將軟質粉末狀固體塞入一個孔中並完全填滿該孔。例如在鋁熱法中，將鋁粉末和金屬氧化物填塞坩堝。(名詞爲 tamping)

**heavy chemical** a chemical in great demand for industrial (p.157) processes and thus manufactured (p.157) in large quantities. Such chemicals are often not very pure (p.20). Examples of heavy chemicals include sulphuric acid, nitric acid, lime and sodium carbonate. See *fine chemicals (p.20)*.

重化學品 指工業 (第 157 頁) 過程中需要量很大、因而大量製造 (第 157 頁) 的化學品。造種化學品往往不是很純 (第 20 頁)。硫酸、硝酸、石灰和碳酸鈉等重化學品的例子。參見"精細化學品"(第 20 頁)。

mixture of aluminium powder and metal oxide
鋁粉和金屬氧化物的混合物

piece of magnesium (to start the reaction)
鎂塊（起始反應）

crucible
坩鍋

powder tamped down
向下填塞的粉末

plug to release molten metal
插入放出熔化的金屬

**Thermit process**
鋁熱法

**ceramic objects**
陶瓷器品

**hydrocarbon** (*n*) an organic (p.55) compound containing (p.55) only the elements carbon and hydrogen; all hydrocarbons are covalent (p.136) compounds. Hydrocarbons include alkanes (↓), alkenes (↓), alkynes (p.174) and benzene (p.179) compounds.

**series** [2] (*n*) a group of organic (p.55) compounds which all have similar chemical properties; physical properties showing a regular change with an increasing number of carbon atoms; can be prepared (p.43) by similar chemical methods; described by a general formula. The compounds are homologous (↓).

**homologous** (*adj*) describes structures which are alike without being exactly the same. In homologous series (↑), all the compounds have a general formula (p.181), the same functional group (p.185) and a gradation of properties (p.9), e.g. the alkanes (↓) form an homologous series; the carboxylic acids form an homologous series, each acid having the functional group – COOH. **homologue** (*n*).

**alkane** (*n*) a hydrocarbon (↑) with the general formula (p.181) $C_n H_{2n+2}$. The alkanes form an homologous (↑) series; the first four members are: $CH_4$ (n = 1); $C_2H_6$; $C_3H_8$; $C_4H_{10}$ (n = 4). Alkanes are either straight chain (p.182) or branched chain compounds. The first four members of the alkanes are methane, ethane, propane and butane; thereafter they have a Greek number prefix for the number of carbon atoms, followed by -ane, e.g. hexane, $C_6H_{14}$. The alkanes have properties similar to methane (↓), becoming less reactive (p.62) with an increasing number of carbon atoms, and passing from gases through liquids to solids for members with a large number of carbon atoms. They are saturated (p.185) organic compounds.

**methane** (*n*) an alkane (↑) of formula $CH_4$. It is an odourless (p.15), colourless (p.15), flammable (p.21) gas. The hydrogen atom can be replaced by halogens (p.117), otherwise methane is unreactive (p.62).

---

烴：碳氫化合物（名）只含（第 55 頁）碳和氫兩種元素的有機（第 55 頁）化合物。一切烴類都是共價（第 136 頁）化合物。烴類包括烷屬烴（↓）、烯屬烴（↓）、炔屬烴（第 174 頁）和苯（第 179 頁）化合物。

系列（名）指具有相似化學性質的一組有機（第 55 頁）化合物。其物理性質隨碳原子數增加而有規律地變化。可以用相似的化學方法製備（第 43 頁）出這些化合物。可用一個通式描述這些化合物。故屬同系（↓）化合物。

同系的（形）描述相似而非完全相同的結構。同系系列（↑）中的所有化合物都有一個通式（第 181 頁），都有相同的官能基（第 185 頁）和一種性質漸變化的性質（第 9 頁）。例如烷屬烴（↓）成爲一個同系系列；羧酸成爲一個同系系列，其中每一種羧酸都含有 – COOH 官能基。（名詞爲 homologue）

烷屬烴（名）通式（第 181 頁）爲 $C_n H_{2n+2}$ 的烴（↑）。烷屬烴爲一個同系（↑）系列；前四個烷屬烴爲：$CH_4$ (n = 1)、$C_2H_6$、$C_3H_8$ 和 $C_4H_{10}$ (n = 4)。烷屬烴可爲直鏈（第 182 頁）或支鏈化合物。前四個烷屬烴英文名稱爲甲烷、乙烷、丙烷和丁烷，在其後的烷屬烴則以希臘數目字頭表示碳原子數目，詞尾則爲 -ane。例如己烷 hexane (C₆H₁₄)，烷屬烴的性質與甲烷（↓）相似，但隨碳原子數增加而逐漸變化較不易反應（第 62 頁），並由氣體變成液體，碳原子數大的烷屬烴則爲固體。烷屬烴是飽和（第 185 頁）的有機化合物。

甲烷（名）化學式爲 $CH_4$ 的烷屬烴（↑）。甲烷是無臭味（第 15 頁）、無色（第 15 頁）的可燃（第 21 頁）氣體。甲烷中的氫原子可爲鹵素（第 117 頁）取代，未被取代的甲烷性質不活潑（第 62 頁）。

---

**homologous series of alkanes 烷屬烴的同系系列**

| formula 化學式 | melting point 熔點 °C | boiling point 沸點 °C |
|---|---|---|
| $CH_4$ | −183 | −162 |
| $C_2H_6$ | −184 | −89 |
| $C_3H_8$ | −188 | −42 |
| $C_4H_{10}$ | −138 | −1 |
| $C_5H_{12}$ | −130 | 36 |
| $C_6H_{14}$ | −95 | 69 |
| $C_7H_{16}$ | −91 | 98 |
| $C_8H_{18}$ | −57 | 126 |

formula 結構式
hydrogen atom 氫原子
carbon atom 碳原子

methane 甲烷

**ethane** (*n*) the second member of the alkanes (↑), formula $C_2H_6$. Its properties are similar to those of methane (↑), but it is less reactive, has a higher boiling point and a greater density. **paraffin²** (*n*) traditional name (p.44) for an alkane (↑).

乙烷（名）烷屬烴（↑）同系物中的第二個化合物，化學式爲 $C_2H_6$，性質與甲烷（↑）相似，但較甲烷不易反應、沸點及密度都比甲烷高。

**鏈烷烴** 烷屬烴（↑）的慣用名（第 44 頁）。

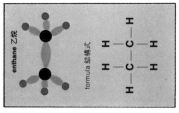

enthane 乙烷

formula 結構式

○ hydrogen atom 氫原子
● carbon atom 碳原子

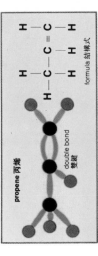

propene 丙烯

double bond 雙鍵

formula 結構式

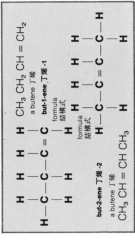

formula 結構式

but-1-ene 丁烯 -1
a butene 丁烯
$CH_3 CH_2 CH = CH_2$

formula 結構式

but-2-ene 丁烯 -2
a butene 丁烯
$CH_3 CH = CH CH_3$

**alkene** (*n*) a hydrocarbon (↑) with the general formula (p.181) $C_nH_{2n}$. The alkenes form an homologous (↑) series; the first three members are either straight chain (p.182) or branched chain compounds. The first four members of the alkenes are: ethene, propene, butene and pentene; thereafter they have a Greek number prefix for the number of carbon atoms, followed by *-ene*, e.g. hexene, $C_6H_{12}$. The alkenes have properties similar to ethene (p.174), becoming less reactive with an increasing number of carbon atoms, and passing from gases through liquids to solids for members with a large number of carbon atoms. They are unsaturated (p.185) organic compounds.

**olefine** or **olefin** (*n*) the traditional name (p.44) for an alkene (↑).

烯屬烴（名）通式（第 181 頁）爲 $C_nH_{2n}$ 的烴（↑）。烯屬烴爲一個同系（↑）系列，前三個烯屬烴爲乙烯 (n = 2)、$C_3H_6$、$C_4H_8$ (n = 4)。烯屬烴可爲直鏈 (第 182 頁) 也可爲支鏈的化合物。系列的前四個烯屬烴英文名稱爲乙烯、丙烯、丁烯和戊烯，其後的烯屬烴均用希臘數目字頭表示原子的數目，其詞尾則爲 *-ene*，例如己烯 ($C_6H_{12}$)、烯屬烴的性質與乙烯（第 174 頁）相似，但隨碳原子數目的增加而逐漸變成較不易反應、並由氣體到液體、碳原子數目大的烯屬烴則爲固體。烯屬烴都是不飽和的（第 185 頁）有機化合物。

**鏈烯烴** 烯屬烴（↑）的英文慣用名（第 44 頁）。

**ethene** (*n*) an alkene of formula $C_2H_4$. It is a colourless (p.15) gas with a sweet smell. The two carbon atoms are joined by a double bond (p.181) which makes ethene reactive (p.62). It undergoes (p.213) addition reactions (p.188), polymerization (p.207) and combustion (p.58).

**ethylene** (*n*) traditional name (p.44) for ethene (↑).

**alkyne** (*n*) a hydrocarbon (p.172) with the general formula (p.181) $C_nH_{2n-2}$. The alkynes form an homologous (p.172) series; the first three members are: $C_2H_2$ (n = 2), $C_3H_4$, $C_4H_6$ (n = 4). Alkynes are either straight chain (p.182) or branched chain compounds. The first three members of the alkynes are ethyne, propyne and butyne; thereafter they have a Greek number prefix for the number of carbon atoms, followed by *-yne*, e.g. hexyne, $C_6H_{10}$. The alkynes have properties similar to ethyne (↓), becoming less reactive with an increasing number of carbon atoms. They are unsaturated (p.185) organic compounds.

**acetylenes** (*n.pl.*) traditional name (p.44) for the alkynes (↑).

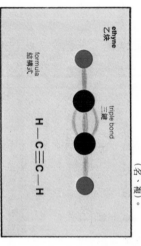

ethyne 乙炔

triple bond 三鍵

formula 結構式

H—C≡C—H

**ethyne** (*n*) an alkyne (↑) of formula $C_2H_2$. It is a colourless (p.15) gas with a sweet smell. Ethyne burns with a bright, white flame (p.58) and is used for lighting. The two carbon atoms are joined by a triple bond (p.181) which makes ethyne very reactive (p.62). Ethyne undergoes (p.213) addition reactions (p.188), and polymerization (p.207). With alkali metals (p.117), acetylides (p.49) are formed.

**acetylene** (*n*) traditional name (p.44) for ethyne.

乙烯（名）化學式為 $C_2H_4$ 的烯屬烴，為一種具有甜味的無色（第 15 頁）氣體，其兩個碳原子由雙鍵（第 181 頁）相連，雙鍵使乙烯易起反應（第 62 頁）。乙烯可經受（第 213 頁）加成反應（第 188 頁）、綜合作用（第 207 頁）和燃燒（第 58 頁）作用。

乙烯（↑）的英文慣用名（第 44 頁）。

炔屬烴（名）通式為 $C_nH_{2n-2}$ 的烴（第 172 頁）。炔屬烴為一同系列（第 172 頁）。前三個炔屬烴分別為乙炔 $C_2H_2$ (n = 2)、$C_3H_4$、$C_4H_6$ (n = 4)。炔屬烴可為直鏈（第 182 頁）也可為支鏈化合物。炔屬烴的前三個為乙炔、丙炔、丁炔，其後的炔屬烴其詞尾則為 *-yne*，例如己炔的英文名為 hexyne $C_6H_{10}$。炔屬烴的性質與乙炔（↓）相似，但隨碳原子數增加而變成較不易起反應。炔屬烴是不飽和的（第 185 頁）化合物。

炔屬烴（↑）的英文慣用名（第 44 頁）為 acetylenes

乙炔（名）化學式為 $C_2H_2$ 的炔屬烴（↑）。乙炔是一種帶有甜味的無色（第 15 頁）氣體，無色燃燒時發出明亮的白色火焰（第 58 頁），故常用於照明。其兩個碳原子由三鍵（第 181 頁）相連，使乙炔極易反應（第 62 頁）。乙炔可經受（第213 頁）加成反應（第 188 頁）和綜合作用（第 207 頁）。與鹼金屬（第 117 頁）取代則生成乙炔化物（第 49 頁）。

電石氣（名）乙炔的慣用名（第 44 頁）。

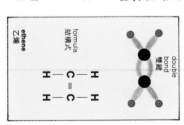

ethene 乙烯

double bond 雙鍵

formula 結構式

H H
 \\ /
  C = C
 /     \\
H       H

**alcohol¹** (*n*) an organic (p.55) compound which contains one or more hydroxyl (p.185) groups. Alcohols can be formed from alkanes (p.172) by substituting (p.188) a hydroxyl group for a hydrogen atom; they form homologous series (p.172) corresponding (p.233) to the alkanes. Alcohols are classified as primary, secondary or tertiary alcohols depending on the position of the hydroxyl group. For primary alcohols, the group is – $CH_2OH$; for secondary alcohols it is ⟩ CHOH; for tertiary alcohols it is ≥ COH.

Alcohols are also classified according to the number of hydroxyl groups in the molecule:

monohydric alcohol  $CH_3CH_2OH$  ethanol

dihydric alcohol  $CH_2OH$ / $CH_2OH$  ethane-1,2-diol (glycol)

trihydric alcohol  $CH_2OH$ / CHOH / $CH_2OH$  propane-1,2,3-triol (glycerol)

Alcohols react with alkali metals (p.117) evolving (p.40) hydrogen and forming an alkoxide; they burn readily and are oxidized to aldehydes (↓), ketones (↓) or carboxylic acids (p.176) depending on the alcohol and the strength of the oxidizing agent. With organic acids alcohols form esters (p.177). The hydroxyl group can be replaced by a halogen (p.117).

**alcohol²** (*n*) trivial name (p.44) for ethanol.

**aldehyde** (*n*) an organic (p.55) compound which contains the functional group – CHO. An aldehyde is formed as a first oxidation product (p.62) from the corresponding (p.233) alcohol (↑), e.g. ethanol is oxidized to ethanal. The name of an aldehyde is derived (p.106) from the corresponding alcohol by changing –ol to -al. Aldehydes are reduced to alcohols; they form addition (p.188) compounds with sodium hydrogen sulphite, hydrogen cyanide and other compounds.

醇（名）含有一個或多個羥基（第185頁）的有機（第55頁）化合物。醇類係由烷屬烴（第172頁）分子中的一個氫原子為一個羥基取代（第188頁）而形成，醇類是與烷屬烴相應（第233頁）的同系列（第172頁）。醇類按其羥基所在之位置分成（伯醇（一級醇）、仲醇（二級醇）和叔醇（三級醇）。伯醇含 –$CH_2OH$ 基、仲醇含 ⟩ CHOH 基、叔醇含 ≥ COH 基。醇亦可按其分子中含的羥基數目分類，如：

一元醇 $CH_3CH_2OH$ 乙醇

二元醇 $CH_2OH$ / $CH_2OH$ 乙二醇-1,2（乙二醇）

三元醇 $CH_2OH$ / CHOH / $CH_2OH$ 丙三醇-1,2,3（丙三醇）

醇與鹼金屬（第117頁）反應放出（第40頁）氫氣並生成醇鹽；醇類容易燃燒並可氧化成醛類（↓）、酮類（↓）或羧酸類（第176頁），這取決於醇和氧化劑的強度。醇與有機酸反應生成酯（第177頁）。醇的羥基可為鹵素所取代（第117頁）。

酒精（名）為乙醇之俗名（第44頁）。

醛（名）含有 –CHO 官能基的有機（第55頁）化合物。醛係由相應的（第233頁）醇（↑）氧化形成的第一個氧化產物（第62頁）。例如乙醇氧化成乙醛。醛的英文名稱是由相應的醇將其詞尾 -ol 改變為 -al 而得出（第106頁）。醛可還原成醇，也可與亞硫酸氫鈉、氰化鹽及其他化合物反應加成（第188頁）化合物。

propan-2-ol
丙醇-2
formula 結構式
a secondary alcohol 仲醇
$CH_3$ — CHOH

2 methylpropan-2-ol
二甲基丙醇-2
formula 結構式
a tertiary alcohol 叔醇
$CH_3$—C—OH
$CH_3$
$CH_3$

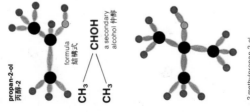

ethanol 乙醇
a primary alcohol 伯醇
formula 結構式

ethanal 乙醛
an aldehyde 乙醛
formula 結構式
$CH_3$ CHO

● carbon atom 碳原子
●● hydrogen atom 氫原子
● oxygen atom 氧原子

**ketone** (n) an organic (p.55) compound containing the functional group — CO. Ketones are prepared (p.43) from the corresponding secondary alcohol (p.175) by oxidation. They are good solvents for organic compounds and are less reactive than aldehydes (p.175), although their reactions are similar to those of aldehydes.

**carboxylic acid** an organic (p.55) compound containing the functional group — COOH. Carboxylic acids are prepared (p.43) from the corresponding alcohol (p.175) by complete oxidation; they are named from that alcohol, e.g. ethanol on oxidation forms ethanoic acid, the ending -ol is replaced by -oic. The

CH₃CH₂OH —(O)→ CH₃COOH

ethanoic acid 乙酸

carboxylic acids are all weak acids (p.45). With alcohols, they form esters (↓); the hydroxyl group (p.185) can be replaced by a halogen (p.117); dehydration forms an acid anhydride (↓). The carboxylic acids form an homologous series (p.172).

**dicarboxylic acid** an organic (p.55) compound with two functional groups of — COOH. The simplest member is ethane di-oic acid (traditional name is ethane di-oic acid oxalic acid). The dicarboxylic acids are stronger acids than the carboxylic acids; they are dibasic (p.46).

**acid anhydride** an organic (p.55) compound prepared from the corresponding (p.233) acid by dehydration or by the action of an acyl chloride (↓) on the sodium salt of the acid.

**acyl chloride** an organic (p.55) compound containing the functional group — COCl; it is the functional group — COOH with the hydroxyl group replaced (p.68) by chlorine. Acyl chlorides are prepared from the corresponding carboxylic acid (↑) by the action of phosphorus pentachloride. They are very reactive (p.62) and react with water to give the corresponding carboxylic acid.

**酮**(名) 含 > CO 官能基的有機(第 55 頁)化合物,可由相應的仲醇(第 175 頁)氧化製備(第 43 頁)而得。酮是有機化合物的良好溶劑。酮的反應和醛相似,但不及醛(第 175 頁)易反應。

**羧酸** 含 — COOH 官能基的有機(第 175 頁)化合物。羧酸可由相應的醇(第 175 頁)完全氧化製備(第 43 頁)而得,羧酸由其醇而得名。例如:乙醇氧化生成乙酸。乙酸的英文名稱是用詞尾 -oic 代替乙醇英文名詞尾的 -ol。羧酸類都是弱酸(↓);羧酸與醇反應生成酯(↓);羧酸的羥基(第 185 頁)可被鹵素(第 117 頁)取代;脫水則生成酸酐(↓)。羧酸類成為一個同系系列(第 172 頁)。

**二羧酸** 含兩個 — COOH 官能基的有機(第 55 頁)化合物。乙二酸(草酸)的慣用名(第 44 頁),是最簡單的二羧酸。二羧酸的酸性比羧酸強;二羧酸都是二元的(第 46 頁)。

**酸酐** 由相應的(第 233 頁)酸脫水而得的一種有機(第 55 頁)化合物。

**醯基氯:醯基氯** 含 — COCl 官能基的有機(第 55 頁)化合物;— COCl 基係 — COOH 官能基中的羥基被氯置換(第 68 頁)生成的(↓)與乙醯氯的羥基。醯基氯可由相應的羧酸(↑)與五氯化磷反應製得。醯基氯極易起反應(第 62 頁)並可與水反應生成相應的羧酸。

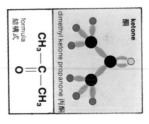

ketone
formula 結構式
dimethyl ketone propanone 丙酮

CH₃—C—CH₃
    ‖
    O

ethanoic acid
formula 結構式
carboxylic acid 羧酸

CH₃—C=O
      OH

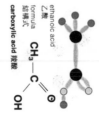

ethanedioic acid
oxalic acid
formula 結構式
dicarboxylic acid 二羧酸

COOH
COOH

ethanoyl chloride
acetyl chloride 乙醯氯
formula 結構式
acyl chloride

CH₃COCl

● carbon atom 碳原子
● hydrogen atom 氫原子
○ oxygen atom 氧原子
○ chlorine atom 氯原子

酯(名) 醇(第175頁)與羧酸(↑)反應生成的有機(第55頁)化合物。羧酸(第180頁)或任何芳香(第180頁)基都可置換(第68頁)酸分子中的氫(第64頁)，反應是可逆的(第213頁)，例如

CH₃CH₂OH + CH₃COOH
⇌ CH₃COOC₂H₅ + H₂O
乙醇 ＋ 乙酸 ⇌ 乙酸乙酯 ＋ 水

酯與無機(第55頁)酸反應也可生成酯。酯類具有如花、果的芳香(第22頁)氣味；酯類不易起反應但可進行(第185頁)水解作用(第66頁)。

脂肪(名) 為甘油三醇-1,2,3(即丙三醇)(↑)與不同羧酸(↑)反應生成的脂(↑)。脂肪所含的酸為飽和的(第185頁)直鏈(第182頁)化合物；而油中所含的酸為不飽和的(第185頁)直鏈化合物。棕櫚酸 (C₁₅H₃₁COOH)、硬脂酸 (C₁₇H₃₅COOH) 及月桂酸 (C₁₁H₂₃COOH) 是動物和植物體中存在的最重要的酸。脂肪的結構如圖示。

醚(名) 由兩個烴(第172頁)基，可爲烷基(第180頁)或芳香基(第180頁)和一個氧原子相連接而成的有機(第55頁)化合物。醚易取自其函素和烷烴。例如乙烷、溴丙烷、碘乙烷(第21頁)，但又不易起反應(第62頁)。

鹵烷(名) 由一個烴(第172頁)的一個氫原子被鹵素(第117頁)原子置換(第68頁)所成的有機(第55頁)化合物。鹵烷的名稱取自其函素和烷烴。例如氯乙烷、溴丙烷、碘乙烷。鹵烷可由醇(第175頁)和二氯一氧化硫 SOCl₂(即亞硫醯氯)起作用而製得。鹵烷類極易反應，可用於許多種有機製劑(第43頁)。

醯胺(名) 含 －CONH₂ 官能基的有機(第55頁)化合物。例如乙醯胺(CH₃CONH₂)。醯胺由相應的羧酸(↑)命名，例如由丙酸可形成丙醯胺。醯胺可由相應之羧酸之銨鹽脫水而得(第66頁)。

---

**ester** (*n*) an organic (p.55) compound formed when an alcohol (p.175) reacts with a carboxylic acid (↑). The hydrogen of the acid is replaced (p.68) by an alkyl (p.180) or any aryl (p.180) group. The reaction is reversible (p.64), e.g.

CH₃CH₂OH + CH₃COOH ⇌ CH₃COOC₂H₅ + H₂O
ethanol + ethanoic acid ⇌ ethyl ethanoate + water

An ester can also be formed with an inorganic (p.55) acid. Esters have fragrant (p.22) odours and are the odours of flowers and fruit; they are not very reactive, but undergo (p.213) hydrolysis (p.66).

**fat** (*n*) an ester (↑) of propane-1,2,3-triol (glycerol) and different carboxylic acids (↑). In fats, the acids are saturated (p.185), straight chain (p.182) compounds. In oils, the acids are unsaturated (p.185), straight chain compounds. The important acids in animal and plant fats are palmitic acid (C₁₅H₃₁COOH), stearic acid (C₁₇H₃₅COOH) and lauric acid (C₁₁H₂₃COOH).The structure of a fat is shown in the diagram opposite.

**ether** (*n*) an organic (p.55) compound with two hydrocarbon (p.172) groups, either alkyl (p.180) or aryl (p.180) joined to one oxygen atom. Ethers are very flammable (p.21) but otherwise are unreactive (p.62).

**alkyl halide** an organic (p.55) compound formed when one hydrogen atom of an alkane (p.172) is replaced (p.68) by a halogen (p.117) atom. They are named from the halogen and the alkane, e.g. chloroethane, bromopropane, iodoethane. Alkyl chlorides are prepared from alcohols (p.175) by the action of sulphur dichloride oxide (thionyl chloride, SOCl₂). They are very reactive, and used in many organic preparations (p.43).

**amide** (*n*) an organic (p.55) compound containing the functional group –CONH₂, e.g. ethanamide CH₃CONH₂. Amides are named from the corresponding carboxylic acid (↑), e.g. propanoic acid forms propanamide. They are formed by dehydrating (p.66) the ammonium salt of the corresponding carboxylic acid.

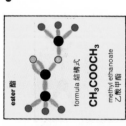

ester 酯
formula 結構式
**CH₃COOCH₃**
methyl ethanoate 乙酸甲酯

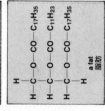

ether (dimethyl ether) 醚（二甲醚）
methoxymethane 甲氧基甲烷
formula 結構式
**CH₃—O—CH₃**

```
    H
    |
H - C - O - CO - C₁₇H₃₅
    |
H - C - O - CO - C₁₁H₂₃
    |
H - C - O - CO - C₁₇H₃₅
    |
    H
```
a fat 脂肪

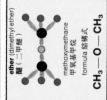

alkyl halide 鹵烷
chloroethane 氯乙烷
formula 結構式
**CH₃CH₂Cl**

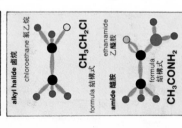

amide 醯胺
ethanamide 乙醯胺
formula 結構式
**CH₃CONH₂**

**amine** (*n*) an organic (p.55) compound containing the functional group – NH₂. Amines are formed by the reaction between ammonia and an alkyl halide (p.177). The amines can be classified by the number of hydrogen atoms in ammonia that are replaced by alkyl (p.180) or aryl groups, e.g.

| | |
|---|---|
| Primary amines | R–NH₂ methylamine CH₃NH₂ |
| Secondary amines | R₂–NH dimethylamine (CH₃)₂NH |
| Tertiary amines | R₃–N trimethylamine (CH₃)₃N |

The amines are colourless (p.15) gases or liquids with a strong fishy odour; they are soluble in water forming weak bases, which form salts with inorganic (p.55) acids, e.g. methylamine reacts with hydrochloric acid to form methylammonium chloride (CH₃NH₃)⁺ Cl⁻.

**nitrile** (*n*) an organic (p.55) compound containing the functional group – CN. The nitriles are named after the corresponding (p.233) alkane with the same number of carbon atoms, e.g. CH₃CH₂CN is propanonitrile. Nitriles are prepared by the action of sodium cyanide (NaCN) on an alkyl halide (p.177). They undergo (p.213) hydrolysis forming first an amide (p.177), then a carboxylic acid (p.176); they undergo reduction (p.193) to an amine (↑).

**isocyanide** (*n*) an organic (p.55) compound containing the group – NC; they are given names such as isocyanoethane (C₂H₅NC). Isocyanides are prepared by the action of silver cyanide (AgCN) on the corresponding (p.233) alkyl halide (p.177). They are colourless (p.15) liquids with a bad odour and are highly poisonous. Isocyanides are hydrolyzed (p.190) to secondary amines.

**amino acid** a carboxylic acid (p.176) with one hydrogen atom replaced by an amino group (– NH₂), e.g. NH₂CH₂COOH, aminoethanoic acid, with a trivial name (p.44) of glycine. Amino acids in solution form both cations (p.125) and anions (p.125) and thus are amphoteric (p.46) in their reactions; they behave as carboxylic acids and as amines. Amino acids are the constituents (p.54) of proteins (p.209); they are capable of forming long chains by condensation (p.191).

胺（名） 含 – NH₂ 官能基的有機化合物。胺可按照氨和鹵烷（第 177 頁）反應生成。胺可按照氨分子中被烷基（第 180 頁）或芳基所取代的氫原子數目作分類。例如：

| | |
|---|---|
| 伯胺 | R–NH₂ 甲胺 CH₃NH₂ |
| 仲胺 | R₂–NH 二甲胺 (CH₃)₂NH |
| 叔胺 | R₃–N 三甲胺 (CH₃)₃N |

胺是有強烈魚腥味的無色（第 15 頁）氣體或液體。可溶於水生成弱鹼，與無機（第 55 頁）酸作用生成鹽。例如甲胺與鹽酸反應生成氨甲氯銨 (CH₃NH₃)⁺ Cl⁻。

腈（名） 含 – CN 官能基的有機（第 55 頁）化合物。腈以含同數碳原子之相應（第 233 頁）烷經命名。例如 CH₃CH₂CN 為丙腈。腈可由氰化鈉 (NaCN) 與鹵烷（第 177 頁）作用而製得，是一種有鹵代氮氨味的無色化合物（第 177 頁）。再生成羥酸（第 176 頁）受還原作用（第 193 頁）成氨胺（↑）。

異腈：胩（名） 含 – NC 基的有機（第 55 頁）化合物；異腈酸賦予種種名稱。例如異氰基乙烷 (C₂H₅NC)。異腈可由氰化銀 (AgCN) 和相應的（第 233 頁）鹵烷（第 177 頁）作用而得。是一種有劇烈氣味而且劇毒的無色（第 15 頁）液體。異腈水解（第 190 頁）成仲胺。

氨基酸（名） 由氨基 (–NH₂) 取代羧酸（第 176 頁）中的一個氫原子而成。例如氨基乙酸 (NH₂CH₂COOH)。其俗用名名（第 44 頁）為甘氨酸。氨基酸在溶液中生成陽離子和陰離子（第 125 頁），故在反應中呈兩性（第 46 頁）：氨基酸起羧酸和胺的作用（第 125 頁）。氨基酸是蛋白質（第 209 頁）的成分（第 54 頁）。能構築縮合反應（第 191 頁）形成長鏈。

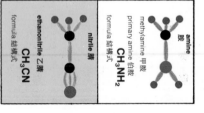

**amine 胺**
primary amine 伯胺
methylamine 甲胺
**CH₃NH₂**
formula 結構式

● carbon atom 碳原子
● hydrogen atom 氫原子
○ oxygen atom 氧原子
● nitrogen atom 氮原子

**nitrile 腈**
ethanonitrile 乙腈
**CH₃CN**
formula 結構式

**isocyanide 異氰酸酯**
methyl isocyanide 甲胩
**CH₃NC**
formula 結構式

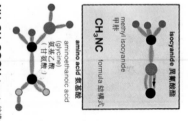

**amino acid 氨基酸**
aminoethanoic acid 氨基乙酸
(glycine) （甘氨酸）
**NH₂CH₂COOH**
formula 結構式

**benzene** (*n*) a hydrocarbon (p.172) of formula $C_6H_6$. The six carbon atoms are combined (p.64) in a ring structure, with all six bonds of equal length and equal activity. The bonds are neither single bonds (p.181) nor double bonds, but have a chemical character peculiar to themselves; this property is called **aromaticity**. Benzene is the first member of two homologous series (p.172); in one, chains (p.182) of carbon atoms are joined to the benzene ring; in the other, two or more benzene rings are combined. The formula of benzene is shown as a hexagon with a circle inside it; the circle represents 6 electrons which do not take part in any one bond, and are known as **delocalized electrons**. The delocalized electrons give benzene its aromatic character. Benzene is a colourless (p.15) liquid with a pleasant (aromatic) smell, is flammable (p.21), a good solvent for organic compounds and is chemically unreactive.

**naphthalene** (*n*) a hydrocarbon (p.172) of formula $C_{10}H_8$. The ten carbon atoms are combined (p.64) in two benzene (↑) ring structures. Naphthalene is a white solid with a strong odour; it is a major (p.226) constituent (p.54) of coal tar and is chemically reactive (p.62). It is a homologue (p.172) of benzene.

**aromatic** (*adj*) describes a compound containing (p.55) a benzene (↑) ring. The compound has chemical properties similar to those of benzene.

**aliphatic** (*adj*) describes a compound consisting of straight chains (p.182) and branched chains, e.g. as in the alkanes, alkenes, and alkynes. Aliphatic compounds include alicyclic (↓) compounds.

**alicyclic** (*adj*) describes a compound consisting of single bonds, or double bonds, with the carbon atoms combined in a ring. Such compounds are aliphatic (↑) and not aromatic (↑). e.g. cyclobutane is an alicyclic compound.

**heterocyclic** (*adj*) describes a compound with an aromatic (↑) or aliphatic (↑) ring structure which includes at least one atom that is not carbon, e.g. pyridine $C_5H_5N$.

苯（名）　化學式為 $C_6H_6$ 的一種烴（第 64 頁）成一個環狀結構（第 172 頁）。其 6 個碳原子結合（第 64 頁）成一個環狀結構中，6 個鍵的鍵長度相等，反應活性相同，這些鍵既非單鍵（第 181 頁）也非雙鍵，而有其本身特有的化學特性；此種屬性稱為芳香性。苯是兩個同系列（第 172 頁）中的頭一個化合物，其中第一個系列是各碳原子的鍵（第 182 頁）都連接到苯環上；而另一個系列是兩個或多個苯環結合起來。苯的結構式以其內有一個圓圈的六角形來表示，圓圈表示 6 個電子不參加任何一個鍵，這些電子通稱為非定域電子。非定域電子使苯具有芳香特性。苯是有令人愉快芳香氣味的一種無色的（第 15 頁）液體，它可燃（第 21 頁），是有機化合物的良好溶劑，且不易起化學反應。

萘（名）　化學式為 $C_{10}H_8$ 的一種烴（第 172 頁），其 10 個碳原子結合（第 64 頁）在兩個苯（↑）環結構中。萘是有強烈氣味的一種白色固體，為煤焦油的主要（第 226 頁）成分（第 54 頁），萘是易反應的（第 62 頁），為苯的一種同系物（第 172 頁）。

芳族的（形）　描述含有（第 55 頁）苯（↑）環的一種化合物。其化學性質與苯相似。

脂（肪）族的（形）　描述含有直鏈（第 182 頁）和支鏈的一種化合物，例如烷屬烴、烯屬烴和炔屬烴。脂族化合物包括脂環族（↓）化合物。

脂環族的（形）　描述含有單鍵或雙鍵，其碳原子結合成環的一種化合物。此種化合物為脂族的（↑）而非芳族的（↑）。例如環丁烷是一種脂環族化合物。

雜環族的（形）　描述含有芳族（↑）或脂族（↑）環狀結構，而其環狀結構中至少有一個原子不是碳原子的一種化合物，例如吡啶（$C_5H_5N$）。

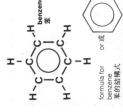

formula for benzene 苯的結構式
benzene 苯
or 或

a homologue with a straight chain 具直鏈的同系物
ethyl benzene 乙苯
straight chain 直鏈
$CH_2CH_3$

aromatic compound 芳族化合物

naphthalene 萘
aromatic compound 芳族化合物

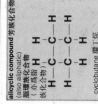

alicyclic compound 芳環化合物 (also aliphatic) 脂環族化合物 (亦為脂族)
cyclobutane 環丁烷
H—C—C—H

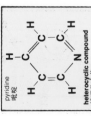

pyridine 吡啶
heterocyclic compound 雜環化合物

**alkyl** (*adj*) describes an organic (p.55) group formed when one atom of hydrogen is taken from an alkane (p.172); examples of alkyl groups are: methyl (CH₃), ethyl (C₂H₅); pentyl (C₅H₁₁). The general formula of an alkyl group is $C_nH_{2n+1}$. **alkylation** (*n*).

**acyl** (*adj*) describes an organic (p.55) group formed when an organic hydroxyl group (—OH) is taken from a carboxylic acid (p.176). *See* **acyl chloride** (p.176).

**aryl** (*adj*) an aromatic (p.179) group formed when one atom of hydrogen is taken from an aromatic hydrocarbon (p.172). **arylation** (*n*).

**phenol** (*n*) an aromatic (p.179) compound of formula C₆H₅OH. It is a white crystalline (p.15) solid with a characteristic odour (p.15); it reacts (p.62) as a weak acid (p.45), forming **phenates**. Phenol is obtained from coal tar; it is used as a common antiseptic and is also important in the plastics (p.210) industry. Homologues (p.172) of phenol are called **phenols**.

phenate 苯酚盐

OH + NaOH → ONa + H₂O

sodium phenate 苯酚鈉

**nitro compound** a combination of an aryl (↑) group and a nitro group (—NO₂) (p.186). Nitro compounds are important because (a) aromatic (p.179) hydrocarbons are easily nitrated by concentrated nitric acid, and (b) they are chemically reactive (p.62). Both nitro and dinitro compounds can be prepared. Alkyl nitro compounds also exist (p.213) but are less important.

**diazonium salt** a combination of an aryl (↑) group with the azo group — N⁺≡N which forms a salt with an inorganic anion (p.125), e.g. (C₆H₅ — N⁺≡N) Cl⁻ which is benzenediazonium chloride. Diazonium salts are prepared from primary aromatic (p.179) amines by the action of nitrous acid. The salts are unstable unless kept at a temperature below 0°C. They are important because many dyes (p.162) are manufactured from diazonium salts.

烷基 (形) 指一個烷屬烴（第 172 頁）分子失去一個氫原子所形成的有機（第 55 頁）基團。例如：甲基（CH₃）、乙基（C₂H₅）、戊基（C₅H₁₁）為烷基。烷基之通式為 $C_nH_{2n+1}$。（名詞為 alkylation）

醯基 (形) 指一個羧酸（第 176 頁）去一個羥基（—OH）所形成的有機基團。參見"醯氯"（第 176 頁）。

芳香基 (形) 指一個芳香烴（第 172 頁）分子失去一個氫原子所形成的芳香基團。（名詞為 arylation）

酚 (名) 化學式為 C₆H₅OH 的芳香族（第 179 頁）化合物。酚為一種有特殊氣味的白色結晶（第 15 頁）固體；它作為弱酸（第 45 頁）反應（第 62 頁）生成酚鹽。酚可由煤焦油中取得。它是一種常用的防腐、殺菌劑，在塑膠製品（第 210 頁）工業上具有重要性。酚的同系物（第 172 頁）稱為酚類。

硝基化合物 由芳香基（↑）和硝基（—NO₂）（第 186 頁）結合而成的化合物。硝基化合物為重要之化合物，因為（a）芳香（第 179 頁）烴易用濃硝酸進行硝化處理；(b) 硝基化合物是易化學反應的（第 62 頁）。可以製備得硝基化合物和二硝基化合物。也存在有（第 213 頁）烷基硝基化合物，但這種化合物不甚重要。

重氮鹽 芳香基（↑）和偶氮基（—N⁺≡N）結合而成的化合物。和一個無機陰離子（第 125 頁）結合而形成的鹽。例如氯化苯基重氮基（C₆H₅ — N⁺≡N）Cl⁻。重氮鹽可從伯芳香（第 179 頁）胺和亞硝酸作用製得。這些鹽在0°C以下才能保持穩定。這些鹽是重要的化合物，由重氮鹽可製得許多種染料（第 162 頁）。

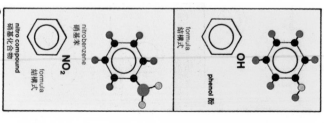

carbon atom 碳原子
hydrogen atom 氫原子
oxygen atom 氧原子
nitrogen atom 氮原子
chlorine atom 氯原子

nitro compound 硝基化合物
nitrobenzene 硝基苯
NO₂
formula 結構式

phenol 酚
OH
formula 結構式

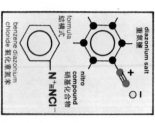

diazonium salt 重氮鹽
benzene diazonium chloride 氯化重氮苯
N⁺≡NCl⁻
formula 結構式

**empirical formula** a formula (p.78) which gives the simplest ratio of elements in a compound, e.g. the empirical formula of ethanoic acid is $CH_2O$.

**molecular formula** a formula (p.78) which shows the number of atoms of each element in a molecule of a compound. This formula is found from the relative molecular mass of a compound. For example, $CH_2O$ is the empirical formula (↑) of ethanoic acid and of glucose (p.206). Their respective (p.233) relative molecular masses are 60 and 180. The relative molecular mass of $CH_2O$ is 30, hence ethanoic acid is $C_2H_4O_2$ and glucose is $C_6H_{12}O_6$. These are molecular formulae.

**structural formula** a formula (p.78) which shows how the atoms in a molecule are grouped together, e.g. the structural formula of ethanoic acid is $CH_3COOH$. A structural formula of an organic (p.55) compound indicates (p.38) how the compound will react.

**graphic formula** a formula (p.78) which shows how the atoms in a molecule are oriented (p.93) in space, relative (p.232) to each other.

**general formula** a formula (p.78) which allows the formula for any member of an homologous series (p.172) to be written, e.g. $C_nH_{2n}$ is the general formula for alkenes. Putting n = 2,3,4, etc. gives the formula for members of the series.

**single bond** a covalent bond (p.136), it is oriented (p.93) in space and allows atoms to turn relative (p.232) to each other.

**double bond** two covalent bonds (p.136) joining together two atoms; each covalent bond is formed by two atoms sharing two electrons. The double bond between two carbon atoms is important, it prevents the atoms turning about the bond and also can cause geometrical isomerism (p.184). A double bond makes an organic (p.55) compound reactive (p.62).

**triple bond** three covalent bonds (p.136) joining together two atoms; each covalent bond is formed by two atoms sharing two electrons. The triple bond between two carbon atoms is important. It has the same effect as a double bond (↑); but is much more reactive.

實驗式　表示化合物分子中各元素間最簡單比例的化學式（第 78 頁）。例如乙酸的實驗式爲 $CH_2O$。

分子式　表示化合物分子中所含元素之原子數的化學式（第 78 頁）。分子式係由該化合物的相對分子質量確定。例如乙酸和葡萄糖（第 206 頁）的實驗式（↑）都是 $CH_2O$，相對分子質量分別（第 233 頁）爲 60 和 180，而 $CH_2O$ 的相對分子量爲 30，因此乙酸之分子式爲 $C_2H_4O_2$，葡萄糖之分子式爲 $C_6H_{12}O_6$。

結構式　表示原子在分子中如何結合在一起的一種化學式（第 78 頁）。有機（第 55 頁）化合物的結構式爲 $CH_3COOH$。有機（第 55 頁）化合物的結構式表明（第 38 頁）該化合物將如何起反應。

圖解式　表示分子所含各原子在空間相對於（第 232 頁）其原子取向（第 93 頁）的化學式（第 78 頁）。

通式　能寫出同系列（第 172 頁）中任何化合物的化學式之分子式（第 78 頁）。例如烯類烴的通式爲 $C_nH_{2n}$，取 n 爲 2、3、4⋯⋯就可寫出系列中各個化合物的化學式。

單鍵　是一種共價鍵（第 136 頁），它在空間取向（第 93 頁），單鍵的原子彼此間能相對地（第 232 頁）旋轉。

雙鍵　使兩個原子連接在一起的兩個共價鍵（第 136 頁），每一個共價鍵都由共用兩個電子的兩個原子所形成。兩個碳原子之間的雙鍵極重要，它阻止原子繞鍵旋轉，並可產生幾何異構現象（第 184 頁）。雙鍵使有機（第 55 頁）化合物易發（第 62 頁）。

三鍵　使兩個原子連接在一起的三個共價鍵（第 136 頁），每一個共價鍵都由共用兩個電子的兩個原子所形成。兩個碳原子之間的三鍵極重要，其效應與雙鍵（↑）相同，但更易反應。

**C:H:O = 1:2:1**
by mass 按質量比
乙酸的實驗式
**empirical formula**
$CH_2O$

for ethanoic acid
相對分子量爲 60
relative molecular
乙酸
mass for ethanoic
acid is 60

**molecular formula**
分子式
$C_2H_4O_2$

**structural formula** 結構式
$CH_3COOH$

● carbon atom 碳原子
● hydrogen atom 氫原子
● oxygen atom 氧原子

ethanoic acid 乙酸

**graphic formula** 圖解式

**chain** (*n*) a structure in which each carbon atom is joined to the next carbon atom to form a chain, or line, of atoms. A chain may be straight (↓) or branched (↓).

**straight chain** a chain of carbon atoms in which any one carbon atom is not joined to more than two other carbon atoms, e.g. as in hexane.

**branched chain** a chain of carbon atoms in which one or more carbon atoms may be joined to 3 or 4 other carbon atoms. *See diagram.*

**cyclic chain** a chain of carbon atoms forming a circle in which the carbon atoms may be joined by either single or double bonds. The resulting (p.39) compound has an aliphatic (p.179) nature.

**ring chain** a chain of carbon atoms which can be cyclic (↑) or can be a benzene (p.179) ring with neither single nor double bonds.

**isomerism** (*n*) the property of having the same molecular formula (p.181), but different structural formulae, e.g. ethanol and methoxymethane (dimethyl ether) have the same molecular formula, but their structural formulae are CH₃CH₂OH and CH₃O CH₃ respectively (p.233) There are different kinds of isomerism. Isomerism is very common in organic (p.55) compounds. **isomer** (*n*), **isomeric** (*adj*).

**isomer** (*n*) one of two, or more, compounds which exhibit (p.221) isomerism (↑), e.g. ethanol and methoxymethane are isomers.

**structural isomerism** isomerism (↑) exhibited by two, or more, organic (p.55) compounds which have the same molecular formulae, but their different structures give them different physical or chemical properties, e.g. ethanol and methoxymethane are structural isomers; a branched chain and a straight chain alkane (p.172) with the same number of carbon atoms are isomers of each other. *See diagram.*

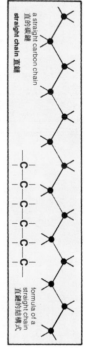

a straight carbon chain
直的碳鏈
**straight chain** 直鏈

formula of a
straight chain
直鏈的結構式

branched carbon chain 分支的碳鏈

formula for a
branched carbon chain*
支鏈的結構式
**branched chain** 支鏈

**鏈**（名）各個碳原子與相鄰碳原子相連連成一列原子的一種結構。鏈可以為直鏈（↓）或支鏈（↓）。

**直鏈** 任何一個碳原子最多連接另外兩個碳原子的一條碳鏈。例如己烷是直鏈的。

**支鏈** 一個或幾個碳原子可以和三或四個其他碳原子相連的碳原子鏈。（見圖）。

**環形鏈** 碳原子之間既可以靠單鍵或雙鍵相連而形成的環形（↑）或苯（第 179 頁）環形的鏈。

**環狀鏈** 各個碳原子可以和三或四個其他碳原子相連成一個環的碳原子鏈。所形成的（第 39 頁）化合物具有脂族（第 179 頁）化合物的性質。

**同分異構現象**（名）具有分子式（第 181 頁）相同而結構式不同的性質。例如乙醇和甲氧基甲烷（即二甲醚）的分子式相同但結構式分別為 CH₃CH₂OH 和 CH₃O CH₃。同分異構現象有不同的型式，在有機（第 55 頁）化合物中，同分異構現象極為普遍。（名 詞 isomer，形容詞 isomeric）

**同分異構體**（名）顯現（第 221 頁）同分異構現象（↑）的兩種或多種化合物之一。例如乙醇和甲氧基甲烷為同分異構體。

**結構同分異構現象** 兩種或多種有機（第 55 頁）化合物所顯現的同分異構現象（↑）。這些化合物的分子式相同，但由於結構式不同而具有不同的物理和化學性質。例如，乙醇和甲氧基甲烷是結構同分異構體；碳原子數相同的支鏈烷烴和直鏈烷烴是彼此的同分異構體（見圖）。

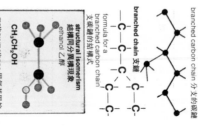

ethanol 乙醇
**CH₃CH₂OH**

methoxymethane 甲氧基甲烷
**CH₃ — O — CH₃**

two compounds
exhibiting structural
isomerism
顯示結構同分異構現象的兩種化合物
**structural isomerism**
結構同分異構現象

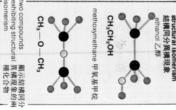

methyl
propane
甲基丙烷

two compounds
exhibiting structural
isomerism

butane 丁烷

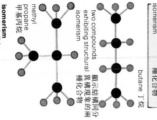

**isomerism**
同分異構現象

**stereoisomerism** (*n*) isomerism that results from a different arrangement in space of the atoms in a molecule. The structural formula does not show the different arrangements which are possible, but the graphic formula of such compounds can show the difference. Stereoisomerism results from the tetrahedral (p.83) directions of the four covalent bonds of carbon. There are two kinds of stereoisomerism: optical isomerism (↓) and geometrical isomerism (p.184).

**optical isomerism** optical isomerism exists when a molecule is not identical (p.233) with its image in a mirror. Two spatial (p.211) arrangements of the atoms are thus possible. The two structures are chemically identical, but one structure rotates (p.218) the plane of polarized light to the left and the other structure rotates it to the right; otherwise their physical properties are the same, e.g. lactic acid exhibits (p.221) optical isomerism.

**enantiomorph** (*n*) (1) a molecule exhibiting (p.221) optical isomerism (↑) which is one of two or more isomeric (↑) forms. (2) one of two crystalline forms when each crystal is the mirror image of the other. **enantiomorphic** (*ad*).

**asymmetric carbon atom** a carbon atom which has four different atoms or groups attached to it. It is the most common cause of optical isomerism (↑), e.g. lactic acid.

**racemate** (*n*) a physical mixture of two enantiomorphs (↑), one with rotation of the plane of polarized light to the right, and the other to the left. The result is optically inactive. When optically active compounds are synthesized, a racemate is produced. **racemic** (*ad*).

立體異構現象 (名) 分子所含原子的不同空間排列所產生之異構現象。結構式不能表示出那些可能的不同排列，但這些化合物的圖解形式就可顯示出其差別。立體異構現象產生於碳的四個共價鍵的四面體（第 83 頁）方向。有兩類立體異構現象：光學異構現象（↓）和幾何異構現象（第 184 頁）。

光學異構現象：旋光異構現象 當一個分子與其鏡像不完全一樣（第 233 頁）時就會存在光學異構現象。因此原子有兩種可能的空間排列（第 211 頁）排列，但其中一種結構使偏光面向左旋轉另一結構使之向右旋轉，否則他們的物理性質就相同，例如乳酸顯示出（第 221 頁）光學異構現象。

對映體：鏡像物 (1) 顯示（第 221 頁）有光學異構現象（↑）的一個分子，它為兩個或多個異構現象（↑）形式中之一種。(2) 兩個晶體互為鏡像物，兩個晶形中之一。（形容詞為 enantiomorphic）

不對稱碳原子 指有四個不同原子或基團與之相連接的碳原子。它是光學異構現象（↑）的最普通起因，例見乳酸。

外消旋物：外消旋 (名) 兩個對映物（↑）的物理混合物，其中之一使偏光面右旋，另一則使之左旋，結果不旋光。當兩種光學化合物合成時，一種外消旋混合物便產生了。（形容詞為 racemic）。

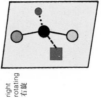

right
rotating
右旋

optically
active forms
of lactic acid
乳酸之旋光形式
enantiomorph
對映體

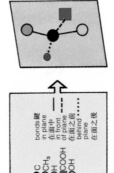

bonds 鍵
— in plane
任面中
--- in front
of plane
任面之前
····· behind
plane
任面之後

C
CH₃
H
COOH
OH

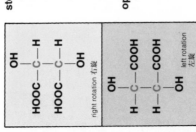

2, 3 dihydroxybutanedioic acid
2,3 二羥基丁二酸
tartaric acid 酒石酸

right rotation 右旋

left rotation
左旋

(two asymmetric
carbon atoms in
the molecule) C
（分子中有兩個不對
稱碳原子）C

optical
isomerism
光學異構現象

CH₃
H—C—OH
COOH

C asymmetric
carbon atom
不對稱碳原子

lactic acid 乳酸

**geometrical isomerism** geometrical isomerism results from two conditions: (1) two carbon atoms joined by a double bond; (2) each carbon atom must have two different atoms or groups joined to it. A double bond prevents rotation (p.218) of the carbon atoms, so two spatial (p.211) arrangements of the atoms are possible. Geometrical isomers have very different physical properties, e.g. but-2-ene has one isomer with a melting point of −139°C and the other −106°C. Their chemical properties tend to be the same, but certain differences also exist, e.g. the difference between maleic and fumaric acids, which are isomers. **geometrical isomer** (*n*).

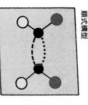

cis-configuration
順式構型

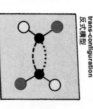

trans-configuration
反式構型

○ carbon atom 碳原子
● hydrogen atom 氫原子
○ chlorine atom 氯原子
bonds 鍵
in plane 在面中
in front of plane 在面之前
behind plane 在面之後
**geometrical isomers**
幾何異構體

**cis-configuration** (*n*) a geometrical isomer (↑) with two like groups on the same side of the double bond. *See diagram.*

**trans-configuration** (*n*) a geometrical isomer (↑) with two like groups on opposite sides of the double bond. *See diagram.*

**tautomerism** (*n*) a state of dynamic equilibrium (p.150) between spontaneously convertible (p.73) isomers. Generally in the conversion of one isomer to another, a hydrogen atom changes its place in the molecular structure, e.g. as in ethyl 3-oxobutanoate (ethyl acetoacetate). The two isomeric forms in tautomerism are the 'keto' and 'enol' forms, shown in the diagram opposite. **tautomer** (*n*) an individual isomer of tautomerism (↑); the 'keto' or 'enol' form of the tautomeric compound. **tautomeric** (*adj*).

**methyl** (*ad*) the alkyl (p.180) group — CH₃.
**ethyl** (*ad*) the alkyl (p.180) group — C₂H₅.
**propyl** (*ad*) the alkyl (p.180) group — C₃H₇.
**butyl** (*ad*) the alkyl (p.180) group — C₄H₉.
**pentyl** (*ad*) the alkyl (p.180) group — C₅H₁₁.

幾何異構現象 幾何異構現象是因於兩個條件：（1）兩個碳原子由一個雙鍵相連；（2）各個碳原子必須有兩個不同的原子或基團與之相連。雙鍵因碳原子旋轉（第218頁），因此各原子間可能有兩種空間（第211頁）排列。幾何異構體的物理性質極不相同，例如丁烯-2 有兩種異構體，其中之一熔點為 −106°C，另一為 −139°C。兩種異構體的化學性質往往相同，但也存在某些差別，例如順丁烯二酸和反丁烯二酸兩種異構體，兩者之間有差別。（名詞 geometrical isomer 意為幾何異構體）

順式構型：反式構型（名）兩個相同的基鄰在雙鍵同一側的一種幾何異構體（↑）。（見圖）

反式構型（名）兩個相同的基鄰在雙鍵兩側的一種幾何異構體（↑）。（見圖）

互變異構現象（名）自發轉化（第73頁）之異構體間的動態平衡（第150頁）狀態。一般而言，一種異構體轉化成另一個異構體時，一個氫原子改變其在分子結構中的位置，如在3-氧代丁酸乙酯（即乙醯乙酸乙酯）。互變異構現象中，此兩種異構體為"酮"式和"烯醇"式，如圖所示。

互變異構體（名）互變異構現象（↑）中的個別異構體：互變異構體的"酮"式或"烯醇"式。

甲基的（形）指烷基（第180頁）基團 — CH₃。
乙基的（形）指烷基（第180頁）基團 — C₂H₅。
丙基的（形）指烷基（第180頁）基團 — C₃H₇。
丁基的（形）指烷基（第180頁）基團 — C₄H₉。
戊基的（形）指烷基（第180頁）基團 — C₅H₁₁。

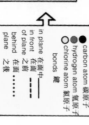

**tautomerism**
互變異構現象

CICH ═ CHCl
1,2 dichloroethene
1,2 —二氯乙烯

ethyl 3 oxobutanoate
3-氧化丁酸乙酯
(乙醯乙酸乙酯)

CH₃
C ═ O
H C H
COOC₂H₅
'keto' form
"酮"式

dynamic equilibrium
動態平衡

CH₃
C — O — H
HC
COOC₂H₅
'enol' form
"烯醇"式

H mobile hydrogen atom
流動氫原子

**saturated² (adj)** describes a carbon chain in which there are only single bonds connecting individual carbon atoms. A saturated chain in un-reactive, as alkanes (p.172) are. **saturation (n)**.

飽和的（形） 描述碳鏈中各個碳原子只靠單鍵相連。飽和鍵不易反應，如烷類（第 172 頁）即是。（名詞爲 saturation）

**unsaturated² (adj)** describes a carbon chain in which there is at least one double or triple bond. An unsaturated chain is reactive, particularly for addition reactions (p.188).

不飽和的（形） 描述碳鏈中至少含有一個雙鍵或三鍵。不飽和鍵易反應，尤其易起加成反應（第 188 頁）。

**polyunsaturated (adj)** describes a carboxylic acid containing more than two double bonds.

多不飽和的（形） 描述羧酸分子中含兩個以上的雙鍵。

**functional group** an atom, or group of atoms, which gives an organic (p.55) compound its reactivity. A functional group is joined to an alkyl (p.180) or aryl group (p.172), which forms the homologous series (p.172) and the functional group gives the chemical characteristics. A long carbon chain with many carbon atoms reduces (p.219) the chemical activity of the functional group. Examples of functional groups are: hydroxyl, carboxyl, amino, chloro. In general formulae an alkyl group is represented by R and an aryl group by Ar or Ph, e.g. RCOOH is a general formula for a carboxylic acid; ArNO₂ is an aromatic (p.179) nitro compound.

官能團；功能團 使有機（第 55 頁）化合物具有反應性的原子和原子團。官能團與烷基（第 180 頁）或芳基相連構成一個同系列（第 172 頁）並使化合物具有其化學特性。含碳原子數多的長碳鏈會降低（第 219 頁）官能團的化學活性。例如羥基、羧基、氨基、氯基都是官能團。通式中用 R 表示烷基、Ar 或 Ph 表示芳基。例如羧酸之通式爲 RCOOH，芳族（第 179 頁）硝基化合物之通式爲 ArNO₂。

**hydroxyl group** the functional group (↑) —OH; it occurs (p.154) in alcohols (p.175) and phenols (p.180). Combined with alkyl groups, the hydroxyl group reacts with the hydrogen of acids to form water. Combined with aryl groups, the hydrogen of the hydroxyl group forms hydrogen ions, making the phenols weakly acidic.

羥基 —OH 官能團（↑），存在（第 154 頁）於醇（第 175 頁）和酚（第 180 頁）分子中。和烷基結合的羥基可與酸中的氫反應生成水；在酚中和芳基結合的羥基，其中的氫可形成氫離子，使酚呈現弱酸性。

**monohydric (adj)** describes an alcohol with one hydroxyl group. See alcohol (p.175).

一羥基的（形） 描述醇分子中含一個羥基。參見"醇"（第 175 頁）。

**dihydric (adj)** describes an alcohol with two hydroxyl groups. See alcohol (p.175).

二羥基的（形） 描述醇分子中含兩個羥基。參見"醇"（第 175 頁）。

**trihydric (adj)** describes an alcohol with three hydroxyl groups. See alcohol (p.175).

三羥基的（形） 描述醇分子中含三個羥基。參見"醇"（第 175 頁）。

**carbonyl group** the functional group (↑) = CO; it occurs in aldehydes (p.175) and ketones (p.176). The carbonyl group has a double bond (p.181) between the carbon and oxygen. This double bond undergoes (p.213) addition reactions (p.188). A hydrogen atom joined to the carbon atom of the carbonyl group makes it more reactive, as in aldehydes.

羰基；碳醯基 分子中的 =CO 官能團（↑）。羰基存在於醛（第 175 頁）和酮（第 181 頁）的碳和氧間同一個雙鍵（第 181 頁）。此雙鍵可經歷（第 213 頁）加成反應（第 188 頁）。氫原子與羰基之碳原子相連使羰基更易反應，如在醛中所見。

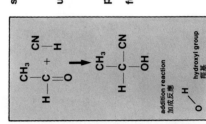

addition reaction
加成反應

CH₃          CN
H—C   +   H
O

→

CH₃
H—C—CN
OH

H
O  hydroxyl group
羥基

dehydration 脫水

RОН → R'COOR + HOH
R'COOH          ester 酯

an aliphatic reaction
脂族反應

an aromatic reaction
芳族反應

ArOH⁺ ⇌ ArO⁻ + H⁺

carbonyl group 羰基

C=O

**carboxyl group** the functional group (p. 185)
—COOH. It forms weak acids and by neutral-
ization it forms salts, e.g. sodium ethanoate.
With alcohols (p. 175) it forms esters (p. 177).

**sulphonate group** the functional group (p. 185)
—SO₂.OH. It occurs almost entirely combined
with aryl groups (p. 180). Aromatic (p. 179)
sulphonic acids are prepared by the action of
concentrated sulphuric acid on benzene (p. 179)
and its homologues; they form salts with alkalis,
and the salts are used as detergents (p. 171).

**nitro group** the functional group (p. 185) —NO₂.
Aliphatic (p. 179) nitro compounds exhibit
(p. 221) tautomerism (p. 184). They are of less
importance than the aromatic (p. 179) nitro
compounds. Benzene and its homologues are
converted to nitro compounds by nitration
(p. 193). Aromatic nitro compounds are
reduced (p. 70) to amines in several stages;
some intermediate compounds are useful
industrially. *See equations opposite.*

**amino group** the functional group (p. 185)
—NH₂. It occurs in primary amines, both
aliphatic (p. 179) and aromatic; in the amides
as part of the amido (p. 179) group; and in the
amino acids. The amino group causes amines
and amino acids to have a basic (p. 46) reaction.

**amido group** the functional group (p. 185)
—CONH₂. It occurs in amides and causes
them to be amphoteric (p. 46) in nature. The
group is dehydrated to form a nitrile (p. 178).

**azo group** the functional group (p. 185)
—N=N—. It occurs in diazo compounds
(p. 180) and is a highly reactive group.

**cyano group** the functional group (p. 185)
—N≡N—. It occurs in the nitriles (p. 178).
It occurs in diazo compounds
(p. 180) and is a highly reactive group. The cyano
group is hydrolyzed (p. 66) to the carboxyl
group and reduced (p. 70) to a primary amine

---

**羧基** —COOH 官能團（第 185 頁），它形成弱酸
並以中和作用方式生成鹽，例如乙酸鈉。與
醇（第 175 頁）作用生成酯（第 177 頁）。

**磺酸基** —SO₂.OH 官能團（第 185 頁），它幾乎都是
與芳基（第 180 頁）結合而存在。芳族（第 179
頁）磺酸可由濃硫酸和苯（第 179 頁）及其同
系物作用製得；芳族磺酸與鹼作用生成鹽，
這些鹽可用作洗滌劑（第 171 頁）。

**硝基** —NO₂ 官能團（第 185 頁），脂族（第 179
頁）硝基化合物顯現（第 221 頁）互變異構現
象（第 184 頁）。其重要性不如芳族（第 179
頁）硝基化合物。苯及其同系物可經硝化作
用（第 193 頁）轉化成硝基化合物。芳族硝基
化合物分幾步還原（第 70 頁）成胺。其中
有一些是工業上很有用的中間化合物。參見
反應式。

**氨基** —NH₂ 官能團（第 185 頁），它存在於脂族
（第 179 頁）和芳族伯胺（第 180 頁）、在醯胺
中作為醯胺基（第一部分、也存在於氨基酸中
以及氨基酸中。氨基使胺和氨基酸具有鹼性（第 46 頁）反應。

**醯胺基：醯胺基** —CONH₂ 官能團（第 185 頁），
它存在於醯胺中並使醯胺具有兩性的（第 46
頁）性質。醯胺基脫水則形成腈（第 178
頁）。

**偶氮基** —N=N— 官能團（第 180 頁），是一個高易反
應的重氮化合物（第 180 頁）中，是一個高易反
應的基團。

**氰基** —C≡N 官能團（第 185 頁），它存在於
腈（第 178 頁）中。氰基（第 185 頁），它存在於
重氮化合物（第 180 頁）中，是一高易反應的
基。氰基可水解（第 66 頁）成羧基，還原成脂
酸及還原（第 70 頁）為伯胺。

---

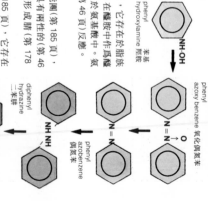

reduction of
aromatic nitro compound

| | |
|---|---|
| NO₂ | nitrobenzene 硝基苯 |
| NO | nitrosobenzene 亞硝基苯 |
| NH.OH | phenyl hydroxylamine 苯基羥胺 |
| NH₂ | phenylamine (aniline) 苯胺 |

方族硝基化
合物的還原

硝基 — nitrobenzene 硝基苯

亞硝基 — nitrosobenzene 亞硝基苯

苯基羥胺 — phenyl hydroxylamine

氨基 — phenylamine (aniline) 苯胺

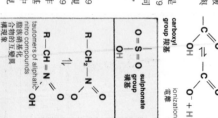

phenyl azoxybenzene 氧化偶氮苯

phenyl azobenzene 偶氮苯

diphenyl hydrazine 二苯肼

偶氮基 — N=N— 

---

**carboxyl group** 羧基

$$-C\overset{\displaystyle O}{\underset{\displaystyle OH}{\diagup}}$$

ionization 電離

$$-C\overset{\displaystyle O}{\underset{\displaystyle O^-}{\diagup}} + H^+$$

**sulphonate group** 磺酸基

$$O=S=O$$
$$OH$$

**nitro compound** 硝基化合物

tautomers of aliphatic
nitro compounds
脂族硝基化
合物的互變異
構現象

$$R-CH_2-N\overset{\displaystyle O}{\underset{\displaystyle O}{\diagdown}}$$

$$\Updownarrow$$

$$R-CH=N\overset{\displaystyle O}{\underset{\displaystyle OH}{\diagdown}}$$

**chloro group** the functional atom (p.185) –Cl. It occurs in the alkyl chlorides (p.177) and the aromatic chloro compounds. The alkyl chlorides are reactive, taking part in substitution (p.188) and elimination (p.189) reactions. The aromatic chloro compounds are relatively (p.232) inert (p.19). The presence of a nitro group (p.186) in the aromatic hydrocarbon increases the reactivity of the chloro group.

**bromo group** the functional atom (p.185) –Br. It occurs in the alkyl bromides, e.g. bromoethane CH₃CH₂Br, and in the aromatic (p.179) bromo compounds. The alkyl and aromatic bromo compounds are more reactive than the corresponding (p.233) chloro compounds, e.g. bromobenzene is more reactive than chlorobenzene.

**iodo group** the functional atom (p.185) –I. It occurs in the alkyl (p.180) iodides, e.g. iodoethane CH₃CH₂I, and in the aromatic (p.179) iodo compounds, e.g. iodobenzene C₆H₅I. The alkyl and aromatic iodo compounds are more reactive than the corresponding bromo compounds (↑).

**chromophore** (n) any chemical group which causes a compound to have a distinctive (p.224) colour. In synthetic (p.200) organic dyes (p.162), such groups as the azo group cause the colour of a compound. **chromophoric** (adj).

**auxochrome** (n) a chemical group which deepens the colour of a compound, where red is taken as the deepest colour of the spectrum and violet as the least deep, e.g. the addition of an auxochrome can change a blue substance to a green substance. In addition, an auxochrome intensifies (p.230) the colour. The amino group (↑) is an example of an auxochrome. **auxochromic** (adj).

**leuco base** a compound formed from a dye (p.162) by reduction (p.70); it is colourless. An insoluble dye is converted to a leuco base which is solubic in alkalis; the alkaline solution is used for dyeing and the dye colour returned by oxidation, e.g. indigo is converted to a leuco base and the colour returned by subsequent oxidation.

**leuco compound** a name for leuco base (↑).

---

**氯基** 一Cl 官能原子 (第 185 頁)，存在於烷基氯 (第 177 頁) 和芳族氯代化合物中。烷基氯基是 易反應的，可參與取代反應 (第 188 頁) 和消除 (第 189 頁) 反應。芳族氯代化合物相對地 (第 232 頁) 是惰性的 (第 19 頁)。芳族烴中的硝基 (第 186 頁) 可增強氯基的活性。

**溴基** 一Br 官能原子 (第 185 頁)，存在於烷基 溴，例如溴乙烷 CH₃CH₂Br 中，也存在於芳 族 (第 179 頁) 溴代化合物中。烷基溴、芳族溴 代化合物都比相應的 (第 233 頁) 氯代化合物 易反應，例如溴代苯比氯代苯易反應。

**碘基** 一I 官能原子 (第 185 頁)，存在於烷基 (第 180 頁) 碘，例如碘乙烷 CH₃CH₂I 中，也存在 於芳族 (第 179 頁) 碘代化合物，如碘代苯 C₆H₅I。烷基、芳族碘代化合物比相應的溴化 合物 (↑) 易反應。

**發色團：發色基** (名) 使某種化合物具有特殊 (第 224 頁) 顏色的任何化學基團。在合成的 (第 200 頁) 有機染料 (第 162 頁) 中，像偶氮 基這類基團可使化合物產生顏色。(形容詞 爲 chromophoric)

**助色基：助色團** (名) 加深化合物顏色的一種化學基 團，其中將紅色取作色譜中最深之顏色、紫 色作爲最淺之顏色。例如加入助色團可使藍 色物質變爲綠色物質。此外，助色團可強化 (第 230 頁) 顏色。氨基 (↑) 爲助色團之一 例。(形容詞爲 auxochromic)

**無色母體** 利用還原作用 (第 70 頁) 自一種染料 (第 162 頁) 生成的化合物，它是無色的。不 溶性染料可以轉變爲可溶於鹼液的無色母體，鹼性溶液可用於染色並藉氧化作用回復染料 的顏色。例如靛藍變成無色母體並藉隨後 的氧化作用回復其顏色。

**隱色化合物** 無色母體 (↑) 的另一名稱。

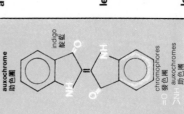

bromoethane
溴乙烷
CH₃CH₂

iodobenzene
碘代苯

auxochrome
助色團

indigo
靛藍

chromophores
發色團

auxochromes
助色團

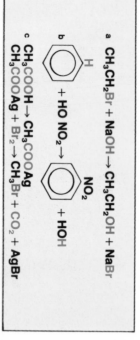

a CH₃CH₂Br + NaOH → CH₃CH₂OH + NaBr

b 
$$+ HO\,NO_2 \rightarrow + HOH$$

(benzene ring with H) + HO NO₂ → (benzene ring with NO₂) + HOH

c CH₃COOH → CH₃COOAg
CH₃COOAg + Br₂ → CH₃Br + CO₂ + AgBr

**substitution** (n) a process in which an atom or a functional group in an organic (p.55) compound is replaced, directly or indirectly, by another atom or functional group. Substitution is one of the most important organic reactions. Examples of substitution are: (a) substituting a hydroxyl group for a halogen in an aliphatic (p.179) compound; (b) substituting a nitro group (p.186) for a hydrogen atom in benzene (p.179); (c) substituting a bromine atom for a carboxyl group in an aliphatic compound. **substitute** (v), substitutional (adj).

**addition** (n) a process in which two substances react to form only one substance. In organic reactions, addition usually takes place across a double bond (p.181), e.g. hydrogen chloride adds across the double bond of ethene.

**addition 加成**
$$\begin{array}{ccc}
CH_2 & & CH_2Cl \\
\| & + \ H \rightarrow & | \\
CH_2 \ Cl & & CH_3
\end{array}$$

**hydrogenation** (n) an addition process (↑) in which hydrogen is added to a molecule in the presence of a suitable catalyst (often finely divided nickel) and at a suitable temperature and pressure. Hydrogenation converts unsaturated (p.185) compounds into saturated compounds. **hydrogenate** (v).

**hydrogenation 加氫**
$$\begin{array}{ccc}
CH_2 & & CH_3 \\
\| & + \ H_2 \rightarrow & | \\
CH_2 & & CH_3
\end{array}$$

取代作用：置換作用（名） 有機（第55頁）化合物中的一個原子或官能團代替另一個原子或官能團代替的過程。取代作用是最重要的有機反應之一。取代作用的例子有：(a) 羥基取代脂族（第179頁）化合物中的鹵素；(b) 硝基（第186頁）取代苯（第179頁）中的氫原子；(c) 溴原子取代脂族化合物中的羧基。（動詞爲 substitute，形容詞爲 substitutional）

加成作用（名） 兩種物質反應只生成一種物質的過程。在有機反應中，加成作用通常是發生在一個雙鍵（第181頁）的兩端。例如，氯化氫加入乙烯的雙鍵的兩端。（形容詞爲 additional）

氫化作用（名） 在有合適催化劑（通常爲磨碎之鎳粉）及合適的溫度和壓力下，氫化作用使不飽和的（第185頁）化合物變化成飽和的化合物。（動詞爲 hydrogenate）

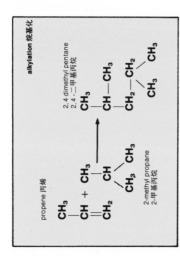

**alkylation 烷基化**

propene 丙烯

$$CH_2=CH-CH_3 + CH_3-CH\overset{CH_3}{\underset{CH_3}{}} \longrightarrow \overset{CH_3}{CH_3-CH-CH_3}$$

2-methyl propane
2-甲基丙烷

$$CH-CH_3 \overset{CH_3}{} \quad CH_2-CH_2 \quad CH_3 \quad CH_3$$

2, 4 dimethyl pentane
2, 4-二甲基戊烷

**alkylation** (*n*) an addition process (↑) in which an alkane (p.172) undergoes (p.213) addition across the double bond of an alkene (p.173) to form a branched chain (p.182) alkane, e.g. propene and 2-methylpropane form 2,4,dimethylpentane. Alkylation is also the substitution (↑) of an alkyl (p.180) group in place of a hydrogen atom. *See Friedel-Crafts reaction (p.199).*

**elimination** (*n*) the removal (p.215) of atoms from a molecule; this usually results in the formation of a double bond, e.g. the removal of the elements of hydrogen and a halide (p.50) from an alkyl halide (p.177) to form an alkene. This is done by the action of a hot, concentrated solution of potassium hydroxide in ethanol (called alcoholic potash), e.g. bromoethane is converted to ethene. Other types of elimination include dehydrogenation (p.190) and dehydration (p.190).

烷基化（名） 烷烴（第 172 頁）在烯烴（第 173 頁）雙鍵的兩端進行（第 213 頁）加成作用生成支鏈（第 182 頁）烷烴的加成過程（↑）。例如丙烯和 2- 甲基丙烷加成生成 2, 4- 二甲基戊烷。烷基化也是烷基（第 180 頁）取代（↑）氫原子位置的作用。參見 " 弗立得 - 克拉夫脫反應 "（第 199 頁）。

消除（名） 由一個分子中除去一個雙鍵。例如，從一個鹵烷（第 177 頁）分子中消除氫和鹵素（第 50 頁）元素而形成一個烯烴分子。這是在熱的氫氧化鉀乙醇濃溶液（稱為鹼性鉀鹼）中進行的。例如，溴乙烷轉化成乙烯。其他的消除型式包括脫氫作用（第 190 頁）和脫水作用（第 190 頁）。

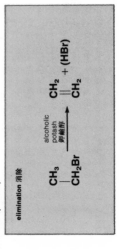

**elimination 消除**

$$CH_3-CH_2Br \overset{\text{alcoholic potash}}{\underset{\text{鉀鹼醇}}{\longrightarrow}} CH_2=CH_2 + (HBr)$$

**dehydrogenation** (*n*) the removal (p.215) of combined hydrogen from an organic (p.55) compound; the process uses a suitable catalyst (e.g. aluminium and chromium oxides) at a high temperature (e.g. 600°C). For example, butane is dehydrogenated to butene. This is an elimination (p.189) process. **dehydrogenate** (*v*).

**dehydration** (*n*) the removal (p.215) of the elements of water from an organic (p.55) compound. For example, (a) the action of excess concentrated sulphuric acid on ethanol produces ethoxyethane (diethyl ether); (b) the action of phosphorus pentoxide on ethanamide produces ethanenitrile. This is an elimination (p.189) process. **dehydrate** (*v*).

**hydrolysis²** (*n*) the reverse process of esterification (↓) and of condensation (↓), and also the substitution of a hydroxyl group into a compound. Hydrolysis of esters is catalyzed by hydrogen ions in acids or by hydroxyl ions in alkalis. Condensation products can be hydrolyzed by boiling with dilute acids. Water alone hydrolyzes alkyl halides (p.177). **hydrolyze** (*v*).

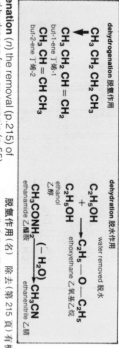

dehydrogenation 脫氫作用

$$CH_3\ CH_2\ CH_2\ CH_3$$
↓
$$CH_3\ CH_2\ CH = CH_2$$
but-1-ene 丁烯-1
$$CH_3\ CH = CH\ CH_3$$
but-2-ene 丁烯-2

dehydration 脫水作用   water removed 脫水

$$C_2H_5OH$$
+
$$C_2H_5OH$$
ethanol
乙醇

$$\longrightarrow C_2H_5-O-C_2H_5$$
ethoxyethane 乙氧基乙烷

$$CH_3CONH_2 \xrightarrow{\ (-H_2O)\ } CH_3CN$$
ethanamide 乙醯胺        ethanenitrile 乙腈

**脫氫作用** (名) 除去 (第 215 頁) 有機 (第 55 頁) 化合物中的化合氫；此過程是在高溫 (例如 600°C) 下使用一種適當的催化劑 (例如氧化鋁和氧化鉻) 進行的。例如丁烷脫氫成為丁烯。這是一種消除 (第 189 頁) 作用過程。(動詞為 dehydrogenate)

**脫水作用** (名) 除去 (第 215 頁) 化合物之水的元素。例如：(a) 過量之濃硫酸對乙醇作用生成乙氧基乙烷 (即二乙醚)；(b) 五氧化二磷對乙醯胺作用生成乙腈。這是一種消除 (第 189 頁) 作用過程。(動詞為 dehydrate)

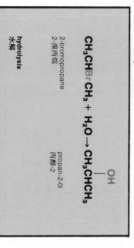

hydrolysis
水解

$$CH_3CHBrCH_3 + H_2O \longrightarrow CH_3CHCH_3$$
$$|$$
$$OH$$

2-bromopropane
2-溴丙烷

propan-2-ol
丙醇-2

**水解作用** (名) 酯化作用 (↓) 和縮合作用 (↓) 之逆過程，也是羥基取代入化合物之作用。酯中之氫離子以及鹼中之羥離子可催化酯的水解作用。縮合的產物可用稀鹽酸煮沸可水解。自能水解鹵烷類 (第 177 頁)。(動詞為 hydrolyze)

**condensation**[3] (*n*) a reaction in which two
organic (p.55) compounds combine to form
one compound with the elimination of water,
hydrogen chloride or ammonia. The most
common compound eliminated is water.
Examples of condensation are (a) aldehydes
and hydroxylamine condensing to aldoximes,
e.g. ethanal oxime $CH_3CH=NOH$; (b)
polymerization (p.207) with condensation, as in
the urea-methanal condensation
polymerization.

縮合作用 (名) 兩種有機 (第 55 頁) 化合物以消
除水、氯化氫或氨的方式結合生成為水的一種縮合作
用的反應。最常消除的化合物為水。縮合作
用的例子有：(a) 醛和羥胺縮合成醛肟，例
如乙醛肟 ($CH_3CH = NOH$)；(b) 以縮合作用
方式聚合 (第 207 頁)，如同脲、甲醛的縮聚
作用。

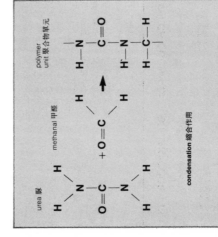

urea 脲　　methanal 甲醛　　polymer
unit 聚合物單元

**condensation 縮合作用**

**esterification** (*n*) the process of converting an
acid into an ester (p.177). The reaction is
reversible, so the water formed in the reaction
is removed by concentrated sulphuric acid to
move the equilibrium mixture (p.150) in the
direction of the ester. For example, the
esterification of ethanoic acid with methanol:
$CH_3COOH + H\ CH_2OH \rightleftharpoons CH_3COOCH_3 + H_2O$
This is a form of condensation (↑).

酯化作用 (名) 酸轉化成酯 (第 177 頁) 的過程。
反應是可逆的，故用濃硫酸除去反應中生成
的水，以使平衡混合物 (第 150 頁) 向酯化的一
邊移動。例如乙酸與甲醇的酯化作用：
$CH_3COOH + HCH_2OH$
$\rightleftharpoons CH_3COOCH_3 + H_2O$
這是縮合作用 (↑) 的一種形式。

**saponification** (n) hydrolysis (p.190) of an ester using boiling aqueous sodium or potassium hydroxide. The alkali combines with and removes (p.215) the acid from the products, and shifts the equilibrium mixture (p.150) towards the hydrolyzed product. **saponify** (v).

**halogenation** (n) the process of substituting (p.188) a halogen (p.117) atom for a hydrogen atom in a molecule of an organic (p.55) compound, e.g. the halogenation of phenol (p.180).

**chlorination** (n) halogenation (↑) when chlorine is the halogen. Chlorinating agents for aliphatic (p.179) compounds include phosphorus penta-chloride, phosphorus trichloride and sulphur dichloride oxide (thionyl chloride, SOCl₂). For aromatic (p.179) compounds, a 'halogen car-rier', such as aluminium chloride or powdered iron, is used and chlorine is passed through at room temperature, causing substitution in the benzene ring (p.179). **chlorinate** (v).

**bromination** (n) halogenation (↑) when bromine is the halogen. For aliphatic (p.179) compounds, red phosphorus and bromine are used as brominating agents. For aromatic (p.179) compounds, a 'halogen carrier' is used as for chlorination (↑). **brominate** (v).

---

$$CH_3COOCH_3 + NaOH \rightarrow CH_3OH + CH_3COONa$$

methyl ethanoate —— methanol + sodium ethanoate
乙酸甲酯 甲醇 乙酸鈉

**saponification**
皂化作用

---

$$CH_3CH_2COOH + Cl_2 \rightarrow CH_3CHClCOOH + HCl$$

propanoic acid 2-chloro propanoic acid
丙酸 2-氯丙酸

**chlorination**
氯化作用

$$CH_3 CO CH_3 + Br_2 \rightarrow CH_3COCH_2Br + HBr$$

propanone bromopropanone 溴丙酮
丙酮

**bromination**
溴化作用

---

**皂化作用** (名) 用沸騰使氫氧化鈉或氫氧化鉀水溶液使酯水解 (第 190 頁)。鹼與酯類結合並除去 (第 215 頁) 產物中的酸，使平衡中的混合物向水解產物一邊移動。(動詞為 saponify)

**鹵化作用** (名) 鹵素 (第 117 頁) 原子取代 (第 188 頁) 有機 (第 55 頁) 化合物分子中之氫原子的過程，例如酚的鹵化作用。

**氯化作用** (名) 鹵素為氯之鹵化作用 (↑)。脂族 (第 179 頁) 化合物用的氯化劑包括五氯化磷、三氯化磷和氯氧化硫 (即亞硫醯氯，SOCl₂)。芳族 (第 179 頁) 化合物則用氯化鋁或鐵粉之類 " 鹵素載體 "，在室溫下讓氯氣通過而引致苯環 (第 179 頁) 發生取代作用。(動詞為 chlorinate)

**溴化作用** (名) 鹵素為溴之鹵化作用 (↑)。脂族 (第 179 頁) 化合物用紅磷和溴作為溴化劑。芳族 (第 179 頁) 化合物的氯化作用 (↑) 則用 " 鹵素載體 "。(動詞為 brominate)

**iodination** (*n*) halogenation (↑) when iodine is the halogen. For aliphatic (p.179) compounds, red phosphorus and iodine are used as iodinating agents. For aromatic (p.179) compounds, iodine in the presence of mercury (II) oxide is used.

碘化作用（名） 鹵素爲碘之鹵化作用（↑）。脂族（第 179 頁）化合物用紅磷和碘作爲碘化劑，芳族（第 179 頁）化合物則在氧化汞 (II) 存在下用碘作碘化劑。

$$2 \langle \bigcirc \rangle + I_2 \longrightarrow 2 \langle \bigcirc \rangle I$$

iodination 碘化作用

**reduction**[2] (*n*) the reduction of organic (p.55) compounds is either hydrogenation (p.188) or the removal of oxygen, e.g. the reduction of ethanoic acid to ethanol or to ethane.

還原作用（名） 有機（第 55 頁）化合物的還原作用是指氫化作用（第 188 頁）或者指脫氧，例如乙酸還原爲乙醇或還原爲乙烷。

reduction 還原作用

$$CH_3COOH \xrightarrow{LiAlH_4} CH_3CH_2OH$$

$$CH_3COOH \xrightarrow{HI} CH_3 CH_3$$

$$\begin{array}{c} CH_2 \\ \| \\ CH_2 \end{array} + H_2 \longrightarrow \begin{array}{c} CH_3 \\ | \\ CH_3 \end{array}$$

reduction (hydrogenation)
還原作用（氫化作用）

**nitration** (*n*) the substitution (p.188) of a nitro group (p.186) in an organic (p.55) compound, using concentrated nitric acid. With aliphatic (p.179) compounds, a high temperature is needed. With aromatic (p.179) compounds, the reaction takes place at much lower temperatures; concentrated sulphuric acid is used to quicken the reaction.

硝化作用（名） 用濃硝酸，使硝基（第 186 頁）在有機（第 55 頁）化合物中所起的取代作用（第 188 頁）。在脂族（第 179 頁）化合物必須用高溫，而結合在芳族（第 179 頁）化合物時，反應可在更低的溫度下進行，可用濃硫酸以加速反應。

**sulphonation** (*n*) the substitution (p.188) of a sulphonate group (p.186) in place of a hydrogen atom in an organic (p.55) compound. Aromatic (p.179) compounds undergo (p.213) sulphonation readily, but aliphatic (p.179) hydrocarbons are extremely difficult to sulphonate. **sulphonate** (*v*).

磺化作用（名） 以磺基的（第 186 頁）取代作用（第 188 頁）替換有機（第 55 頁）化合物之氫原子。芳族（第 179 頁）化合物易經受（第 213 頁）磺化作用，而脂族（第 179 頁）烴則極難磺化。（動詞爲 sulphonate）

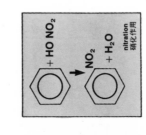

sulphonation
磺化作用

benzene 苯

oleum 發烟硫酸

SO₂OH

benzenesulphonic
acid 苯磺酸

+ HO NO₂

NO₂

+ H₂O

nitration
硝化作用

**methylation** (*n*) a process in which a methyl group (p.184) is substituted for a hydrogen atom in an organic compound. Methylating agents include dimethyl sulphate, diazomethane and iodomethane. **methylate** (*v*).

**cracking** (*n*) an industrial (p.157) process in which hydrocarbons from petroleum fractions (p.160) are strongly heated under pressure; large molecules are broken up into smaller molecules, e.g. the higher fractions (p.202) are converted to petrol and kerosene. **crack** (*v*).

**fermentation** (*n*) the decomposition of carbohydrates (p.205) caused by enzymatic (p.72) action; the enzymes are produced by yeasts or by bacteria. **ferment** (*v*).

**ortho-para-directing** (*adj*) a functional group (p.185) in a benzene ring (p.179) directs (p.233) the position of further substitution (p.188) in the ring. Some functional groups direct towards the *ortho* and *para* positions, and a second substitution will take place in these positions. An *ortho-para-directing* group activates (p.21) the benzene ring, so the second substitution is faster. The following groups are *ortho-para-directing*: alkyl, amino, hydroxyl, halogeno, methoxy.

**meta-directing** (*adj*) a functional group in a benzene ring (p.233) the position of further substitution in the ring. See *ortho-para directing* (↑). Some groups direct towards the *meta* position, these reduce (p.219) the activity of the ring so that the second substitution is slower and more difficult. The following groups are *meta-directing*: nitro, sulphonate, carboxyl, aldehyde, cyano.

$$C_2H_5COOH + CH_2N_2 \rightarrow C_2H_5COOCH_3 + N_2$$

diazobenzene
重氮苯

methylation
甲基化作用

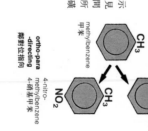

meta-directing
間位指向

NO₂ → NO₂ / NO₂

1,3 dinitro benzene
1,3-二硝基苯

甲基化作用（名）甲基（第 184 頁）取代有機化合物中氫原子的過程。甲基化劑包括硫酸二甲酯、重氮甲烷和碘化甲烷。（動詞為 methylate）

裂解（名）將從石油餾出物（第 160 頁）取得的烴在壓力下強烈加熱的一種工業（第 157 頁）過程；在此過程中大的分子被斷裂成小的分子。例如，將較高沸點之餾出物（第 202 頁）轉化成汽油和煤油。（動詞為 crack）

發酵（名）由酶的（第 72 頁）作用引起碳水化合物（第 205 頁）的分解下用；酶係由酵母或由細菌所產生。（動詞為 ferment）

鄰對位指向：鄰對位定向（形）苯環（第 179 頁）中的官能團（第 185 頁）指示（第 233 頁）環中下一步取代（第 188 頁）的位置。某一些官能團指向鄰位和對位，因而第二個取代位置將發生在這個位置。鄰對位指向基使苯環活化（第 21 頁），因而第二個取代基更快速。烷基、氨基、羥基、鹵代基、甲氧基都是鄰對位指向基。

間位指向：間位定向（形）苯環中的官能團指示下一步取代的位置。參見"鄰對位指向"(↑)。某些基團指向"間位"，這會降低（第 219 頁）苯環的活性，所以第二個取代速度較困難。硝基、磺基、羧基、醛基、氰基都局間位指向基。

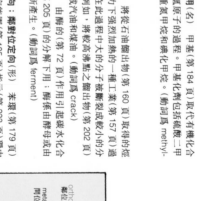

reference point
參攷點

ortho 鄰位
meta 間位
para 對位

2-nitro-methylbenzene
2-硝基甲苯

4-nitro-methylbenzene
4-硝基甲苯

methylbenzene
甲苯

ortho-para-directing
鄰對位指向

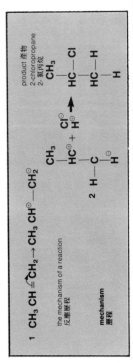

$$1 \quad CH_3 \; CH \rightleftharpoons CH_2 \rightarrow CH_3 \; CH^{\oplus} — CH_2^{\ominus}$$

the mechanism of a reaction
反應歷程

**mechanism**
歷程

$$CH_3 \qquad\qquad CH_3$$
$$HC^{\oplus} — C \qquad Cl^{\ominus} \rightarrow HC — Cl$$
$$\qquad\quad H^{\oplus} \qquad\quad H^{\oplus} + H^{\oplus}$$
$$2 \; H — C \qquad\qquad HC — H$$
$$\qquad\quad H \qquad\qquad\qquad H$$

product 產物
2-chloropropane
2- 氯丙烷

**mechanism** (*n*) the mechanism of a reaction gives an explanation of the way in which a reaction takes place. Reactivity occurs (p.63) because points of electron excess (p.230) or electron deficiency (p.232) appear in an organic (p.55) molecule and chemical attack takes place at these points. The reaction mechanism of the addition of hydrogen chloride to propene, shown above, explains how the reaction takes place and why the particular product is obtained.

**ozonolysis** (*n*) the process of passing ozonized oxygen $(O_2 + O_3)$ through a solution of an alkene containing zinc dust and ethanoic acid. An ozonide, is first formed by ozone adding across the double bond. This decomposes to form two compounds with a carbonyl group (p.185); these compounds may be aldehydes or ketones. Ozonolysis is used to determine (p.222) the position of a double bond in an unsaturated (p.185) molecule, by identifying (p.225) the products.

**歷程；機理；機制** (名) 反應歷程是對反應進行方式的一種解釋。由於在一個有機 (第 55 頁) 分子中出現電子過剩 (第 230 頁) 或電子不足 (第 232 頁) 的位置，並在這些位置發生化學反應，因而出現反應性。氯化氫加成於丙烯的反應歷程如上圖所示，它解釋了反應是怎樣發生以及之所以獲得特定產物的原因。

**臭氧分解** (名) 讓臭氧化的氧氣 $(O_2 + O_3)$ 流過一種含有鋅粉和乙酸的烯烴溶液的方法。首先是臭氧加入雙鍵的兩端形成臭氧化物 (見圖)，接著臭氧化物分解成含羰基 (第 185 頁) 的兩個化合物，此化合物可爲醛或酮。臭氧分解常用於確定 (第 222 頁) 不飽和 (第 185 頁) 分子中雙鍵的位置以便驗證 (第 225 頁) 產物。

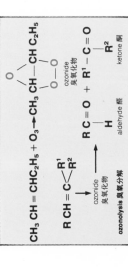

$$CH_3 \; CH = CHC_2H_5 + O_3 \rightarrow CH_3 \; CH \underset{O}{\overset{O}{\underset{\displaystyle O}{\diagup\!\!\!\!\diagdown}}} CH \; C_2H_5$$

ozonide
臭氧化物

$$R \; CH = C{\diagup R^1 \atop \diagdown R^2} \xrightarrow{\qquad} RC = O + R^1 — C = O$$
$$\qquad\qquad\qquad\quad\; H \qquad\qquad R^2$$

ozonide
臭氧化物

aldehyde 醛 ketone 酮

**ozonolysis 臭氧分解**

**Schiff's reagent** a solution containing red fuchsine dye decolorized (p.73) by sulphur dioxide. When an aldehyde is added to Schiff's reagent, the red, or pink, colour returns. Ketones have no effect on the reagent, so the two kinds of compound can be distinguished (p.224).

**Fehling's test** a solution containing copper (II) sulphate, sodium potassium tartrate and sodium hydroxide is added to a solution to test for a sugar (p.205). If a red precipitate appears when the solution is boiled, a reducing sugar is present (p.206) is present (p.217). A red precipitate also indicates that an aldehyde (p.175) is present.

**Benedict's test** a solution containing copper (II) sulphate, sodium citrate and sodium carbonate is added to a solution to test for a sugar (p.205). If a red precipitate appears when the solution is boiled, a reducing sugar (p.206) is present (p.217). A red precipitate also indicates that an aldehyde (p.175) is present.

**biuret test** sodium hydroxide solution is added to a substance, followed by one or two drops of 1% copper (II) sulphate solution. A violet colour shows that protein (p.209) is present.

**iodine test** iodine solution is added to a substance. A deep blue colour indicates that starch (p.207) is present.

**Lassaigne test** a test to identify (p.225) the elements in an organic compound. The compound is fused with sodium metal in a test-tube. The tube is then broken and an aqueous solution of its contents formed. Carbon and nitrogen form sodium cyanide, tested for by iron (II) sulphate. Sulphur forms sodium sulphide, tested for as hydrogen sulphide. A halogen forms a sodium halide, tested for by silver nitrate. Thus C, N, S, Cl, Br, I are tested for.

$$CH_3CH_2CH_2Br + CH_3CH_2CH_2Br + 2Na \longrightarrow CH_3CH_2CH_2CH_2CH_2CH_3 + 2NaBr$$

bromopropane 溴丙烷    hexane 己烷    Wurtz reaction 伍茲反應

席夫試劑 用二氧化硫使紅色的品紅染料脫色（第73頁）。當醛加入席夫試劑時，可使之回復紅色或粉紅色。酮則對試劑無影響。因此可以區別（第224頁）出這兩類化合物。

斐林試驗 在一種溶液中加入含硫酸銅(II)、酒石酸鉀鈉和氫氧化鈉的溶液用以檢驗糖（第205頁）的試驗。如溶液煮沸時出現紅色沉澱，則表明存在有（第217頁）還原糖（第206頁）。紅色沉澱物也表示存在有醛（第175頁）。

本尼迪特試驗 在一種溶液中加入含硫酸銅(II)、檸檬酸鈉和碳酸鈉的溶液用以檢驗糖（第205頁）的試驗。如溶液煮沸時出現紅色沉澱物（第217頁）有還原糖（第206頁），表明存在有醛（第175頁）。

縮二脲試驗：雙縮脲試驗 淋氫氧化鈉溶液加入某種物質中，再加一、二滴1%硫酸銅(II)溶液。出現紫色表示存在有蛋白質（第209頁）。

碘試法 淋碘溶液加入某種物質中，出現深藍色表示存在有澱粉（第207頁）。

拉薩涅試驗 鑑別（第225頁）某種有機化合物所含元素的試驗。將化合物與金屬鈉放在試管內熔化。然後將試管打破並做成其內容物的一種水溶液。碳和氫形成氰化鈉，可用硫酸亞鐵(II)作檢驗。硫生成硫化鈉，可用硫化氫作檢驗。鹵素形成一種鹵化鈉，可用硝酸銀作檢驗。因此可檢驗出C、N、S、Cl、Br和I。

**Wurtz reaction** the production of an alkane (p.172) from an alkyl halide (p.177) by the action of metallic sodium, using an inert (p.19) solvent, e.g.

$$2C_2H_5Br + 2Na \rightarrow C_2H_5{-}C_2H_5 + 2NaBr$$

The general reaction is:

$$2RX + 2Na \rightarrow R\text{-}R + 2NaX.$$

**Fittig reaction** the production of a benzene homologue (p.172) from an aromatic (p.179) halide by the action of metallic sodium in ether.

**Wurtz-Fittig reaction** the production of an alkyl benzene from an aliphatic halide and an aromatic halide by the action of metallic sodium in ether. By-products are also formed.

伍爾茲反應 使用一種惰性（第 19 頁）溶劑，藉金屬鈉的作用，自一種鹵烷（第 177 頁）生產烷烴（第 172 頁）的反應。

$$2C_2H_5Br + 2Na \longrightarrow C_2H_5{-}C_2H_5 + 2NaBr$$

反應通式爲：

$$2BX + 2Na \longrightarrow R\text{-}R + 2NaX$$

菲提希反應 在醚中藉金屬鈉的作用，自某一種芳族（第 179 頁）鹵化物生產苯的同系物（第 172 頁）。

伍茲－菲提希反應 在醚中藉金屬鈉的作用，自某一種脂族鹵化物和芳族鹵化物生產烷基苯。也生成副產物。

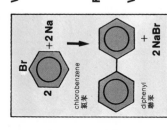

chlorobenzene 氯苯
diphenyl 聯苯

**Fitting reaction**
菲提希反應

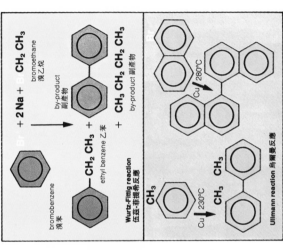

bromobenzene 溴苯
by-product 副產物
ethyl benzene 乙苯
bromoethane 溴乙烷
by-product 副產物

**Wurtz-Fittig reaction**
伍茲-菲提希反應

**Ullmann reaction**
烏爾曼反應

**Ullmann reaction** the production of higher aromatic (p.179) homologues from a bromo or iodo compound by the action of copper powder. *See equation below.* The general reaction is $2ArX \rightarrow Ar\text{-}Ar$ where Ar is an aryl (p.180) group and X is iodine or bromine.

烏爾曼反應 藉銅粉的作用自溴代或碘化合物生產高級芳族（第 179 頁）同系物。反應通式爲 $2ArX \rightarrow Ar\text{-}Ar$，式中 Ar 爲芳基（第 180 頁），X 爲碘或溴。

**Williamson's synthesis** a process for preparing simple or mixed ethers (p.177) from an alkyl halide (p.177) and the sodium derivative (p.200) of an alcohol, e.g.

$$C_2H_5I + CH_3ONa \rightarrow C_2H_5 - O - CH_3 + NaI$$

Methoxyethane ($CH_3OC_2H_5$) is a mixed ether.

**Grignard reagent** a reagent which is prepared by the action of metallic magnesium on an alkyl halide in ether. An alkyl magnesium halide is formed, and this is the reagent, e.g.

$$CH_3CH_2Br + Mg \rightarrow CH_3CH_2MgBr$$

A Grignard reagent undergoes (p.213) many different reactions and the different reagents that can be prepared are used in the synthesis of many organic compounds. The symbol R is used in a Grignard reagent to represent any alkyl group (p.180) and the symbol X to represent a halogen (p.117). Grignard reagents can also be prepared using aryl (p.180) halides, e.g. $C_6H_5Br + Mg \rightarrow C_6H_5MgBr$. Only bromo and iodo compounds can be used for aryl halides.

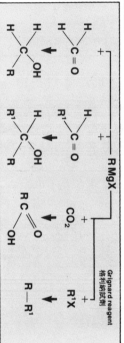

**Sandmeyer reaction** the use of copper (I) chloride in concentrated hydrochloric acid to convert an aqueous diazonium (p.180) chloride to a chloro-compound. A similar reaction takes place with copper (I) bromide and a diazonium bromide. A copper salt is unnecessary for the corresponding iodo compound.

**Gattermann reaction**[1] the use of copper powder to convert a diazonium salt to a halogeno compound, e.g. $C_6H_5 - N^+ \equiv N Cl^- \rightarrow C_6H_5Cl$. The yield (p.159) from this reaction is not as good as for the Sandmeyer reaction (↑).

**威廉遜合成法** 自鹵烷（第 177 頁）和醇的鈉衍生物（第 200 頁）製備簡單醚和混合醚（第 177 頁）的方法。例如

$$C_2H_5I + CH_3ONa \rightarrow C_2H_5 - O - CH_3 + NaI$$

甲氧基乙烷（$CH_3 - O - C_2H_5$）是一種混合醚。

**格利納鎂試劑** 由金屬鎂在醚中與鹵烷基起作用製得的一種試劑。可以製得格氏烷基試劑即烷基鹵化鎂。例如：

$$CH_3CH_2Br + Mg \rightarrow CH_3CH_2MgBr$$

格利納鎂試劑可進行（第 213 頁）許多不同的反應而製成不同的反應物。用以製得的許多有機化合物。格利納鎂試劑中符號 R 表示任何烷基（第 180 頁），符號 X 表示鹵素（第 117 頁）。也可用芳基（第 180 頁）鹵製備格利納鎂試劑。例如：

$$C_6H_5Br + Mg \rightarrow C_6H_5MgBr$$

只能用溴代和碘代化合物作為芳基鹵。

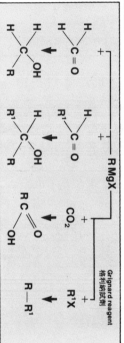

**Grignard reagent 格利納試劑**

**賈特曼反應** 使用銅粉將重氮鹽轉化成鹵代化合物。例如，水的氯化重氮鹽（第 180 頁）轉化成一種氯代化合物。用溴化銅（I）和溴化重氮鹽也可進行類似反應。相應的碘代化合物不需用銅鹽。

**山德梅耶反應** 在濃鹽酸中使用氯化銅（II）將含水的氯化重氮鹽（第 180 頁）轉化成一種氯代化合物，例如，$C_6H_5 - N^+ \equiv N Cl^- \rightarrow C_6H_5Cl$，此反應之產率（第 159 頁）比山德梅耶反應（↑）低。

**Sandmeyer reaction 山德梅耶反應**

**Friedel-Crafts reaction** a process for the alkylation (p.189) of benzene rings (p.179). Alkyl halides (p.177) react with aromatic hydrocarbons,in the presence of anhydrous aluminium chloride.

**Reimer-Tiemann reaction** a reaction between phenols (p.180), aqueous sodium hydroxide and trichloromethane (chloroform), which puts an aldehyde group (p.175) on the benzene ring. The reaction does not produce a high yield (p.159) but is the easiest method of putting an aldehyde group on a benzene ring The substitution generally takes place in the *ortho*-position.

**Gattermann reaction**[2] the combination of phenols with hydrogen chloride or hydrogen cyanide to give a product which is hydrolyzed by water to an aldehyde.

**Cannizzaro reaction** aldehydes (p.175) with no hydrogen atom on the second carbon atom in the aliphatic group, when treated (p.38) with cold concentrated aqueous sodium hydroxide, are both oxidized and reduced (p.70) at the same time; e.g. with methanal the reaction is:

$2HCHO + NaOH \rightarrow HCH_2OH + HCOONa$

The reaction also takes place with aromatic (p.179) aldehydes.

弗瑞德 - 克來福特反應 苯環(第179頁)的過程。鹵烷類(第177頁)和芳烴在無水氯化鋁催化下發生反應。

雷默 - 蒂曼反應 酚類(第180頁)、氫氧化鈉水溶液和三氯甲烷(即氯仿)之間的反應。此反應將酚將醛基(第175頁)代入苯環。雖然反應的產率(第159頁)不高，但卻是將醛基代入苯環的最簡便方法。一般是在鄰位發生此取代作用。

賈特曼反應 酚類與氯化氫或氰化氫合生成的產物與水的水解成醛的反應。

康尼查羅反應 脂族基團中的第二位碳原子上不含氫原子的醛(第175頁)，當用冷的濃氫氧化鈉水溶液處理(第38頁)時，同時發生氧化和還原(第70頁)反應。例如用甲醛時，反應為：
$2HCHO + NaOH \rightarrow HCH_2OH + HCOONa$
用芳族(第179頁)醛也可發生此反應。

Reimer-Tiemann reaction
雷默·蒂曼反應

$OH + CHCl_3 + 3 NaOH \rightarrow$

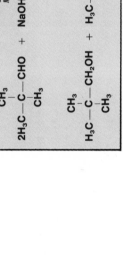

Cannizzaro reaction
康尼查羅反應

$$2H_3C-C-CHO + NaOH \rightarrow H_3C-C-CH_2OH + H_3C-C-COONa$$

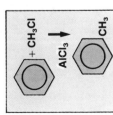

Friedel-Crafts reaction
弗瑞德·克來福特反應

$+ CH_3Cl \xrightarrow{AlCl_3} CH_3$

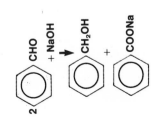

Gattermann reaction
賈特曼反應

**aldol addition** a reaction in which aldehydes and ketones form addition compounds across the double bond (p.181) of the carbonyl group (p.185). The original reaction was:

$$CH_3CHO + CH_3CHO \rightleftharpoons$$
$$CH_3CH(OH)CH_2CHO \text{ (aldol)}$$

The reacting molecules need not be identical, i.e. the addition product is formed only in small quantities. A trace (p.20) of potassium hydroxide catalyzes (p.72) the reaction.

**Kolbe electrolytic reaction** a process for preparing alkanes (p.172) by the electrolysis (p.122) of the sodium salts of carboxylic acids. The carboxylic group must be at the end of a carbon chain. A cold concentrated aqueous solution of a salt is electrolyzed using platinum electrodes.

$$CH_3COONa \rightarrow CH_3{-}CH_3 + CO_2 + NaOH$$

Only alkanes with an even number of carbon atoms are prepared by this reaction.

**derivative** (n) a compound made by substitution of one or more atoms, or groups of atoms, in an original substance. Because the compound is a derivative of an original substance, it has the same kind of structure. For example, taking propane as an original substance, then its derivatives include propanol, propanal, propanoic acid and chloropropane. Derivatives are used to determine structure or composition. If the original structure is known, then the structure of the derivative is known. Ozonolysis (p.195) is frequently used to prepare derivatives. The derivatives are identified and from that the original structure of an alkene, including the position of the double bond (p.181), is found. **derive** (v).

**synthesis** (n) (1) making a compound by chemical means from its elements. (2) making a compound by a series of chemical processes. e.g the synthesis of ascorbic acid (vitamin C). A simple synthesis is shown in the equations opposite. **synthetic** (adj). **synthesize** (v).

**synthetic** (adj) describes any compound made by synthesis (↑) from simpler compounds. A synthetic substance is one that generally replaces a naturally occurring (p.154) substance.

醛醇加成作用　醛和酮在聚基（第 185 頁）雙鍵（第 181 頁）的兩端形成成化合物的反應。初始反應爲：

$$CH_3CHO + CH_3CHO$$
$$\rightleftharpoons CH_3CH(OH)CH_2CHO \text{ (醛醇)}$$

反應中的分子無需相同。平衡基本上是向反應的左邊，亦即只生成少量加成產物，可用氫氧化鉀催化（第 72 頁）此反應。

科爾伯電解反應　將羧酸的鈉鹽電解（第 122 頁）以製備烷屬烴（第 172 頁）的一種方法。羧基必須是在碳鏈的一端。利用鉑電極電解這些鹽的冷而濃縮的鹽水溶液。

$$CH_3COONa \longrightarrow CH_3{-}CH_3 + CO_2$$
$$+ NaOH$$

此反應只能製得含偶數碳原子的烷屬烴。

衍生物（名）　在某種原始物質中取代一個或多個原子或某原子團而製成的化合物。因爲此物化合物是原始物質的衍生物，故具有相同的結構形式。例如，取丙烷作爲一種原始物質，則其衍生物包括丙醇、丙醛、丙酸和氯丙烷。衍生物可用於確定結構和組成。如已知其原始結構，則也知其衍生物的結構。臭氧解（第 195 頁）常用於製備各種衍生物。鑒別出各種衍生物由此鑒定一種烯烴的原始結構，包括其雙鍵（第 181 頁）的位置。（動詞爲 derive）

合成法（名）　(1)用化學方法由其組成元素製備一種化合物；(2)通過一系列化學過程製造一種化合物。例如：抗壞血酸（維生素 c）的合成。右面的反應式表示簡單的合成法。（形容詞爲 synthetic，動詞爲 synthesize）

合成的（形）　描述任何化合物係以合成法（↑）於較簡單化合物製成者。合成的物質通常用以於取代一種天然存在（第 154 頁）的物質。

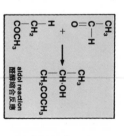

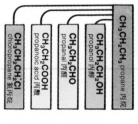

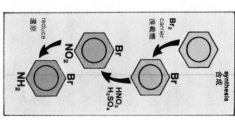

**aldol reaction** 醛醇綜合反應

**derivatives** 衍生物

CH₃CH₂CH₃ propane 丙烷

CH₃CH₂CH₂OH propanol 丙醇

CH₃CH₂CHO propanal 丙醛

CH₃CH₂COOH propanoic acid 丙酸

CH₃CH₂CH₂Cl chloropropane 氯丙烷

**synthesis** 合成

Br₂ carrier 溴載體

reduce 還原

重整：重組 (名) 直鏈烷屬烴 (第 172 頁) 或脂環族化合物 (第 179 頁) 脫氫 (第 190 頁) 生成芳族烴或支鏈烴的過程。

回流的 (形) 液體或氣體逆原來的 (第 220 頁) 方向流回。(名詞爲 reflux，動詞爲 reflux)

回流冷凝器 裝在燒瓶上方的一種冷凝器 (第 28 頁)，可使燒瓶加熱時產生的蒸汽冷凝並流回燒瓶中。此作用可防止 (第 216 頁) 燒瓶煮乾並使有機反應物在被加熱時有充足 (第 231 頁) 時間進行反應，因爲反應物不會從燒瓶逸出。

蒸餾釜：蒸餾器 (名) 工業上用於蒸餾 (第 33 頁) 之一種設備 (第 23 頁)，通常用金屬製成，可產生 (第 62 頁) 大量蒸餾液 (↓)。

餾出液：蒸餾液 (名) 從蒸餾 (第 33 頁) 過程冷凝所取得的液體。

分餾 用一支分餾柱 (↓) 將蒸餾瓶 (第 28 頁) 瓶頸加長 (第 213 頁) 進行蒸餾 (第 33 頁) 的方法。將燒瓶出來的蒸汽冷凝並流回燒瓶中。到達蒸餾柱頂部的蒸汽才能進入冷凝器 (第 28 頁) 並作爲餾出液 (↑) 收集。此法用於分離沸點接近的液體。分餾柱可有高度不同的幾個出口，以便分離和收集在不同溫度下冷凝 (第 11 頁) 的不同蒸汽。工業上 (第 157 頁) 上也採用這種蒸餾方式，特別是石油 (第 160 頁) 的蒸餾。

分餾柱：分餾管 蒸餾瓶瓶頸的加長 (第 213 頁) 部分，在柱頂端有一個出口 (第 215 頁) 供蒸汽通入冷凝器 (第 28 頁)，並使之在一特定溫度下冷凝成液體。參見 "分餾" (↑)。

**reforming** (*n*) a process in which straight chain alkanes (p.172) or alicyclics (p.179) are dehydrogenated (p.190) and form aromatic or branched chain hydrocarbons.

**reflux** (*ad*) of a liquid or gas, the flowing back in an opposite direction from its original (p.220) direction. **reflux** (*n*), **reflux** (*v*).

**reflux condenser** a condenser (p.28) fitted above a flask so that the vapour, formed by heating the flask, is condensed and flows back into the flask. This action prevents (p.216) the flask boiling dry and allows organic reactants sufficient (p.231) time to react while being heated, as the reactants do not escape from the flask.

**still** (*n*) an apparatus (p.23) used in industry for distillation (p.33). It is generally made of metal and produces (p.62) large quantities of distillate (↓).

**distillate** (*n*) the liquid obtained as the result of condensation from a process of distillation (p.33).

**fractional distillation** a process of distillation (p.33) in which the neck of a distillation flask (p.28) is extended (p.213) by a fractionating column (↓). Vapour from the flask condenses and falls back into the flask. Only vapour reaching the top of the column passes to a condenser (p.28) and is collected as a distillate (↑). This process is used to separate liquids which have boiling points close together. A column can have exits at different heights, allowing vapours which condense (p.11) at different temperatures to be separated and collected. This type of fractional distillation is used in industry (p.157), especially in the distillation of petroleum (p.160).

**fractionating column** an extension (p.213) of the neck of a distillation flask with an exit (p.215) at the top of the column, to pass the vapour to a condenser (p.28). An industrial fractionating column has several exits to take away vapour condensing to a liquid at a particular temperature. See *fractional distillation* (↑).

**refluxing** 回流

no liquid or vapour escapes
無蒸汽或液體逸出

**reflux** 回流

reflux condenser
回流冷凝器

cold water
冷水

vapour condenses and flows back into flask
蒸汽冷凝並流回燒瓶中

organic liquids heated reaction takes place
有機液體加熱時發生反應

heat 加熱

flask 燒瓶

thermometer 溫度計

to condenser and distillation
往冷凝器和蒸餾

**fractionating column** 分餾柱

**fractional distillation** 分餾作用

glass beads 玻璃珠

mixture of liquids with boiling points close together
沸點接近的液體混合物

flask 燒瓶

heat 加熱

**fraction** (n) the distillate (p.201) collected at a particular temperature from a fractionating column (p.201), e.g. in the distillation of petroleum, four main fractions are obtained. These fractions can be fractionally distilled again to obtain a better separation (p.34) of the different liquids.

餾分：餾出物（名） 在某一特定溫度下從分餾柱（第 201 頁）收集得的餾出液體（第 201 頁）。例如石油蒸餾時獲得四個主要餾分，這些餾分可以再次分餾以使不同的液體更好地分離（第 34 頁）。

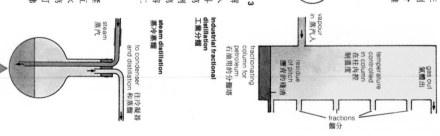

**liquid extraction** the removal (p.215) of a substance, dissolved in one solvent, by using a second solvent. The second solvent is added to the original (p.220) solution in a separating funnel (p.27). The two solvents must be immiscible (p.18). The solute (p.86) dissolves in both solvents and the two liquids are separated (p.34) by a separating funnel; e.g. ethanol is extracted from ethoxyethane (diethyl ether) by adding a concentrated solution of sodium chloride, and running off the aqueous layer in which the ethanol is dissolved.

液體萃取 轉使用第二溶劑（第 215 頁）溶解於一種溶劑中的某種物質。將第二溶劑加入一個盛有原始（第 220 頁）溶液的分液漏斗（第 27 頁）中。溶質（第 86 頁）溶解在這兩種溶劑中，用一個分液漏斗將此兩種液體分離（第 34 頁），用一個分液漏斗將此兩種液體分離（第 34 頁）。例如藉加入濃氯化鈉溶液並放出溶解有乙醇的水層，從而自乙氧基乙烷（即二乙醚）萃取得乙醇。

original solution 原始溶液

liquid extraction 液體萃取

solvent 溶劑

substances dissolved in solvent 溶解在溶劑中的物質

**steam distillation** steam is blown through the heated mixture of the products of an organic (p.55) reaction. If a liquid product of the reaction is immiscible (p.18) with water, then the vapour of the product together with steam passes into a condenser and the condensed liquid (p.201) contains water and the condensed liquid, e.g. phenylamine (aniline) can be steam distilled. The advantage of steam distillation is that the product is obtained at a temperature below its boiling point.

蒸汽蒸餾 使水蒸汽吹過有機（第 55 頁）反應產物的熱混合合物。如某反應的液體產物和水互不溶混（第 18 頁），那麼將該產物的蒸汽和水蒸汽一起通入一個冷凝器中，而餾出液（第 201 頁）中含有水和冷凝的液體。例如苯胺可在低於沸點的溫度下獲得產物。蒸汽蒸餾的優點是可在低於沸點的溫度下獲得產物。

**vacuum distillation** distillation is carried out (p.157) at a low pressure. This method (p.221) of distillation is used with substances that decompose (p.65) at temperatures below their boiling points. At a lower pressure, a substance boils at a temperature below its boiling point.

真空蒸餾：減壓蒸餾 在低壓力下進行（第 157 頁）的蒸餾。分解（第 65 頁）溫度低於其沸點的物質可用此種蒸餾方法（第 221 頁）。較低壓力使物質在低於其沸點的溫度下沸騰。

vapour in 蒸汽入

gas out 氣體出

temperature controlled in column 在柱內控制溫度

residue of pitch 瀝青的殘渣

fractionating column for petroleum 石油用的分餾柱

**industrial fractional distillation** 工業分餾

fractions 餾分

steam 蒸汽

heat 加熱

to condenser and distillation 往冷凝器和蒸餾

**steam distillation** 蒸汽蒸餾

**dry distillation** the heating of a solid to make it give off (p.41) a vapour (p.11); the vapour is condensed to a liquid. For example, the dry distillation of solid calcium ethanoate (acetate) produces propanone (acetone), which is a volatile (p.18) liquid.

**destructive distillation** the heating of solid or liquid organic (p.55) materials to a temperature which is high enough to decompose (p.65) the material so that a residue and a distillate (p.201) are produced; a gas may be produced as well. Coal undergoes (p.213) destructive distillation. The residue is coke (p.162); the distillate consists of tar (p.162); and coal gas is also produced. Wood can also be destructively distilled.

**spirit** (*n*) (1) a liquid consisting mainly of ethanol (ethyl alcohol) obtained by the distillation (p.33) of fermented (p.194) fruit, potatoes, or cereals. (2) any volatile (p.18) liquid obtained by distillation from natural products such as wood and petroleum; used for combustion (p.58), especially in internal combustion engines. (3) any solvent for fat, gums, paints.

**methylated spirit** a mixture of ethanol (ethyl alcohol) and methanol (methyl alcohol), usually with added colouring material to show the mixture is poisonous. Methylated spirit is used as a fuel (p.160) and as a solvent (p.86).

**absolute alcohol** ethanol (ethyl alcohol) from which all traces (p.20) of water have been removed. This is done by adding small quantities of calcium oxide (quicklime) which removes the water from 96% ethanol obtained by distillation.

to condenser-distillate 收集送住冷凝器的餾出液
solid 固體
heat 加熱
▲ **dry distillation 乾餾**

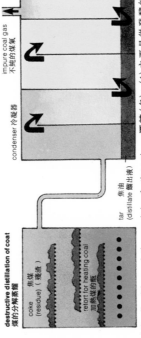

**destructive distillation of coal**
煤的分解蒸餾

coke (residue) 焦煤（殘渣）
retort for heating coal 加熱煤的瓶
impure coal gas 不純的煤氣
condenser 冷凝器
tar 焦油（distillate 餾出液）

**乾餾** 加熱一種固體（使之發出（第 41 頁）蒸汽（第 11 頁）；此蒸汽冷凝成液體。例如，固體的乙酸鈣（即醋酸鈣）乾餾產生丙酮、丙酮係一種揮發性（第 18 頁）液體。

**分解蒸餾** 將固體或液體有機（第 55 頁）材料加熱至高溫（使之分解（第 65 頁）產生殘渣和餾出液（第 201 頁），並產生氣體。煤經受（第 213 頁）分解蒸餾產生殘渣為焦碳（第 156 頁）。餾出液由焦油（第 162 頁）組成，同時產生煤氣。木材也可分解蒸餾。

**酒精**（名）(1) 主要是從發酵的（第 194 頁）水果、馬鈴薯或穀類蒸餾（第 33 頁）取得之乙醇（即乙基醇）組成的一種液體。(2) 藉蒸餾作用從木材和石油之類天然產物取得的任何一種揮發性（第 18 頁）液體，可供燃燒（第 58 頁）之用，尤其是供內燃機用；(3) 供脂肪、樹膠、油漆用的任何一種溶劑。

**甲基化酒精** 乙醇（即乙基醇）和甲醇（即甲基醇）的混合物，通常都加入著色材料以表示此種混合物有毒。甲基化酒精可用作燃料（第 160 頁）和作溶劑（第 86 頁）。

**無水酒精；絕對酒精** 所含微量的（第 20 頁）水已全部除去的乙醇（即乙基醇）。在蒸餾取得的 96% 乙醇中加入少量氧化鈣（即生石灰）以除去水分而製得。

| | | |
|---|---|---|
| **oils** 油類 neutral, viscous, ether soluble immiscible with water 中性、黏稠、溶於乙醚、和水不混溶 | | |
| **mineral oils** 礦物油類 volatile hydrocarbons (alkanes, cycloalkanes) 揮發烴烃 ( 烷類、環烷 ) | **fixed oils** 不揮發油類 non-volatile esters of glycerol 不揮發性 甘油脂 | **essential oils** 香精油類 volatile esters 揮發性酯 |

**essential oil** 香精油
oil of wintergreen 冬青油
methyl 2-hydroxybenzoate 2-羥基苯甲酸
COOCH₃ OH

**oil** (n) (1) any substance, obtained from animals, plants or minerals, which is neutral (p.45), viscous, combustible (p.58) and soluble in ethanol or ether (ethoxyethane) but insoluble in water. The three main kinds of oil are fixed oils (↓), mineral oils and essential oils (↓). Mineral oils are petroleum (p.160). (2) a neutral fat (p.177) which is liquid below 20°C, is called an oil; such substances are fixed oils (↓). **oily** (adj).

**essential oil** a volatile (p.18) oil produced from a plant and occurring (p.154) especially in the flowers of a plant. Essential oils give a plant its characteristic odour (p.15).

**fixed oil** a non-volatile oil occurring in plants. Fixed oils are generally edible and are used in cooking, e.g. coconut oil, peanut oil. The fixed oils are esters (p.177) of glycerol (a trihydric alcohol (p.175) and unsaturated (p.185) or polyunsaturated (p.185) carboxylic acids, although they may contain a small proportion (p.76) of saturated (p.185) carboxylic acids.

**hydrocarbon oil** any oil obtained industrially (p.157) from petroleum (p.160).

**flash point** the lowest temperature at which a volatile (p.18) liquid, especially an oil (↑), gives off (p.41) enough vapour to produce a small flame, but not to catch fire, when touched by a small flame or hot object.

**ignition point** the lowest temperature at which a volatile (p.18) liquid, especially an oil, will catch fire, or burst into flames, when touched by a flame or hot object. **ignite** (v).

**explosive** (adj) describes any substance that ignites (p.32) so quickly that it causes an explosion (p.58).

油 (名) (1)從動、植物或礦物取得的任何中性(第45頁)、黏稠(第18頁)物質。油可燃燒(第58頁)及溶於乙醇或乙醚(即乙氧基乙烷)而不溶於水。油主要有三類:不揮發油(↓)、礦物油和香精油(↓)。礦物油為石油(第160頁)。(2)在溫度20°C以下為液體的中性脂肪(第177頁)稱為油,此種物質屬不揮發油(↓)。(形容詞屬 oily)。

香精油 : 必需油脂 某種植物所產的揮發性(第18頁)油。主要存在(第154頁)於植物的花中。香精油使植物具有其特徵氣味(第15頁)。

不揮發油 : 固定油 植物體中存在的一種非揮發性的油。不揮發油通常可供食用,例如椰子油、花生油。不揮發油為甘油(即三元醇(第175頁)及不飽和(第185頁)或多不飽和(第185頁)羧酸的酯(第177頁),但可能含有小比例(第76頁)的飽和(第185頁)羧酸。

烴油 : 以工業(第157頁)方法從石油(第160頁)取得的任何一種油。

閃點 : 指一種揮發性(第18頁)液體,尤其是油汽,但在小火焰或熱物體接觸時又不致着火的最低溫度。

著火點 : 燃點 指一種揮發性(第18頁)液體,尤其是油,當用火焰或熱物體接觸時會着火或燃着的最低溫度。(動詞屬 ignite)

爆炸性 (形) 描述能快速點燃(第32頁)而引起爆炸(第58頁)的任何物質。

**carbohydrate** (n) an organic (p.55) substance with the general formula $C_x(H_2O)_y$ and a complex molecular structure (p.83). The carbohydrates are divided into two groups, sugars (↓) and polysaccharides (p.207).

**sugar** (n) a colourless, crystalline, water-soluble solid with a sweet taste. Sugars are classified as monosaccharides (↓), disaccharides (p.206), trisaccharides, tetrasaccharides, etc.

**monosaccharide** (n) a sugar (↑) with the formula $C_n(H_2O)_n$, with the most common members of the group having n = 5 for a pentose (↓) or n = 6 for a hexose (↓). The two most common monosaccharides are glucose and fructose (p.206), they are both hexoses (↓). The monosaccharides exhibit stereoisomerism, the isomers are discussed under hexose (↓); monosaccharides are not hydrolyzed to simpler sugars.

**pentose** (n) a monosaccharide (↑) with a molecular formula (p.181) of $C_5H_{10}O_5$. The pentoses can be divided into aldoses (↓) and ketoses (p.206). The most common is ribose.

**hexose** (n) a monosaccharide (↑) with a molecular formula (p.181) of $C_6H_{12}O_6$. The hexoses can be divided into aldoses (↓) and ketoses (p.206). The carbon chains for the two kinds of molecule are shown in the diagram opposite in structural formulae. An aldose (↓) has an aldehyde (p.175) group and a ketose (p.206) has the carbonyl group (p.185) of a ketone (p.176). The hydrogen atoms and hydroxyl groups can have different spatial (p.211) directions, producing stereoisomers (p.183). Asymmetric carbon atoms (p.183) produce optical isomerism (p.183).

**aldose** (n) a monosaccharide (↑) with an aldehyde group (p.175); it is a reducing sugar (p.206). In a hexose, which is also an aldose, called an **aldohexose**, there are four asymmetric carbon atoms (p.183), none of which is combined with the same organic (p.55) group, hence each asymmetric atom will have two enantiomorphs (p.183). This produces 16 possible isomers; only three occur abundantly (p.231) in nature. Aldoses exhibit (p.221) the reactions of aldehydes and alcohols (p.175).

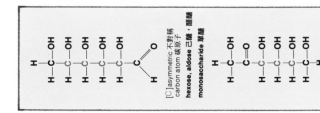

[C]asymmetric 不對稱
carbon atom 碳原子
hexose, aldose 己醣，醛醣
monosaccharide 單醣類

[C]asymmetric 不對稱
carbon atom 碳原子
hexose, ketose 己醣，酮醣
monosaccharide 單醣類

a hexose 己醣
an aldose 一種醛醣
a monosaccharide 單醣類

**ketose** (n) a monosaccharide (p.205) with a carbonyl group (p.185) giving the reactions of a ketone (p.176). A ketose is also a reducing sugar (↓). In a hexose which is also a ketose, called a **ketohexose**, there are three asymmetric carbon atoms (p.183); each asymmetric atom will have two enantiomorphs (p.183). This produces 8 possible isomers. Ketoses exhibit (p.221) the reactions of ketones and alcohols (p.175).

**reducing sugar** a sugar (p.205) which reduces Fehling's solution (p.196) and Benedict's solution (p.196). All monosaccharides are reducing sugars, but only some disaccharides.

**glucose** (n) an aldohexose. It is less sweet than sucrose (↓) and very soluble in water; the naturally occurring form is optically active (p.19), rotating the plane of polarized light to the right, and so is also known as 'dextrose'.

**fructose** (n) a ketohexose. It is the sweetest of all sugars and very soluble in water; the naturally occurring form is highly optically active (p.19), rotating the plane of polarized light to the left.

**disaccharide** (n) a sugar (p.205) which consists of two monosaccharides (p.205) chemically combined. On hydrolysis (p.66) one molecule of a disaccharide produces two molecules of monosaccharides; the monosaccharide molecules can be the same, or different. The most common disaccharides have a molecular formula (p.181) of $C_{12}H_{22}O_{11}$.

**sucrose** (n) a disaccharide (↑); on hydrolysis (p.66) it produces equal proportions of glucose (↑) and fructose (↑). A molecule of sucrose consists of a molecule of glucose and a molecule of fructose combined by a condensation (p.191) reaction. Sucrose is a non-reducing sugar (↓).

**non-reducing sugar** a sugar (p.205) which does not reduce Fehling's solution (p.196) or Benedict's solution (p.196).

**maltose** (n) a disaccharide (↑); on hydrolysis it produces glucose (↑). A molecule of maltose consists of two molecules of glucose combined by a condensation reaction (p.191). Maltose is a reducing sugar.

酮醣：酮醣（名） 含有可起酮（第 176 頁）反應之羰基（第 185 頁）的一種單醣（第 205 頁）。酮醣也是一種還原糖（↓）。己糖也是酮醣，稱為己酮醣，有一個不對稱碳原子（第 183 頁）。每一個不對稱碳原子呈現有兩種對映體（第 183 頁），可產生 8 個可能的異構體。酮醣顯示（第 221 頁）酮和醇（第 175 頁）的反應。

還原醣（名） 能使斐林溶液（第 196 頁）和本尼迪特溶液（第 196 頁）還原的一種糖（第 205 頁）。一切單醣都是還原糖，只有幾種雙醣是還原糖。

葡萄醣（名） 為一種己醛醣。甜味雖不及蔗糖（↓）但極易溶於水。其自然存在形式是光學活性的（第 19 頁），能使偏光面右旋，故稱"右旋糖"。

果醣（名） 為一種己酮醣。係所有糖中極甜者，極易溶於水，其自然存在形式是光學活性的（第 19 頁），能使偏光面左旋。

雙醣（名） 由兩個單醣（第 205 頁）分子化學結合構成的一種糖（第 205 頁）。雙醣水解時（第 66 頁），一個雙醣分子產生兩個單醣分子；所產生的異構單糖分子可以是相同，也可以是不相同的。最普通的雙醣其分子式（第 181 頁）為 $C_{12}H_{22}O_{11}$。

蔗醣（名） 不能使斐林溶液（第 196 頁）或本尼迪特溶液（第 196 頁）還原的一種糖（第 205 頁）。

非還原醣（名） 不能使斐林溶液（第 196 頁）還原的一種糖（第 205 頁）。

麥芽醣（名） 為一種雙醣（↑）；蔗糖水解時生成等化例之葡萄糖（↑）和果糖（↑）。蔗糖分子由一個葡萄糖分子和一個果糖分子縮合而成。蔗糖是一種非還原糖（↓）。

麥芽醣（名） 為一種雙醣（↑）；麥芽糖水解產生葡萄糖（↑）。麥芽糖分子由兩個葡萄糖分子縮合而成。麥芽糖是一種還原糖。

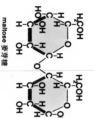

**maltose** 麥芽糖
(two hexoses combined)
**disaccharide** 雙醣
（兩個己醣結合）

**多醣：多糖 (名)** 係一種不甜的碳水化合物（第205頁）。不溶於水，不結晶，大多數多醣都有膠體（第98頁）的性質。多醣係由己醣（第205頁）通過縮合反應（第191頁）生成的，其通式為 $(C_6H_{10}O_5)_n$，其中 n 的數目很大。澱粉（↓）和纖維素（↓）是最常見的多醣。

**starch** (n) a polysaccharide (↑), which on hydrolysis produces glucose (↑). One starch molecule consists of 4000 to 30 000 glucose molecules chemically combined.

**澱粉 (名)** 係一種多醣（↑），水解時生成葡萄糖（↑）。一個澱粉分子由 4000 至 30000 個葡萄糖分子化學結合而成。

**cellulose** (n) a polysaccharide (↑), which on hydrolysis produces glucose (↑). One cellulose molecule consists of about 3000 glucose molecules.

**纖維素 (名)** 係一種多醣（↑），水解時生成葡萄糖（↑）。一個纖維素分子包含約 3000 個葡萄糖分子。

**polymerization** (n) a chemical process in which molecules of the same compound combine together to form a molecule of high relative molecular mass (p.114). The formation of starch from glucose is a form of polymerization. There are two kinds of polymerization, addition polymerization (p.208) and condensation polymerization (p.208).

**聚合作用 (名)** 由同種化合物的分子結合形成相對分子質量（第114頁）之高分子的化學過程。由葡萄糖形成澱粉就是一種聚合作用式。聚合作用分成加聚（第208頁）和縮合聚合（第208頁）兩類。（動詞為 polymerize，名詞為 polymer，形容詞為 polymeric）

**polymerize** (v), **polymer** (n), **polymeric** (adj).

**monomer** (n) a molecule or substance which can be polymerized; it has a molecule of low relative molecular mass (p.114), e.g. chloroethene (vinyl chloride) has a molecular formula (p.181) $C_2H_3Cl$; it is the monomer of polyvinyl chloride (P.V.C.) which contains between 900 to 1300 molecules of the monomer chemically combined.

**單體 (名)** 可被聚合之分子或物質；其分子有低的相對分子質量（第114頁）。例如氯乙烯（乙烯基氯）的分子式為 $C_2H_3Cl$，它為聚氯乙烯（P.V.C.）的單體，聚氯乙烯含有 900 至 1300 個化學結合的單體分子。

**dimer** (n) a molecule or compound, formed by the chemical combination of two simpler molecules, e.g. nitrogen dioxide, $NO_2$, forms a dimer, dinitrogen tetroxide, $N_2O_4$.

**二聚物 (名)** 由兩個較簡單的分子化學結合形成的一個分子或化合物。例如二氧化氮（$NO_2$）形成的二聚物為四氧化二氮（$N_2O_4$）。（名詞為 dimerization，形容詞為 dimeric，動詞為 dimerize）

**dimerization** (n), **dimeric** (adj), **dimerize** (v).

**polymer** (n) a material with a molecule of high relative molecular mass (p.114) formed by polymerization (↑). For example, using the monomer (↑) chloroethene, a polymer is produced with a relative molecular mass of 50 000 – 80 000. The constitutions of the molecules of a polymer vary, it is described as being between a range of relative molecular masses.

**聚合物 (名)** 由聚合作用（↑）生成的一種材料，其分子之相對分子質量（第114頁）極高。例如利用氯乙烯單體（↑）合成的一種聚合物，其相對分子質量為 50000 至 80000。聚合物分子的構造不同，它是被形容在一系列相對分子質量之間。

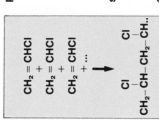

$$CH_2 = CHCl$$
$$+$$
$$CH_2 = CHCl$$
$$+$$
$$CH_2 = CHCl$$
$$+\ldots \longrightarrow$$

$$\begin{array}{ccc} & Cl & Cl \\ & | & | \\ CH_2-CH-CH_2-CH.. \end{array}$$

**polymer** 聚合物

a polymer with 900 to 1300 molecules of the monomer combined
結合 900 至 1300 個單體分子的一種聚合物

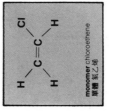

$$\begin{array}{ccc} H & & Cl \\ \diagdown & & \diagup \\ & C=C & \\ \diagup & & \diagdown \\ H & & H \end{array}$$

**monomer** chloroethene
單體 氯乙烯

$$CH_2 = CH_2 + CH_2 = CH_2 + CH_2 = CH_2 + \cdots$$

ethene (monomer)
乙烯（單體）

$$-CH_2 - CH_2 - CH_2 - CH_2 -$$
$$-CH_2 - CH_2 - CH_2 -$$

polythene (polymer)
聚乙烯（聚合物）

**addition polymerization**
**加聚作用**

**addition polymerization** (n) polymerization (p.207) in which monomers (p.207) combine together by an addition reaction, so that the polymer has the same empirical formula (p.181) as the monomer. For example, polyvinyl chloride is a polymer (p.207) with an empirical formula of $(C_2H_3Cl)_n$, where n is between 900 and 1300; the empirical formula of the monomer is $C_2H_3Cl$. Addition polymerization takes place with alkenes (p.173) and their derivatives (p.200). Chloroethene ($C_2H_3Cl$) is the monomer (p.207) of polyvinyl chloride.

**condensation polymerization** (p.207) polymerization (p.207) in which monomers (p.207) combine together in a condensation reaction (p.207). For example, aminoethanoic acid (glycine) undergoes (p.213) condensation polymerization. Most condensation polymerizations are a kind of copolymerization (↓).

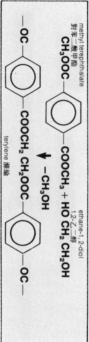

methyl terephthalate
對苯二酸甲酯

$$CH_3OOC - \bigcirc - COOCH_3 + HO\ CH_2\ CH_2\ OH$$

$$-CH_3OH$$

$$-OC - \bigcirc - COOCH_2\ CH_2\ OOC - \bigcirc - OC-$$

ethane-1,2-diol
1,2-乙二醇

terylene 滌綸

copolymerization 共聚作用

(also a condensation polymerization)
（亦為一種縮合聚合）

**copolymerization** (n) polymerization (p.207) using two, or more, monomers (p.207); e.g. the polymer terylene starts with two monomers: dimethyl benzene-1,4-dicarboxylate (methyl terephthalate) and ethane-1,2-diol. Condensation polymerization (↑) takes place with the elimination (p.189) of methanol.

**polythene** (n) a polymer (p.207) which has ethene (p.174) as a monomer (↑). Molecules of ethene in the presence of a catalyst undergo (p.213) addition polymerization (↑). Polythene is inert (p.19), a good electrical insulator, and can be moulded (p.210).

加成聚合作用（名）在第 207 頁結合一些的聚合作用（第 207 頁），因而聚合物之實驗式（第 181 頁）與其單體相同。例如聚氯乙烯是一種聚合物（第 207 頁），其實驗式為 $(C_2H_3Cl)_n$，式中的 n 介於 900 至 1300 之間，其單體之實驗式為 $C_2H_3Cl$。可由飽屬烴（第 173 頁）和它的衍生物（第 200 頁）加聚製得。氯乙烯（$C_2H_3Cl$）為聚氯乙烯之單體（第 207 頁）。

縮聚作用 單體（第 207 頁）在結合反應（第 191 頁）中結合在一起的聚合作用（第 207 頁）。例如胺基乙酸（即甘胺酸）經受（第 213 頁）縮聚作用。縮聚作用多數是一種共聚作用（↓）。

共聚作用（名）使用兩個或多個單體（第 207 頁）的聚合作用（第 207 頁）。例如聚合物滌綸以 1,4-苯二羧酸二甲酯（即對苯二羧酸甲酯）和 1,2-乙二醇兩個單體開始。隨消除（第 189 頁）甲醇而發生縮聚作用（↑）。

聚乙烯（名）以乙烯（第 174 頁）作爲單體（↑）的聚合物（第 207 頁）。乙烯分子在有催化劑的情況下經歷（第 213 頁）加聚作用（↑）。乙烯是惰性的（第 19 頁），是一種良好的絕緣體，可以用於模塑（第 210 頁）。

**peptide** (*n*) a polymer (p.207) formed by the condensation copolymerization (↑) of several amino acids (p.178) joined by a peptide bond (↓). On hydrolysis it produces amino acids.

**peptide bond** a condensation reaction (p.191) between an amino group (p.186) and a carboxyl group (p.186) with the elimination (p.189) of water. In hydrolysis the peptide bond is broken to form the original (p.220) amino acids.

**protein** (*n*) a polymer (p.207) formed from peptides (↑) by condensation polymerization (↑). A protein contains 50 or more amino acids (p.178) joined by peptide bonds (↑). On hydrolysis a protein is first decomposed to peptides and then to amino acids.

**rubber** (*n*) a naturally occurring (p.154) elastic material which is a polymerized (p.207) hydrocarbon (p.172). Destructive distillation (p.203) of natural rubber produces 2-methylbuta-1,3-diene. Synthetic rubber is generally manufactured (p.157) by the copolymerization of phenylethene (styrene) and buta-1,3-diene (butadiene). *See diagram.*

**rubber 橡膠**

$$CH_2 = C - CH = CH_2$$
$$\quad\quad\ |$$
$$\quad\quad CH_3$$

2-methyl buta-1,3-diene
monomer of
natural rubber
2-甲基丁二烯 1,3
天然橡膠之單體

**glasses** (*n,pl.*) amorphous (p.15) solids composed of silicon dioxide ($SiO_2$) with the silicon and oxygen atoms forming a tetrahedral (p.83) structure. Cations of various metals form bonds with the atoms in the tetrahedral structure. Like polymers (p.207) glasses have molecules of very high relative molecular mass (p.114). Soda glass, the normal soft glass, is manufactured (p.157) by fusing together a mixture of sand, limestone and sodium carbonate. Glasses have no fixed melting point (p.12).

肽（名） 若干個氨基酸（第 178 頁）通過肽鍵（↓）相連縮聚（↑）而成的一種聚合物（第 207 頁）。肽水解生成氨基酸。

肽鍵 一個氨基（第 186 頁）和一個羧基（第 186 頁）之間消除（第 189 頁）水之縮合反應（第 191 頁）所形成的鍵。水解時，肽鍵斷裂形成原來的（第 220 頁）氨基酸。

蛋白質（名） 由許多肽（↑）分子縮聚（↑）而成的一種聚合物（第 207 頁）。一個蛋白質分子含有 50 或 50 多個氨基酸（第 178 頁）通過肽鍵（↑）相連。水解時，蛋白質先分解成肽，再分解成氨基酸。

橡膠（名） 係一種天然存在的（第 154 頁）一種彈性材料。它是天然橡膠分解蒸餾（第 203 頁）產生 2-甲基丁二烯-1,3。一般而言，合成橡膠是用苯基乙烯（苯乙烯）和丁二烯-1,3（即丁二烯）共聚製造（第 157 頁）成的。（見圖）

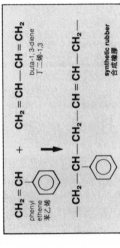

pheryl
ethene
苯乙烯

buta-1,3-diene
丁二烯-1,3

**synthetic rubber
合成橡膠**

玻璃（名，複） 由二氧化矽（$SiO_2$）組成的無定形（第 15 頁）固體，其中的矽和氧原子形成一種四面體構（第 83 頁）結構。各種金屬之陽離子與四面體結構中的原子形成鍵。玻璃與聚合物（第 207 頁）一樣，其分子的相對分子質量（第 114 頁）極高。鈉玻璃，即普通的軟玻璃是由砂、石灰石和碳酸鈉的混合物熔融一起是由砂（第 157 頁）的。玻璃無固定熔點（第 12 頁）。

**plastic²** (*n*) a material manufactured (p.157) by the polymerization (p.207) of organic (p.55) substances. It is usually a hard material, or a threadlike material used for making cloth, and does not appear to be plastic (p.14) in the usual meaning of the word. A plastic material has its shape changed, either when hot or cold, and then retains (p.215) this shape. A manufactured plastic has the property of plasticity at some stage of its manufacture. There are two types of plastics, thermoplastics and thermosetting plastics.

**thermoplastic** (*n*) a kind of plastic which exhibits plasticity when hot. The material is heated and pressed into shape in a mould (↓). On cooling it retains the shape of the mould. On heating again, its shape can be changed by pressure, e.g. polyvinyl chloride (P.V.C.) and polythene (p.208). These materials are soluble in organic solvents.

**thermosetting plastic** a kind of plastic which exhibits plasticity (p.14) when first heated; by heating, the structure of the polymer is changed and bonds are formed between molecules to produce a three-dimensional structure, making the plastic very strong when it has cooled. When the plastic is heated again, its shape cannot be altered, i.e. it has lost its plasticity, e.g. urea-methanal resins and Bakelite.

**mould** (*n*) a vessel into which a hot liquid material is poured to become cold. When cold it solidifies (p.10) in the shape of the mould. Powders can be used instead of liquids and then pressure and heat are applied (p.232) to turn the powder into a solid of the same shape as the mould. This method is used with plastics (↑). **mould** (*v*), **moulding** (*n*).

**plasticizer** (*n*) a substance which is added to a polymer (p.207) to keep it plastic, e.g. a plasticizer is added to polyvinyl chloride to alter its properties. With a small quantity of plasticizer, a strong solid is produced; with a large quantity of plasticizer, an elastic solid is produced.

---

塑膠 (名) 以有機 (第 55 頁) 物質聚合 (第 207 頁) 製成的 (第 157 頁) 一種材料。塑膠通常是一種堅硬材料或線狀材料，用的線狀材料，而並不表現出一般字面意義的塑性 (第 14 頁)。塑性材料在加熱或冷卻時可以改變其形狀，而後又可保持 (第 215 頁) 此形狀。塑性塑膠製品在加工的某個階段具有塑性的性質。塑膠分熱塑性塑膠及熱固性塑膠兩種類型。

熱塑性塑膠 (名) 受熱時顯現 (第 221 頁) 塑性的一類塑膠。此種材料可置於一個模型 (↓) 內加熱並壓製成某種形狀。冷卻之後即保留有模型的形狀。再次加熱時可藉壓力改變其形狀。例如聚氯乙烯 (P.V.C.) 和聚乙烯 (第 208 頁) 即屬此類塑膠。這類材料可溶於機溶劑。

熱固性塑膠 此類塑膠在第一次加熱時顯現塑性的 (第 14 頁)。加熱後可改變聚合物的結構，在分子之間形成鍵而產生一種三向結構，當冷卻之後，即可使塑膠極強硬。此種塑膠再次加熱時已不能改變其形狀，即已失去其塑性。例如尿醛樹脂和電木即屬此類塑膠。

模型 (名) 一種容器，在其中使液態材料冷卻。此材料冷卻之後即保持液態材料倒入變冷。當冷 (第 10 頁) 成模型之形狀，也可用粉料。然後施加 (第 232 頁) 壓力與熱，使粉料轉變與模型相同形狀的固體。此法用於製造塑膠製品 (↓)。名詞為 mould，動詞為 moulding。

塑化劑 (名) 加入聚合物 (第 207 頁) 中使之保持塑性的物質。例如聚氯乙烯加入塑化劑以改變其性質。用少量塑化劑，產生一種強硬之固體，用大量塑化劑則產生其具有彈性之固體。

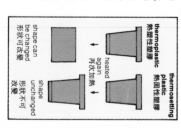

powdered plastic put in mould 粉狀塑料放入模型中

mould heated pressure applied 模型加熱及施加壓力

moulding plastic 模製塑膠製品

moulded cooled 模製成的模型冷卻

moulded plastic 模製成的塑膠製品

**thermoplastic plastic** 熱塑性塑膠

shape can be changed 形狀可改變

heated again 再次加熱

**thermosetting plastic** 熱固性塑膠

shape unchanged 形狀不可改變

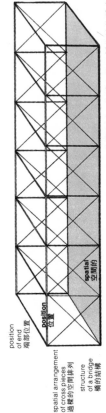

position of end
端部位置

spatial arrangement
of cross pieces
過樑的空間排列

**position**
位置

**spatial**
空間的

structure
of a bridge
橋的結構

**position** (*n*) the position of an object is its place in space in relation (p.232) to other objects. For example, (a) the position of a table in a room is its place in relation to a door, a window, or chairs in the room; (b) the position of an element in the electrochemical series (p.130) shows the relation of that element to the other elements in the series, i.e. whether it has a greater or lesser electrode potential. **position** (*v*).

**spatial** (*adj*) describes an arrangement or a direction or an extent (p.213) in space, e.g. the spatial properties of covalent bonds (p.136) describe the direction and arrangement of the bonds in space. **space** (*n*), **space** (*v*).

**limit**[1] (*n*) the value of a quantity (p.81) beyond which it is generally not possible to go. If a limit can be exceeded, *see diagram*, then different circumstances or different physical laws act on the quantity. A limit can be the largest or smallest possible value, e.g. the limit of solubility of a crystalline substance is the greatest amount that dissolves in boiling water.

**structure**[2] (*n*) the arrangement in space of the connected (p.24) parts of a whole object. For example, (a) the structure of a bridge is the arrangement in space of its different parts; (b) the structure of a molecule of ethanol (p.175) shows the arrangement in space of the atoms in the molecule. **structural** (*adj*).

**structural** (*adj*) describes anything to do with structure, e.g. structural isomerism (p.182) is a kind of isomerism which depends on molecules having different structures.

**construct** (*v*) to make a structure, e.g. to construct a model (p.223) is to make a model with a particular structure. **construction** (*n*).

**slit** (*n*) a long, very narrow hole in a surface.

位置 (名) 物體之位置係指該物體在空間中和其他物體方位的關係 (第 232 頁)。例如：(a) 某子在房內的位置是指該某子在房內的門、窗或椅子的方位的關係；(b) 電化序 (第 130 頁) 中元素位置表示該元素和電化序中其他元素的關係，即該元素是否有較高，或者較低的電極電位。(動詞爲 position)

空間的 (形) 描述在空間中的一種排列或方向或範圍 (第 213 頁)。例如共價鍵 (第 136 頁) 的空間性質是描述這些鍵在空間中的方向和排列。(名詞爲 space，動詞爲 space)

極限 (名) 指通常不可能逾越的一個量 (見圖) (第 81 頁) 的值。如果能超出這個極限 (見圖)，那麼不同的情況或不同的物理定律會影響此量。極限可以是最大的可能值，也可以是最小的可能值。例如，結晶物質的溶解度極限是其在沸水中的最大溶解量。

結構；構造 (名) 一個完整物體的各連接 (第 24 頁) 部件在空間中排列。例如：(a) 一座橋的結構是指橋的不同部件在空間中的排列；(b) 乙醇 (第 175 頁) 分子的結構表示乙醇分子中的各原子在空間中的排列。(形容詞爲 structural)

結構的 (形) 描述任何與結構有關的事物。例如，結構的同分異構現象 (第 182 頁) 是一種異構現象，它取決於分子的不同結構。

構造 (動) 製作一種結構。例如構造一個模型，製作一個模型 (第 223 頁) 是製作一個具有特定結構的模型。(名詞爲 construction)

狹縫；裂縫 (名) 在一個表面內的一個長而極狹窄的孔。

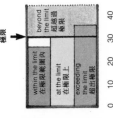

limit
極限

beyond
the limit
超越過
極限

within the limit
在極限範圍內

at the limit
在極限上

exceeding
the limit
超出極限

km/hr
千米／小時

0　10　20　30　40

slit in a surface
在一個表面
中的狹縫

**slit**
狹縫

**system** (*n*) (1) a fixed way of carrying out a process (p.157), e.g. the system of naming chemical substances. (2) a set of objects which obey physical laws, have an effect on each other and form a whole unit (↓), e.g. the substances in an equilibrium mixture (p.150) form a system. **systematic** (*adj*).

**unit** (*n*) (1) a value of a quantity which is accepted as a standard (p.229) for that quantity, e.g. the kilogram is accepted as the unit of mass. (2) a whole thing, made from different parts, which acts as a whole, e.g. a molecule (p.77) of a substance is made of different parts (the atoms) and it acts as a single object, it forms a unit. (3) each member of a series (p.172) is a unit.

**circumstances** (*n.pl.*) everything that may or may not have an effect on an object or a substance, together make the circumstances of that object or substance; e.g. the circumstances of a liquid are: the vessel containing it, the temperature, the atmospheric pressure, etc. Contrast *conditions*: *circumstances* that have an effect on a process involving the object or substance are its *conditions*.

**local** (*adj*) describes anything near an object or substance, e.g. the local conditions or the local circumstances. *Contrast* ambient (p.103) which describes anything surrounding an object.

**general** (*adj*) describes properties, qualities and nature possessed by all members of a set, e.g. the properties which all acids possess are their general properties. *Contrast* the additional properties of any one acid which are its particular properties. **generalize** (*v*), **generalization** (*n*).

**special** (*adj*) describes properties, qualities and nature which are very different for one member of a set and distinguish (p.224) it from all other members of the set.

**universal** (*adj*) (1) describes a statement which is always true and has no exceptions (p.230), e.g. a universal law. such as the attraction between unlike electric charges, has no exceptions. (2) describes a class of substances which have a characteristic action without exception, e.g. a universal solvent for organic substances.

系統：體系（名）（1）實現（第 157 頁）某一過程的命名系統；（2）遵循物理定律，依此互有影響並構成一個完整單元（↓）的各種物質構成，例如平衡混合物（第 150 頁）中的各種物質構成一個系統。（形容詞爲 systematic）

單位：單元（名）（1）人們公認爲某之標準（第 229 頁）的量值，例如千克被公認爲質量的單位；（2）由不同部分構成之作爲一個整體的完整事物，例如一物質的分子（第 77 頁）係由不同部分（原子）所組成的，而分子作爲一個單一、成爲一個單元；（3）系列（第 172 頁）中的各個成員是一個單元。

情況：情形（名）對一件事物體或一種物質鄰近的任何事物，或無影響之一切事物合一起就構成此物體或物質的情況。例如一液體的情況是：盛裝液體之容器、溫度、大氣壓力等。circum-stances（情況）與 conditions（條件）有差別：對涉及物體或物質之過程有影響的情況（cir-cumstances）就是條件（condition）。

局部的（形）描述在一個物體或物質鄰近的任何事物，或無影響之局部條件或局部情況。local 與 ambient（第 103 頁）對比：ambient（外界）係描述一個物體周圍的任何事物。

一般的（形）描述一組中所具有之任何性質、品質和本質，例如一切酸類共有之性質就是一般性質。與之對比：任何一種被所另有的性質則屬此種被獨特的性質。（動詞爲 generalize。名詞爲 generalization。）

特殊的（形）描述對一組中的某一成員而言其性質、品質及本質是極不同，並使之與組中所有其他成員相區別（第 224 頁）。

普遍的：通用的（形）（1）描述永遠爲真實而無例外（第 230 頁）的一種陳述。例如有關電荷相吸這一普遍定律沒有例外；（2）描述無例外地具有特有作用的一類物質，例如有機物質的通用溶劑。

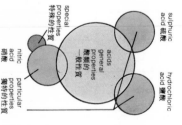

sulphuric acid 硫酸 special

use of general particular special 一般的、獨特的之用法

special properties 特殊的性質

acids general properties 酸類的一般性質

nitric acid 硝酸

hydrochloric acid 鹽酸

particular properties 獨特的性質

**exist** (v) to be, of objects, materials and substances, i.e. to be perceptible (p.42) to sight, hearing, feeling, tasting or smelling. For example, carbon exists in two crystalline forms. See occur (p.154). **existence** (n).

**extend** (v) (1) to take up space between two points or to cover an area, e.g. the coal seam (p.154) extends from a depth of 200 metres to a depth of 250 metres. (2) to take up time between two points in time, e.g. the whole process of separating the mixture by fractional distillation extended from 2.00 p.m. to 4.30 p.m. **extent** (n).

**extent** (n) (1) the space, time or activity between limits, e.g. the extent of the decomposition of phosphorus pentachloride into phosphorus trichloride and chlorine depends on the temperature and pressure (the temperature and pressure make the limits). (2) the space between limits over which something extends (↑). **extensive** (adj), **extensively** (adv).

**accompany** (v) to take place (p.63) or to exist (↑) at the same time, e.g. the reaction between metals and concentrated acids is usually accompanied by the evolution (p.40) of heat. **accompaniment** (n).

**limit²** (v) to make a limit (p.211) for time or activity, e.g. (a) refluxing (p.201) the mixture was limited to half an hour; (b) pressure limits the formation of ammonia from nitrogen and hydrogen.

**effort** (n) the use of force or energy, e.g. an effort is needed to keep the reactants in a chemical process under pressure.

**undergo** (v) to take part in an action or a chemical reaction because of an outside agent (p.63), e.g. iron undergoes rusting in damp air (the damp air is an outside agent causing the rusting).

**overcome** (v) to make an action take place when there is resistance (↓) to the action, e.g. sufficient energy has to be supplied (p.154) to reactants to overcome the energy barrier (p.152).

**resist** (v) to try to prevent an action taking place, e.g. a chemical reaction does not take place until the energy barrier (p.152) is overcome (↑); the energy barrier resists the reaction taking place. **resistance** (n).

---

存在（動）　實際上有物體、材料和物質，可由視覺、聽覺、觸覺、味覺或嗅覺感覺到（第42頁）。例如碳存在兩種晶形。參見 " 發現有 "（第154頁）。(名詞爲 existence)

延伸：延續（動）　(1) 佔據兩點間的空間或者覆蓋一面積。例如煤層（第154頁）從深度200米伸到250米；(2) 佔據兩點間的時間。例如利用分餾方法分離混合物的整個過程從下午兩點延續到下午四點半。(名詞爲 extent)

程度：範圍（名）　(1) 限度之間的空間、時間或活動。例如五氯化磷分解成三氯化磷和氯的程度取決於溫度和壓力（溫度和壓力成爲其限度）；(2) 某些事物延伸限度之間的空間。(形容詞爲 extensive，副詞爲 extensively)

伴隨（動）　同時發生（第63頁）或存在（↑）。例如金屬和濃硫酸發生反應通常伴隨放出（第40頁）熱量。(名詞爲 accompaniment)

限制（動）　造成活動的一個限度（第211頁）。例如：(a) 使混合物的回流（第201頁）限制於半小時；(b) 氮和氫反應形成氨受壓力限制。

努力（名）　力或能量之使用。例如保持化學過程中的反應物在壓力下需要作出努力。

經歷：經受：進行（動）　因一種外界作用物（第63頁）而參加一種作用或化學反應。例如鐵在潮濕空氣中經歷生銹過程（潮濕空氣引起生銹的外界作用物）。

克服（動）　必須克服物提供（↓）下發生。例如：必須反應物提供（第154頁）足夠能量才能克服能壁（第152頁）。

阻止：抵抗（動）　試圖防止發生一種作用。例如：阻止（↑）能壘（第152頁）之後才能進行化學反應：能壘阻止反應發生。(名詞爲 resistance)

**extent**
範圍

the extent of
a coal seam
煤層範圍

upper limit 上限
lower limit 下限
coal 煤
extent 範圍

**overcome**
克服

man overcomes-
resistance-
case moves
人克服阻力使箱子移動

case resists
moving
箱子阻止運動

**resist** 阻止

**cause** (n) a substance, a form of energy, or an event which makes a change, a process, or another event take place. For example, (a) oxygen in the air is a cause of iron rusting; (b) a flame is the cause of many explosions of methane in coal mines; (c) the escape of coal gas was the cause of the explosion. **cause** (v).

**effect** (n) the change, process or event produced by a cause (↑), e.g. heat causes water to boil, the boiling of the water is the effect of the heat.

**effective** (adj) describes anything that produces an effect (↑). For example, (a) the hydroxyl group in an alcohol (p.175) is the effective part of the molecule; (b) propanone (acetone) is a very effective solvent for organic substances (it produces the effect of dissolving the solute).

**effectiveness** (n).

**efficient** (adj) describes a device, a piece of apparatus, or a process which produces a result with little or no waste of energy or material. For example, (a) a fractionating column is an efficient piece of apparatus for separating a liquid mixture. (b) the Solvay process (p.169) is the most efficient way of manufacturing sodium carbonate. **efficiency** (n).

**facilitate** (v) to make an action or a process take place more easily, e.g. a catalyst facilitates the chemical combination of nitrogen and hydrogen in the Haber process (p.170). **facility** (n).

**favour** (v) to provide the conditions (p.103) so that a chemical reaction takes place more easily or more quickly, e.g. high pressure favours the combination of nitrogen and hydrogen to form ammonia, high pressure is the condition, it does not cause the reaction to take place more easily, but the pressure offers less resistance (p.213) to the change. **favourable** (adj).

**terminate** (v) to stop a process, or a chemical reaction, before it has had time to finish, e.g. the distillation was terminated when the flask cracked. **termination** (n).

**duplicate** (v) to carry out the same action twice, e.g. the experiment was duplicated so that the two sets of results could be compared (p.224). **duplicate** (n), **duplicate** (adj).

起因：原因 (名)　引致變化、過程或另一事件發生的一種物質、一種的能量形式或一件事。例如：(a) 空氣中的氧是鐵生銹的起因之一；(b) 火焰是煤礦中許多次甲烷爆炸的起因；(c) 煤氣漏是爆炸的起因。(名詞爲 cause)

效果：影響 (名)　由於某種原因 (↑) 所產生的變化、過程或事件。例如加熱引致水沸騰，水沸騰就是熱的效果。

有效的 (形)　描述任何事物能產生某種效果 (↑)。例如：醇 (第 175 頁) 的羥基是某分子的有效部分；(b) 丙酮是一種極有效的有機物質用的溶劑 (它產生溶解溶質的效果)。(名詞爲 effectiveness)

效率高的 (形)　描述一套裝置、一台儀器或一個過程很少或沒有浪費能量或材料。例如 (a) 分餾塔是分離液體混合物的高效設備；(b) 索爾維法 (第 169 頁) 是效率最高的碳酸鈉製造方法。(名詞爲 efficiency)

促進：使便利 (動)　使一種作用或一個過程更容易發生。例如在哈柏法中，催化劑促進氮和氫的化學結合 (第 170 頁)。(名詞爲 facility)

有利於 (動)　提供各種條件 (第 103 頁) 使化學反應更容易或更快發生。例如高壓力有利於氮和氫化合生成氨；高壓力是條件，它不使反應更易發生，但高壓力對變化提供較低的阻力 (第 213 頁)。(形容詞爲 favourable)

終止 (動)　使一個過程反應在達到完成之前停止下來。例如在燒瓶破裂時終止蒸餾作用。(名詞爲 termination)

重複 (動)　使相同的作用進行兩次。例如重複實驗以便比較 (第 224 頁) 兩組實驗的結果。(名詞爲 duplicate，形容詞爲 duplicate)

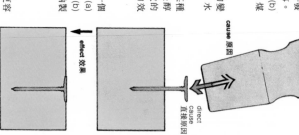

cause 原因

direct cause 直接原因

effect 效果

one key is the duplicate of the other 這一把鑰匙是另一把的複製品

duplicate 複製品

**remove** (v) to take an object, a material, or a substance from a place, e.g. (a) to remove a residue (p.31) from a filter paper; (b) to remove impurities from a metal. **removal** (n), **removable** (adj).

除去 (動) 自一處取去物體、材料或物質，例如：(a) 除去濾紙中的濾渣 (第 31 頁)；(b) 除去金屬中的雜質。(名詞為 removal，形容詞為 removable)

**substitute** (v) to put one object in the place of another object, when the two objects are not similar and do not have the same properties, e.g. to substitute a chlorine atom for a hydrogen atom in an organic compound (the two atoms are not similar and do not have the same properties). **substitution** (n).

取代 (動) 在兩件不同物體無相同性質下，用一件物體代替另一件物體。例如氯原子取代一個有機化合物分子中的氫原子 (氯原子和氫原子是不同的原子，兩者之性質不同)。(名詞為 substitution)

**exchange** (v) to put one object in the place of another object when the two objects are not similar but do have the same use or property, e.g. when a solution of ammonium sulphate passes through soil, calcium sulphate solution passes out of the soil; the ammonium ions have been exchanged for calcium ions (the two ions are not similar, but they have the same properties of being cations (p.125). **exchange** (n).

交換 (動) 當兩件物體不同但用途或性質相同時，將一件物體代替另一件物體。例如硫酸銨溶液流過土壤，硫酸鈣溶液從土壤中流出時，銨離子已為鈣離子 (兩種離子不相同但都具有陽離子 (第 125 頁) 的性質) 所交換。(名詞為 exchange)

**interchange** (v) to put one object in the place of another object when the two objects are identical, e.g. when water is in equilibrium (p.150) with its vapour in a cloud vessel, the molecules in the water interchange with the molecules in the vapour. **interchange** (n).

互換 (動) 在兩件物體完全相同時，將一件物體代替另一物體。例如在雲室中，水與其蒸汽平衡 (第 150 頁) 時，水中的分子與蒸汽中的分子互換。(名詞為 interchange)

**retain** (v) to continue to possess an object, energy, substance or property when conditions try to remove (↑) the object, energy, substance or property. For example, (a) after distillation, the flask always retains some liquid; (b) after purification of pig iron to wrought iron (p.163) the iron still retains some impurities. **retention** (n).

保留 (動) 當種種條件圖除去 (↑) 物體、能量、物質或屬性時，繼續擁有一個物體、能量、物質或屬性。例如：(a) 蒸餾之後，燒瓶內總是保留着一些液體；(b) 生鐵提純成鍛鐵 (第 163 頁) 之後，鐵中仍保留一些雜質。(名詞為 retention)

**revert** (v) to go back to an original (p.220) state, e.g. on heating rhombic sulphur slowly it changes to monoclinic sulphur at 96.5°C. On cooling, monoclinic sulphur slowly reverts to rhombic sulphur.

復原 (動) 回到原先的 (第 220 頁) 狀態。例如在 96.5°C 下將斜方晶硫慢慢加熱，它變成單斜晶硫。而冷卻後，單斜晶硫又復原為斜方晶硫。

**return** (v) to go back to an original (p.220) place, e.g. if an electron is removed from an atom, a positive ion is formed; if an electron returns to the ion, the atom is formed.

返回：回復 (動) 回到原先的一個位置。例如被從一個原子除去一個電子，形成陽離子，電子返回該離子時，又形成原子。

**exit** (n) a place through which something goes out, e.g. in a fractionating column there are several exits, one for each fraction. **exit** (v).

出口 (名) 某些東西經此出去的一個位置。例如，分餾柱有若干個出口，每一出口分別供一種餾分用。(名詞為 exit)

remove
除去

removing an
electric light bulb
除去一隻電燈泡

electron
taken away
電子被取走

return
返回

electrons retained
when other electron
removed 在其他電子
除去時保留着的電子

retain
保留

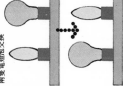

exchanging two
electric light bulbs
兩隻電燈都交換

exchange 交換

**prevent** (v) to cause an action or a process not to take place or to cause or not to happen. An agent is needed as the cause. The presence (↓) of the agent, or the action of the agent, is the cause. The agent can be an object, substance, condition or a person. **prevention** (n).

**interfere** (v) to cause a process to become slower, or to stop, or make it difficult to observe (p.42). The process is usually one that is wanted, e.g. (a) the presence of propane-1,2,3-triol (glycerol) interferes with the decomposition of hydrogen peroxide, the process is slowed down; (b) the presence of a sodium salt interferes with the flame test for other metals (makes the test difficult to observe). **interference** (n).

**counteract** (v) to act against a process so that the process is slowed down or stopped, or made to go in the reverse (↓) direction. The process is usually, but not necessarily, one that is not wanted, e.g. substances are added to rubber to prevent atmospheric oxygen destroying its properties, these substances counteract the effect of atmospheric oxygen. **counteraction** (n).

**reverse** (v) to make a process go in the opposite direction, e.g. heating a liquid causes it to vaporize (p.11), cooling the vapour reverses the process and the liquid is condensed. **reverse** (n), **reverse** (adj).

**tend** (v) to have a possible action or behaviour, with the action taking place slowly, or not taking place if conditions (p.103) are not suitable, e.g. a solution of sodium hydroxide tends to absorb carbon dioxide from the air, this can be counteracted (↑) by using a tight-fitting rubber bung (p.24). **tendency** (n).

**trend** (n) the general direction of change in a set of related facts, e.g. (a) in homologous (p.172) series the trend is for a decrease in activity with an increase in the number of carbon atoms (the general direction of related change); (b) in the set of alkali metals (p.117) the trend is for increasing ease of ionization with increasing atomic number.

防止：阻止（動） 引致一種作用或一個過程不進行或不發生。這需要有一種媒介物作為其起因。媒介物的存在（↓），或媒介物的作用就是起因、物體或物質，條件或某人都可成為媒介物。（名詞為 prevention）

干擾（動） 使一個過程進行得較慢或停止或令其難於觀察（第 42 頁），這個過程通常是所要的過程。例如：(a) 丙三醇 -1, 2, 3（即甘油）的存在會干擾過氧化氫的分解，使過程變慢；(b) 鈉鹽的存在會干擾其他金屬的火焰試驗（令試驗難以觀察）。（名詞為 inter-ference）

抑制（動） 使一個過程進行較慢，或使之朝相反的（↓）方向進行。這種過程通常不是不必要的。但也不一定是不需要之過程。例如：加入一些物質要止大氣氧破壞橡膠之性能，這些物質能抑制大氣氧的影響。（名詞為 counteraction）

使反向（動） 使一個過程向相反方向進行。例如，將液體加熱引致它汽化（第 11 頁），將蒸汽冷卻令過程反向進行，使液體冷凝。（名詞為 reverse，形容詞為 reverse）

傾向於（動） 具有一種可能的作用或進行，但行動是緩慢發生，或條件（第 103 頁）不適合時不發生。例如，氫氧化鈉溶液傾向於吸收空氣中的二氧化碳，以橡膠要止（第 24 頁）鬆緊可抑制（↑）之。（名詞為 tend-ency）

趨向（趨勢）（名） 一組相關事件中變化的一般方向。例如：(a) 在同系（第 172 頁）系列中化學活動性的趨向是隨碳原子數的增加而下降（有關變化的一般方向）；(b) 在一組鹼金屬（第 117 頁）中，其趨向是電離的容易程度隨原子數增加而增加。

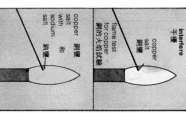

interfere 干擾

copper salt 銅鹽

copper salt with sodium salt 銅鹽 和 鈉鹽

flame test for copper 銅的火焰試驗

colour of sodium flame interferes with flame test for copper 鈉火焰的顏色干擾銅的試驗

**present¹** (*v*) to produce a thought in an observer, e.g. (a) making volatile (p.18) compounds react when heated presents a difficulty to an observer which he overcomes (p.213) by refluxing (p.201) the compounds; (b) finding a use for waste products presents a problem to a manufacturer.

**present²** (*ad j*) describes an object or substance which is in a particular, named place. For example, (a) the iodine test shows whether starch is present in food; (b) the presence of chloride ions (p.123) is tested by silver nitrate solution. **presence** (*n*).

呈現出：引起：提出 (動) 在一位觀察者中產生一種想法。例如：(a) 揮發性 (第 18 頁) 化合物受熱時發生反應使觀察者出現困難，他採取加熱的方法克服了 (第 213 頁) 此困難：(b) 向製造廠家提出尋求利用廢物的難題。

存在的：在場的 (形) 描述一件物質或一種物質是在一個特別提及的地方。例如：(a) 碘試驗顯示食品中是否存在含澱粉：(b) 存在的氯離子 (第 123 頁) 可由硝酸銀溶液檢驗出。(名詞爲 presence (n)。

**recur** (*v*) to occur (p.63) time after time, usually at a definite interval (p.220) of time, e.g. conditions for the pollution of the atmosphere recur every year when the temperature is suitable for the formation of oxides of nitrogen. **flow** (*v*).

重現 (動) 通常以一定時間間隔 (第 220 頁) 反復地出現 (第 63 頁)。例如：每年當溫度適合於形成氮的氧化物時，都重現大氣污染的條件。

**flow** (*v*) to move, of a liquid or gas, along a pipe or over a surface. Electric current and heat flow along a conductor. **flow** (*n*).

流動 (動) 液體或氣體沿管道或在一表面之上移動。電流和熱沿導體流動。(名詞爲 flow)

**obstruct** (*v*) to prevent (↑) the flow of a liquid or gas, or the flow of a stream of particles (p.110), e.g. a deposit of calcium carbonate in a water-pipe obstructs the flow of water. **obstruction** (*n*).

阻礙 (動) 阻止 (↑) 液體或氣體流動。例如，水管中的微粒 (第 110 頁) 流的流動。例如，水管中的碳酸鈣沉積物阻礙水的流動。(名詞爲 obstruction)

**stoppage** (*n*) the state of a flow being stopped, e.g. a stoppage in the delivery tube connecting a distillation flask to a condenser means that the flow of vapour has stopped and an explosion is likely.

阻塞 (動) 流動被停止着的狀態。例如，連接燒瓶和冷凝器的導出管阻塞意謂着蒸汽流已受阻，可能會發生爆炸。

**capture** (*v*) to attract and to hold an object by the use of force. To catch and to keep an escaping object. For example, (a) an atomic nucleus (p.110) captures a neutron (p.110) during bombardment (p.143) by neutrons (the neutron is held in the nucleus by nuclear forces); (b) a steam trap (p.29) captures any water escaping with the steam. **capture** (*n*).

俘獲：收集 (動) 藉力以吸引或保持一個物體。捕捉和保持逃出的物體。例如：(a) 原子核 (第 110 頁) 受中子 (第 110 頁) 轟擊 (第 143 頁) 時俘獲中子 (中子靠核力保持在核中)：(b) 瓣閥 (第 29 頁) 收集隨蒸汽逸出的水分。(名詞爲 capture)

**contact** (*n*) the state of two, or more, objects being in touch with each other. For example, (a) an electrical contact is formed when two conductors (p.122) touch each other; (b) when sulphur dioxide and oxygen are both in contact with a platinum catalyst, the two gases combine (the two gases touch each other and the catalyst). **contact** (*n*).

接觸 (名) 兩個或多個物體彼此相接觸之狀態。例如：(a) 當兩個導體 (第 122 頁) 彼此相觸形成一個電連接器：(b) 二氧化硫和氧同時與鉑催化劑接觸，此兩種氣體結合一起 (兩種氣體依此兩氣體與催化劑觸及)。(名詞爲 contact)

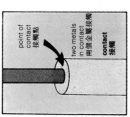

pipe
管子

water flowing
水流動

flow
流動

obstruct
阻礙

obstruction
阻礙

flow
obstructed
流動受阻礙

**obstruct**
**阻礙**

point of
contact
接觸點

two metals
in contact
兩個金屬接觸

contact
接觸

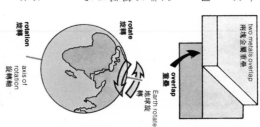

**overlap** (v) to cover part of a flat object by putting another flat object above it, e.g. the pieces of wood on the side of a boat overlap each other. **overlap** (n).

**rotate** (v) to turn, of an object, if the object turns about an axis going through the body, e.g. the Earth rotates about its axis. **rotation** (n).

**vary** (v) to change in detail (p.226) only, of a quantity (p.81), a quality (p.15), or a shape, e.g. (a) atmospheric pressure varies from day to day; (b) the electric current passing through an electrolyte (p.122) varies with the voltage (p.126) applied to the electrodes (the process of electric current being conducted by the electrolyte remains the same, only the detail of the value of the current varies). **variation** (n), **variable** (adj), **variable** (n), **varied** (adj).

**variation** (n) (1) the amount of change which takes place if a quantity or quality varies (↑); e.g. the variation in atmospheric temperature from the highest to the lowest temperature of the day, (2) the action of varying (↑).

**variable**[1] (n) a quantity, such as temperature, pressure, humidity, concentration, etc. which can change in value or can be changed in value, e.g. the vapour pressure (p.103) of a liquid is a variable because it varies (↑) with temperature.

**variable**[2] (adj) describes a quantity, quality, or shape which varies (↑) or can be varied, e.g. atmospheric pressure is variable as it changes from day to day.

**maximum** (n) (maxima n.pl.) the greatest possible, or the greatest recorded (p.39), value of a variable (↑), e.g. (a) 41°C was the maximum recorded for atmospheric temperature in the country; (b) the maximum pressure the chamber will withstand is 200 atmospheres.

**minimum** (n) (minima n.pl.) the least possible, or the lowest recorded, value of a variable (↑), e.g. 630 mm of mercury was the minimum recorded for atmospheric pressure in the country; (b) the minimum pressure recorded in the vacuum distillation was 50 mm of mercury.

重疊 (動) 於一個平的物體上方部分蓋以另一個平的物體。例如，小艇側邊的木板互相重疊。(名詞爲 overlap)

旋轉 (動) 指物體繞穿過該物體之軸轉動。例如，地球繞其軸旋轉。(名詞爲 rotation)

變化：使不同(動) 只在一個細節(第 81 頁)、質的(第 15 頁)或形狀的細節(第 226 頁)方面作改變。例如：(a) 大氣壓每天都不同；(b) 流過電解質(第 122 頁)的電流隨施于電極的電壓(第 126 頁)而變化(傳導電流的過程保持相同，僅電流值的細節不同)。(名詞爲 variation，形容詞爲 variable，名詞爲 varying，varied)

變化量 (名) (1)指質或量變化(↑)時所發生的變化總量。例如：大氣溫度由最高及最低溫度變化；(2)指不同的(↑)作用。

可變量 (名) 重量值可以改變或被改變的量。例如溫度、壓力、濕度等。液體的蒸汽壓力(第 103 頁)是一個可變量，因爲它隨溫度而變化(↑)。

可變的 (形) 描述重、質或形狀可以變化(↑)的值。例如之變化。大氣壓力是可變的，因爲氣壓日日都改變。

最高值 (名) 最大可能的或最高記錄的(第 39 頁)可變量(↑)的值。例如：(a) 41°C 是該國氣溫記錄的最高值；(b) 此容器能耐的最高壓力爲 200 大氣壓。

最低值 (名) 最低可能或最低記錄的可變量(↑)的值。例如：在該國所記錄的最低氣壓值爲 630 mm 水柱；(b) 在真空蒸餾中記錄下的最低壓力爲 50 mm 水柱。(複數爲 minima)

改變 (動) 在量、形狀或條件上作一種變化 (↑)。例如：(a) 將氣體壓力從 1 大氣壓改變成 2 大氣壓；(b) 藉加入催化劑改變可逆反應的條件 (條件的變更)。(名詞爲 alteration)

**alter** (*v*) to make one variation (↑) in a quantity, shape, or condition, e.g. (a) to alter the pressure of a gas from 1 atmosphere to 2 atmospheres; (b) to alter the condition in a reversible reaction by the addition of a catalyst (change of conditions). **alteration** (*n*).

改善：改進 (動) 改變 (↑) 一個過程或一個物體，使之更適合於一種具體用途。例如分餾柱 (第 201 頁) 可加以改良以便能生產兩種餾分 (第 202 頁) 而不是只生產一種餾分；過程和物體都被改善。在改善一個過程或一個物體時，並不改變其用途。(名詞爲 modification)

**modify** (*v*) to alter (↑) a process, or an object, to make it more suitable for a particular purpose, e.g. a fractionating column (p.201) can be modified to produce two fractions (p.202) instead of one (both a process and an object are altered). If its purpose is not changed. **modification** (*n*).

減少 (動) 指數量 (第 81 頁) 或數目變少或數目變少引致一個數量 (第 81 頁) 或數目變少引致；降低反應物 (第 62 頁) 的溫度使以逐漸減低乙醇和乙酸之間的反應速率 (第 149 頁)；(b) 氣體之體積隨施于氣體壓力增加而減少。(名詞爲 decrease)

**decrease** (*v*) to become, or to cause to become, less, of a quantity (p.81) or number. For example, (a) the rate of reaction (p.149) between ethanol and ethanoic acid is decreased by lowering the temperature of the reactants (p.62); (b) the volume of a gas decreases as the pressure on the gas increases. **decrease** (*n*).

縮減；降低 (動) 引致一個數量 (第 81 頁) 或數量之減少。例如：(a) 在電解 (第 122 頁) 過程中，電極上的沉積物隨過程之進行而積聚；(b) 當懸浮體沉降時，沉降物積聚在容器底上。(名詞爲 accumulation)

**reduce** (*v*) to cause a quantity (p.81), or number to become less, e.g. the concentration of the solution was reduced from 2.0M to 1.0M by adding a suitable volume of solvent. To contrast *decrease* (↑) and *reduce*: lowering the temperature by 80°C *decreases* the volume of the gas by 160 cm³. Lowering the temperature of the gas by 80°C *reduces* the volume to 5.30 dm³. *Decreasing* describes a continuous change; *reducing* describes a change from one state, or degree (p.227), to another. **reduction** (*n*).

縮減：降低 (動) 引致一個數量 (第 81 頁) 或數目變少。例如：(a) 與 *reduce* (縮減) 之比較 1.0 M 縮減至 1.0 M。度由 2.0 M 縮減至 1.0 M。氣體溫度下降 80°C，則 *reduce*：氣體溫度下降 80°C，則其體積縮減 (decrease) 160cm³。氣體溫度下降 80°C *decreasing*；氣體溫度下降 80°C 形容一種連續變化；*reducing* (縮減) 形容從一種狀態或程度 (第 227 頁) 變成另一種狀況或程度。(名詞爲 reduction)

積聚 (動) 在一段時間內藉加入而增加數量 (第 81 頁) 或數量。例如：(a) 在電解 (第 122 頁) 過程中、電極上的沉積物隨過程之進行而積聚；(b) 當懸浮體沉降時，沉降物積聚在容器底上。(名詞爲 accumulation)

**accumulate** (*v*) to increase a quantity (p.81) or an amount by addition over a period of time, e.g. (a) during electrolysis (p.122) the deposit on an electrode accumulates as the process goes on; (b) as a suspension settles, the sediment accumulates at the bottom of the vessel. **accumulation** (*n*).

加速 (動) 隨時間推移而增加速度。例如某些催化劑比其他催化劑更能加速反應速度；亦即這些催化劑能增加反應速率。

**accelerate** (*v*) to increase speed with time, e.g. some catalysts accelerate the rate of reaction more than others dp, i.e. they increase the rate of reaction.

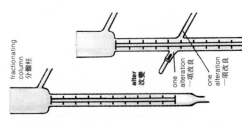

fractionating
column
分餾柱

**alter 改變**

one
alteration
一項改良

one
alteration
一項改良

column modified
to collect two
distillates
改良蒸
餾柱以收
集兩種餾出液

**modify**
**改良**

**event** (n) a change that takes place at a particular time and usually at a particular place, e.g. (a) the appearance of a precipitate is an event; (b) an explosion is an event. **eventual** (adj), **eventually** (adv).

**interval** (n) the distance between two points or the length of time between two events.

**duration** (n) the length of time in which a process takes place, e.g. (a) the duration of a process of distillation; (b) the duration of a chemical reaction. To contrast *duration* and *extent* (p.213): The *duration* of the distillation is one hour, the *extent* of the distillation is from the time of starting to heat the flask until the last drop of distillate is collected, i.e. *duration* gives the time for the process, while *extent* gives the limits of the process. **durable** (adj), **endure** (v).

**precede** (v) to come immediately before an event in time, or before a unit (p.212) in a series (p.172), e.g. number 7 precedes number 8 in the series 5, 6, 7, 8, 9, 10. Numbers 5, 6, 7, are the preceding terms for number 8, while 7 is the preceding term. **preceding** (adj).

**previous** (adj) describes events which come before a named event, or a named time, or the present time, e.g. in a previous experiment, which took place last year, the results were different. To contrast *previous* with *preceding*: the *preceding* (↑) experiment is one immediately before the present experiment; a *previous* experiment is one which took place some time ago. **previously** (adv).

**preliminary** (adj) describes the first stage of a process, e.g. preliminary tests in analysis indicate (p.38) the subsequent (↓) stages that have to be carried out.

**original** (adj) describes the very first event, action or process, e.g. Becquerel carried out the original experiments on radioactivity.

**subsequent** (adj) describes events which come after a named event, or a named time, or the present time, e.g. the results of the present experiment are not sufficient for a conclusion, subsequent experiments may give a better result. **subsequently** (adv).

事件（名）：在一段特定時間以及通常在一個特定地點發生的變化。例如：(a) 出現沉澱物是一個事件；(b) 爆炸是一個事件。（形容詞為 eventual，副詞為 eventually）

間隔（名）：兩點間的距離或兩次事件間的時間長度。

持續時間（名）：過程發生時間之長度。例如：(a) 蒸餾過程的持續時間；(b) 化學反應的持續時間。duration（持續時間）和 extent（程度）的比較（第 213 頁）：蒸餾持續時間（duration）是 1 小時，蒸餾程度（extent）是從開始加熱的時間開始至收集得最後一滴蒸餾出液的時間為止，即持續時間指明過程的時間而程度是指明過程的限度。（形容詞為 durable，動詞為 endure）

（時間、位置）在前的（形）：時間上緊接在一個事件之前，或一個系列（第 172 頁）中的一個單元（第 212 頁）之前。例如：在 5、6、7、8、9、10 系列中，數字 7 位於數字 8 之前，數字 5、6、7 是居前項的各項，而 7 是居前項。（形容詞為 preceding）

（時間或順序上）在前的：早先的（形）描述在一個事件之前，或在一個系列（第 172 頁）中的一次試驗，或先前的（previous）實驗是先發生的事件。例如在去年進行的一次實驗與 preceding（在前的）之比較：前一次實驗是不同的。Previous 前一次（preceding）（↑）實驗是緊接在本次實驗前面作的一次實驗；早先的（previous）實驗是指在某段時間進行的一次實驗。（形容詞為 previously）

初步的（形）：初步試驗指明（第 38 頁）隨後（↓）必須進行的各個試驗階段。分析中的初步試驗階段。（形容詞為 preliminary）

原始的（形）：最初的（形）描述最早的事件、作用或過程。例如貝克勒耳完成有關放射性的原始實驗。

隨後的（形）：繼起的（形）描述跟在一次指定事件之後，或指定時間之後或現在以後的一些事件。例如，目前的實驗結果不足以作結論，（但隨後的實驗也許可以得出更好的結果。（副詞為 subsequently）

interval 間隔

point A A 點
interval 間隔
pint B B 點
interval 間隔
point C C 點

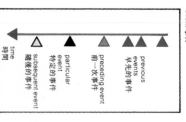

event 事件

previous events 早先的事件

preceding event 前一次的事件

particular event 特定的事件

subsequent event 隨後的事件

time 時間

**simultaneous** (*adj*) describes two or more events which take place at the same time, e.g. the closing of a switch in an electrical circuit and the flow of electric current in the circuit are simultaneous. **simultaneously** (*adv*).

同時發生的 (形) 描述兩個或多個事件在同一時間發生。例如在一個電路中，開關之閉合和電流在電路中之流動是同時發生的。(副詞為 simultaneously)

**immediate** (*adj*) describes one event which is next in time to another event without anything happening between the two events, e.g. precipitation is immediate when silver nitrate solution is added to the solution of a chloride. **immediately** (*adv*).

即刻的；直接的 (形) 描述一次事件跟著另一次事件之後發生，而兩件事件之間沒有發生任何事情。例如以將氧化物溶液加入硝酸銀溶液時即刻刻生成沉澱。(副詞為 immediately)

**order** (*n*) the arrangement of parts of a whole, or of members of a series (p.172), so that a pattern (p.93) is formed; or the arrangement follows particular rules; or the arrangement occurs (p.63) in time. For example, (a) the molecules in a giant structure are arranged in a particular pattern which has a definite order; (b) the stages of distillation follow each other in order; (c) words can be arranged in an alphabetical order. **ordered** (*adj*).

次序 (名) 一個整體的各部分的排列，或一個系列 (第 172 頁) 的各成員的排列，因而構成一種圖案 (第 93 頁)；或排列遵循特定的規則；或排列按時發生 (第 63 頁)。例如：(a) 巨大結構中的分子按列一定次序的特定模式排列；(b) 蒸餾的各階段彼此遵循一定次序；(c) 詞彙可按字母次序排列成。(形容詞為 ordered)

**method** (*n*) a particular way of carrying out a process, e.g. (a) the method of separating a liquid mixture by fractional distillation; (b) copper (II) sulphate can be prepared by digesting copper (II) oxide in sulphuric acid or by digesting copper (II) carbonate in sulphuric acid; these are two different methods.

方法 (名) 完成一個過程的具體途徑。例如：(a) 利用分離作用分離液體混合物的方法；(b) 將氧化銅(II) 放在硫酸中煮解和將硫酸銅(II) 放在硫酸中煮解都可製備得硫酸銅(II)，這是兩種不同的方法。

**control** (*v*) to start or stop a process, to increase or decrease the rate of a reaction; to vary a quantity for a purpose; e.g. to control the manufacture of ammonia by using suitable conditions of temperature and pressure and adding suitable amounts of reactants (p.62). **control** (*n*).

控制 (動) 開始或停止一個過程以加快或降低反應速度；為某種目的而改變一個量。例如利用適當的溫度和壓力條件以及加入適量反應物 (第 62 頁) 以控制氨的製造。(名詞為 control)

**progress** (*v*) of a process, to go successfully from one stage (p.159) to the next stage in the correct order (↑), e.g. as the distillation progressed, the distillate (p.201) accumulated (p.219) in the receiver. **progress** (*n*).

進行 (動) 指過程以正確的次序 (↑) 成功地從一個階段 (第 159 頁) 進入下一個階段。例如隨蒸餾之進行，餾出液 (第 201 頁) 積聚 (第 219 頁) 在接受器中。(名詞為 progress)

**exhibit** (*v*) to start a property (p.9) to be observed when it is not always open to observation (p.42). For example sodium carbonate crystals exhibit efflorescence (p.67), i.e. they can be observed to effloresce only under particular conditions.

顯示；表現 (動) 當一種性質不易被觀察 (第 42 頁) 時，使此種性質 (第 9 頁) 能夠被觀察到。例如，碳酸鈉晶體顯示出風化 (第 67 頁)，也就是說只有在特殊的種種條件下才可觀察到這些晶體發生風化現象。

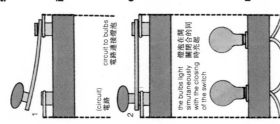

(circuit)
電路

**simultaneous**
同時發生的

1　circuit to bulbs
電路連接燈泡

2　the bulbs light simultaneously with the closing of the switch
燈泡在開關閉合的同時亮起

**statement** (n) a description of related facts in language or in symbols, e.g. a statement of Boyle's law in language or in symbols. The related facts are the mass of the gas, the pressure, its volume and its temperature. **state** (v).

**information** (n) the knowledge given by a statement (↑), e.g. information on the change in volume of a gas under particular conditions can be obtained from a statement of the gas laws (p.109). **inform** (v).

**determine** (v) (1) to find a value of a quantity if more than one measurement has to be made and the final value obtained by calculation, or to find a very accurate value of a quantity by careful measurement, e.g. to determine the boiling point of a substance using a very accurate thermometer and special apparatus. **determination** (n). (2) if one variable (p.218) depends on another variable, then the second variable determines the value of the first variable, e.g. the volume of a gas depends on its pressure, so its pressure determines its volume.

**deduce** (v) to come to a conclusion (p.43) using known facts, e.g. fact 1: substance X neutralizes an alkali; fact 2: acids neutralize alkalis; deduce that substance X is an acid. **deduction** (n).

**verify** (v) to carry out an experiment and to record results using materials and substances, when similar experiments have been carried out previously (p.220), to show that the previous results, and the deductions (↑) from the results, were true. For example, to carry out an experiment to verify the statement (↑) of Boyle's law. **verification** (n).

**assume** (v) to take a statement (↑) or a fact, without verification (↑) to be true, e.g. in a calculation on relative vapour density (p.12) we assume the gas laws. **assumption** (n).

**clarify** (v) to make a statement (↑) clear enough to understand it, e.g. a statement about the direction of the covalent bonds of a carbon atom is clarified by making a drawing of the atom and bonds. **clarification** (n).

陳述：報告（名） 以語言或以符號作有關事實的闡述。例如波義耳定律以語言或以符號的闡述。其有關事實是氣體的質量、壓力、體積和溫度。（動詞爲 state）

資料：信息（名） 陳述（↑）中所提供的知識。例如，從氣體定律（第 109 頁）的陳述中可獲得在特定條件下氣體體積變化的資料。（動詞爲 inform）

測定：決定（動）（1）作出一次以上測量，以求一個量的值再由計算獲得最後值或須經仔細的測量找出一個極準確的值。例如，利用極準確的溫度計和特殊儀器測出一種物質的沸點。名詞爲 determination 意爲測定。（2）如一個可變量第一個可變量取決於另一個可變量，那麼第二個可變量確定第一個可變量的值。例如氣體之體積決定於它所受的壓力，故壓力決定其體積。

推論：推斷（動） 利用已知事實得出結論（第 43 頁）。例如事實 1：X 物質能中和一種鹼，事實 2：酸類能中和鹼類；因此，推論 X 物質是一種酸。（名詞爲 deduction）

證實（動） 當預先（第 220 頁）已完成一些類似的實驗，爲了指出先前的實驗結果以及由這些結果作出的推論（↑）是真實實際，而利用材料和物質進行一項實驗及記錄下實驗結果。例如進行一項實驗以證實波義耳定律的陳述（↑）。（名詞爲 verification）

假定（動） 未經證實（↑）是真實而作的一個陳述（↑）或一件事實。例如在有關相對蒸汽密度（第 12 頁）的計算中，我們假定了氣體定律。（名詞爲 assumption）

闡明（動） 作出一條足夠清楚、可理解的陳述（↑）。例如從碳原子共價鍵出發的圖，以闡明有關碳原子共價鍵方向之陳述。（名詞爲 clarification）

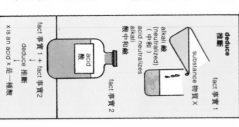

deduce 推斷

substance 物質 X

alkali 鹼
(neutralized)（中和）

acid neutralizes alkali 酸中和鹼

acid 酸

fact 事實 1

fact 事實 2

fact 事實 1 + fact 事實 2

deduce 推斷

x is an acid x 是一種酸

**model** (*n*) a device (p.23) which represents (↓) an object and allows a person to think about the object when the object itself cannot be observed (p.42). A **scale model** is an exact (p.79) representation with each part a known fraction, in size, of the object, e.g. a scale model of a house. An **adequate model** has sufficient (p.231) detail for a particular purpose, e.g. an adequate model of an atom describes the nucleus (p.110) and extra-nuclear electrons (p.110). An **analogue model** uses only similar characteristics, e.g. a permeable membrane is represented as a sieve with holes.

**represent** (*v*) to make a drawing which is like an object and allows a person to remember or to think about the object, e.g. the drawings of apparatus on page 23 represent each piece of apparatus; they are not the same as pictures of the apparatus, but the representation allows a person to imagine the actual piece of apparatus. **representation** (*n*).

**apparent** (*adj*) describes anything that appears to be correct when observed (p.42) by the senses, but can be shown by experiment, or is known, not to be correct, e.g. the apparent r.m.m. of ethanoic acid, when dissolved in some solvents, is twice the actual value, i.e. the measurement does not give a correct value.

**recapitulate** (*v*) to state (↑) facts a second time, usually in a different way, in order to clarify (↑) the information.

**refer** (*v*) to bring facts, statements, or information to a person's attention, e.g. when discussing electrovalency it is useful to refer to the atomic structure of elements, i.e. to direct a person's attention to the information on atomic structure.

**random** (*n*) describes an event, a process or a state which exhibits (p.221) no order (p.221), e.g. it is not possible to predict (p.85) the path that the molecules of a gas will follow.

**probability** (*n*) the mathematical fraction which describes chance, e.g. when a die is thrown, there is an equal chance that any one of the six faces will appear; the probability of any one particular face appearing is 1/6.

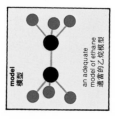

**model**
模型
an adequate
model of ethane
適當的乙烷模型

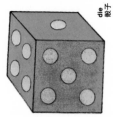

die
骰子

模型（名）　代表（↓）一件物體的裝置（第 23 頁）。當人們不可能觀察（第 42 頁）此物體本身時，此裝置使人們想像出這件物體。比例模型是一種正確的（第 79 頁）表示法，模型各個部分的尺寸才都爲該物體按比例縮小的一知分數，例如房屋的模型是按比例縮小的。適當模型足夠（第 231 頁）詳細，可用於一種特定用途，例如可用一個原子的適當模型描述原子核（第 110 頁）和核外的電子（第 110 頁）。模擬模型只使用相似的特徵，例如用一種帶有許多孔的篩來表示可滲透性膜。

代表：象徵（名）　畫一幅像一種物體的圖畫，使人們可記住該想象該物體。例如第 23 頁上的儀器圖代表各件儀器，這些圖並不等同於儀器之照片，但表示法卻使人們能想像一件真正的儀器。（名詞爲 representation）

表觀的：外顯的（形）　描述某種事物憑感覺觀察（第 42 頁）時好像是正確的，但卻可以通過實驗揭示，或被認爲此觀察不正確，例如乙酸溶於某些溶劑中時，測得乙酸的表觀相對分子質量爲實際值的兩倍，亦即該測量沒有得出正確的值。

扼要重述（動）　第二次陳述（↑）事實，通常是採用不同的方式，以便闡明（↑）該資料。

提及：參考（動）　使事實、陳述或資料引起人們注意。例如在討論電價時，提及元素的原子結構是有用的，即是說可引導人們去注意有關原子結構的資料。

隨機（名）　無規則（名）　指一次事件、一個過程或一種狀態顯示出（第 221 頁）無次序（第 221頁），如不可能預測（第 85 頁）氣體分子流動的路徑。

概率：或然率（名）　描述機會的數學分數，例如在拋擲骰子時，六個面中的任何一個面出現之機會均相等，任何一個特定面出現的概率都是 1/6。

**compare** (v) to list and discuss the similar (p. 233) properties and discuss the similar objects, or the similar characteristics of two processes, e.g. to *compare* sodium *to* potassium. If both similar and different properties or characteristics are discussed then one thing is *compared with* another. **comparison** (*n*), **comparable** (*adj*).

**contrast¹** (v) to list and discuss the different properties and characteristics of two objects or the different characteristics of two processes. **contrast** (*n*), **contrasting** (*adj*).

**contrast²** (*n*) (1) a list of the differences between two objects or processes; (2) the existence of differences between two objects or processes.

**distinguish** (v) to recognize differences in detail between two substances, two objects or two kinds of radiation (p. 138) when the substances, objects or characteristics are similar in most properties or characteristics. For example, (a) the properties of ethanol and propanol are very similar but they can be distinguished by their boiling points; (b) a person can *distinguish between* kerosene and petrol; he can *distinguish* kerosene *from* petrol. **distinction** (*n*), **distinguishable** (*adj*).

**distinction** (*n*) a difference between two substances, two objects or two kinds of radiation (p. 138) which allows the substances, objects or radiations to be distinguished (↑), e.g. distinctions can be clear or slight; a distinction can be made or can be drawn.

**distinctive** (*adj*) describes a property or characteristic which allows a substance, an object or a kind of radiation (p. 138) to be distinguished (↑) easily from all others, e.g. chlorine has a distinctive odour (p. 15) which allows it to be easily recognized.

**distinct** (*adj*) describes an object or a kind of radiation (p. 138) which possesses at least one distinctive (↑) property or characteristic, e.g. (a) copper is distinct among all metals because of its colour; (b) chlorine is distinct amongst gases because of its odour.

---

比較 (動) 列出並討論兩個物體的相似 (第 233 頁) 性質和特徵或兩個過程的相似特徵。例如將鈉與鉀作比較。比較某件事物時，既可論其相似的性質或特徵，也可論其不同的性質或特徵。(名詞為 comparison，形容詞為 comparable)

對比 (動) 列出並討論兩個物體的不同性質和特徵或兩個過程的不同特徵。(名詞或形容詞為 contrast，形容詞為 contrasting)

對比 (名) (1) 兩個物體或過程之間的差異的列表；(2) 兩個物體或過程之間差異的存在。

辨別：分別 (動) 在物質、物體或放射線 (第 138 頁) 的大部分性質或特徵都很相似的情況下，認出兩種物質、兩件物體或兩種放射線 (第 138 頁) 之間的詳細差異。例如：(a) 乙醇和丙醇的性質極相似，但可由其沸點的不同辨別出；(b) 人們可以辨別煤油和汽油，也可從汽油中辨別出煤油。(名詞為 distinction，形容詞為 distinguishable，distinct)

區別 (名) 兩種物質、兩件物體之間 (第 138 頁) 的差異，它使物質、物體或放射線可被辨別 (↑)。例如，區別可有明顯和輕微之分；可以作出區別或指出區別。

有區別的 (形) 描述使某種物質、物體或某種放射線 (第 138 頁) 能容易辨別 (↑) 出的一種性質或特徵。例如，氯有獨特的氣味 (第 15 頁)，容易認出。

(性質、種類) 不同的 (形) 描述一件物體或一種放射線 (第 138 頁) 至少具有一種特殊的 (↑) 性質或特徵，例如：(a) 銅因其顏色而與其他金屬不同；(b) 氯因氣味而與其他氣體不同。

---

A

B

① **compare A to B**
　A, B similar shape
　A, B same colour

② A, B similar shape
　A, B same colour

③ **compare A with B**
　A, B similar shape
　A, B similar shape

④ A, B same colour
　A is big
　B is small

⑤ **contrast A with B**
　A is big
　B is small

⑥ A is big
　B is small

**distinguish**
辨別

C

D

C, D same shape
same colour
C, D only the small black circle
(small detail)
C, D distinguished by small black circle

① **A 與 B 比較**
　A、B 的形狀相似
　A、B 的顏色相同

② A、B 的形狀相似
　A、B 的顏色相同

③ **A 同 B 比較**
　A、B 的形狀相似
　A、B 的形狀相似

④ A、B 的顏色相同
　A 大
　B 小

⑤ **A 同 B 對比**
　A 大
　B 小

⑥ A 大
　B 小

區分：分化 (動) 指出一種物質、一件物品或一種放射線之所以與其他相似物體或其他相似放射性不同的理由。例如波義耳利用元素不能分解而化合物可以分解的說法，將元素和化合物相區分。(名詞爲 differentiation)

檢測：探測 (動) 技巧地完成各種特殊試驗以發現一種物質、一件物體或一種放射性的存在。例如：(a) 檢測銀試樣中存在的雜質；(b) 檢測放射源射出的 α 射線。(名詞爲 detection)

鑑定 (動) 將一種物質的試樣、一個過程，或一種放射線比較並觀察其相同的 (第 233 頁) 性質和特徵之後將之說出。例如一種具揮發性且可燃燒的無色液體，其沸點爲 78°C，與乙酸反應生成酯，氧化成乙酸，可與五氯化磷爲乙醇。這些都是乙醇的性質，故將之鑑別爲乙醇。(名詞爲 identification、identify，形容詞爲 identifiable)

鑑別：證實 (名) 鑑別 (↑) 一種物質、過程或放射線作用之方法。

同一性：本性 (名) 物質、過程、或放射線之名稱，或可用於隨後鑑定性質或特徵之描述。例如在貝克勒耳發現放射性源的放射線時，並非立即確定 (↓) 放射線的真實本性。

確定：確立 (動) (1) 使一種新發現的粒子、物質、過程或放射線的本性 (↑) 被人們接受。例如查得韋克以他的實驗結果證實中子的本性。(2) 使學說或假說被人們接受。例如亞佛加德羅假學在人們認識其在測定分子質量上的用處之後才被確立。(名詞爲 establishment)

**differentiate** (v) to give reasons why a substance, an object or a kind of radiation is different from other similar objects or other similar radiations. For example, Boyle differentiated between elements and compounds by saying elements could not be decomposed whereas compounds could be decomposed. **differentiation** (n).

**detect** (v) to discover the presence of a substance, an object, or a kind of radiation by the use of particular tests, skilfully carried out, e.g. (a) to detect the presence of impurities in a specimen of silver; (b) to detect alpha radiation from a radioactive source. **detection** (n).

**identify** (v) to name a specimen of a substance, a process, or a radiation after comparing it to a known substance, process or radiation, and observing that the properties and characteristics are identical (p.233). For example, a colourless liquid is volatile and flammable, it has a boiling point of 78°C, reacts with ethanoic acid to form an ester, is oxidized to ethanoic acid and reacts with phosphorus pentachloride. These are the properties of ethanol, so the liquid is identified as ethanol. **identification** (n), **identify** (n), **identifiable** (ad).

**identification** (n) the process of identifying (↑) a substance, process, or radiation.

**identity** (n) the name of a substance, process or radiation or a description of the properties and characteristics which allow it to be used for identification subsequently, e.g. when Becquerel discovered radiation from radioactive sources, the true identity of the radiation was not immediately established (↓).

**establish** (v) (1) to make the identity (↑) of a newly discovered particle, substance, process or radiation, accepted, e.g. Chadwick established the identity of the neutron by his experimental results. (2) to make a theory or hypothesis accepted, e.g. Avogadro's hypothesis was not established until its usefulness in determining relative molecular masses became known. **establishment** (n).

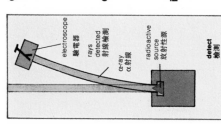

electroscope 驗電器
rays detected 射線檢測
α-ray α射線
radioactive source 放射性源

**detect** 檢測

**detail** (n) a small part of a structure, object, or process; or a fact, which is not important, in a set of facts. For example, (a) in describing the structure of a molecule of ammonia, the angles between the bonds are a detail; (b) in describing the preparation of copper (II) sulphate from copper (II) oxide and sulphuric acid, an experimental detail is that the oxide should be added in small amounts (a part of a process, or a fact in a set of descriptive facts).

**detailed** (adj) describes a statement (p.222) which gives as many details as possible.

**essential** (adj) describes any part of a whole without which the whole loses its identity (p.225). An essential property or characteristic is one that must be possessed for the purpose of identification, e.g. (a) the amino group (p.186) is an essential part of the molecule of an amine (p.178); without it the substance is no longer an amine; (b) the ability to produce hydrogen ions in solution is an essential property of an acid; if a substance does not exhibit this property it cannot be classified (p.120) as an acid. **essentially** (adv).

**major** (adj) describes any part, property, characteristic or fact which is important or is the most important, e.g. (a) the major use of chlorine is in the manufacture of plastics and synthetic rubber (fact); (b) the major source of bromine is sea water. **majority** (n).

**minor** (adj) describes any detail, property, characteristic or fact which is not important or is of lesser importance. For example, (a) a minor use of lead is in lead tetraethyl, added to petrol; (b) a minor detail in the process of distillation is controlling (p.221) the flow of water through the condenser to obtain the best conditions for condensation. **minority** (n).

**definite** (adj) describes a statement, a relation (p.232), a property or a characteristic about which there is no doubt, as previous experimental work has established (p.225) facts, e.g. aluminium oxide has a definite covalent crystalline structure, i.e. the structure has been established (p.225).

---

細節：或在：(a) 關述氨分子的結構所用的鍵間的角度就是一個細節；(b) 在用氧化銅的細節和硫酸酸製備硫酸銅 (II) 的鬆述中，實驗的細節是：氧化物愈少量加入（過程的一部分，或一組描述事實中的一件事實）。

詳細的 (形)　描述一條陳述 (第 222 頁) 提供了盡可能多的細節。

必要的：基本的 (形)　描述一個整體的任何部分：沒有這一部分，整體也就失去其本性 (第 225 頁)。必要的性質或特性是指鑑定目的所必具之性質或特性。例如：(a) 氨基是胺 (第 178 頁) 分子的必要部分，第 186 頁) 是胺 (第 178 頁) 分子的一個氨基是各氨基之物質亦不成為胺：(b) 酸的一種必要性是在溶液中能產生氫離子：不顯示此性質之物質不能歸類 (第 120 頁) 局酸。（副詞局 essentially）

主要的：較大的 (形)　描述任何細節、性質、特徵或事實是重要的或最重要的。例如：(a) 氯的主要用途是製造塑膠和合成橡膠（事實）；(b) 海水是溴的主要來源。（名詞局 majority）

次要的：較少的 (形)　描述任何細節、性質、特徵或事實較次要性較低，較不重要。例如：(a) 鉛的次要用途是：(b) 加入汽油中的四乙基鉛是鉛的次要用途；(b) 控制 (第 221 頁) 流過冷凝器的水流以達到最佳的冷凝條件是蒸餾過程的次要細節。（名詞局式局 minority）

明確的 (形)　描述一條陳述、一種關係 (第 232 頁)、一種性質或特徵不存在疑問，因為早先的實驗作業經已確定 (第 225 頁) 這些事實，例如氧化鋁有明確的巨大共價晶體結構，即兼已證實 (第 225 頁) 此種結構。

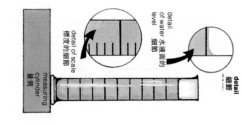

detail of water 水液面的 level 標度的細節

detail of scale 標度的細節

measuring cylinder 量筒

**detail** 細節

度：程度（名）（1）在品質（第 15 頁）的標度上的
等級。例如熱物體的熱程度可分成：微溫、
溫、熱、沸騰、熾熱、白熱、紅熱。程度這個詞描
述一種簡單的測量方法並可應用於數量（第
81 頁）以及應用於質量；（2）一個過程達到的
限度，例如，離子化程度被電解質度量的限
度。這是一種準確的（↓）量度，通常用百分
數或小數表示。

明顯的：可觀的（形）描述量（第 81 頁）或質（第
15 頁）的變化被認為足夠大或足夠重要的意
義。例如：(a) 電解過程中，電極周圍的離
子濃度明顯下降。因此，要攪拌電解質；(b)
二氧化碳在冷水中有相當大的溶解度，這點
在碳循環（第 61 頁）中有重要意義。

準確的（形）描述以最好的儀器（第 23 頁）或器
具（第 23 頁）所作出的測量結果可以利用（第
85 頁）。例如，在移吸管（第 26 頁）所標註的
溫度下使用，移吸管可準確量度液體之體
積。（名詞為 accuracy）

準確度（名）所具之準確（↑）的等級。任何儀器
和標度（第 26 頁）都不可能讀至極準確的
值。故任何測量都不可能絕對正確（第 79
頁）。在提供一個測量結果時必須指出（第
222 頁）其準確的程度（↑）。通常是指出準確
度的範圍，例如亞佛加厥羅常數確定（第
225 頁）之準確值為 (6.02252 ± 0.00028)×10²³
摩爾⁻¹；±0.00028 這個數字表示準確度範
圍，因而是該測量的準確度。

不符合：矛盾（名）當測量值必須相同或者陳
述應該是具有相同意義的兩個測量值或兩
條陳述之間出現明顯的（↑）的差異。例如同
一個溶液的三個滴定結果分別為：21.3 cm³、
22.8 cm³、21.3 cm³，則第二個結果與其他兩
個結果不符合。

**degree** (*n*) (1) a stage on a scale of quality
(p.15), e.g. the degrees of a hot object are:
lukewarm, warm, hot, boiling, red hot, white
hot. A degree describes a simple method of
measurement and can be used for quantities
(p.81) as well as for qualities. (2) the extent
reached by a process, e.g. the degree of
ionization measures the extent to which a
substance is ionized. This is an accurate (↓)
measurement and is usually stated as a
percentage or a decimal fraction.

**appreciable** (*adj*) describes a change in a
quantity (p.81) or a quality (p.15) which is big
enough or important enough to be considered,
e.g. (a) during electrolysis there is an
appreciable lowering of the concentration of
ions around an electrode; because of this, the
electrolyte is stirred; (b) carbon dioxide has an
appreciable solubility in cold water, this is
important in the carbon cycle (p.61).

**accurate** (*adj*) describes a measurement made
by the best instruments (p.23), or apparatus
(p.23), available (p.85), e.g. a pipette (p.26)
gives an accurate measurement of volume of a
liquid at the temperature marked on the
pipette. **accuracy** (*n*).

**accuracy** (*n*) the quality of being accurate (↑). No
measurement can be exact (p.79) as all instruments and scales (p.26) cannot be read to an
exact value. In giving a measurement, the
degree (↑) of accuracy must be stated (p.222).
This is usually done by giving the limits of
accuracy, e.g. the accurate value established
(p.225) for the Avogadro constant is (6.02252 ±
0.00028) × 10²³ mol⁻¹; the figure ± 0.00028
gives the limits of accuracy, and thus the
degree of accuracy of the measurement.

**discrepancy** (*n*) an appreciable (↑) difference
between two measurements or two statements
when the measurements should have been the
same or the statements should have had the
same meaning, e.g. if the results for three titra-
tions of the same solutions are: 21.3 cm³,
22.8 cm³, 21.3 cm³, then there is a discrepancy
between the second result and the other two.

**accurate**
準確的

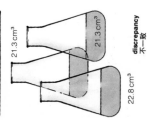

21.3 cm³

22.8 cm³

21.3 cm³

**discrepancy**
不一致

**define** (v) to state clearly, in known words, a description of a term which excludes (p.230) similar terms and allows the terms to be discussed (↓) without misunderstanding. Examples of terms are quantities, units of measurement, categories of objects and substances, properties and characteristics. **definition** (n).

**definition** (n) the result of defining a term. It can also include giving a name to a new idea, such as the naming of the neutron. Examples are: the definition (a) of an ion using the words: atom, electric charge, positive and negative, electron; (b) of electrochemical equivalent using the words: mass, deposit, coulomb. **discussion** (n).

**discuss** (v) to examine a statement and to give reasons, in writing or by speaking, for the statement, or against it, e.g. to discuss the statement that fractional distillation is not necessary to separate two liquids if their boiling points are 10°C apart (there are reasons for and against the statement). **discussion** (n).

**comment** (v) to explain, in writing or by speaking, why a statement is true or untrue, giving reasons for what is said. e.g. to comment on Boyle's law. **comment** (n).

**exemplify** (v) to give examples in order to clarify (p.222) a statement, e.g. the dehydrating properties of concentrated sulphuric acid can be exemplified by its action on sucrose, in which the sugar is converted to carbon. **exemplification** (n).

**complex** (adj) describes a structure, process or system formed from parts connected or combined, in an order that can be recognized, to form a whole, e.g. a polymer (p.207) molecule is complex; it consists of small molecules (the parts) combined into a whole, and the structure of the whole molecule can be recognized from the combination of the parts. **complexity** (n). **complex** (n). **complex** (v).

**simple** (adj) describes anything that is not made from parts, is not complex (↑), is easy to understand, e.g. compare a complex ion with a simple ion. **simplicity** (n).

下定義 (動)：闡釋 (名)　用人們熟悉的詞彙清楚地闡明一個術語之敍述，而此敍述並使這些術語能被討論 (↓) 而不致令人誤解。術語之例如：物體和物質的種類、性質和特性為術語的例子。（名詞爲 definition）

定義 (名)　對一個術語下定義之結果。定義也包括爲新概念取名，如如中子取名。使用原子、電荷、正和負、電子這些詞給離子下定義：(a)使用質量、沉積、庫侖這些名詞給電化當量下之定義。

討論 (動)　審查一條陳述並以文字或說明或反對這條陳述的理由。（名詞爲 discussion）

評論 (動)：批評 (動)　以文字或語言解釋一條陳述之爲真實或不真實，並提出所說之理由。例如評論波義耳定律。（名詞爲 comment）

舉例說明 (動)：例證 (動)　舉一些例子以闡明 (第 222 頁) 一條陳述。例如濃硫酸的脫水性質可舉遭硫酸對蔗糖的作用爲例說明。在此作用中，糖被轉化成碳。（名詞爲 exemplification）

複雜的 (形)　描述由連接的或結合的各部分構成的一個結構、過程或系統，以一種可以識別的次序形成一個整體。例如聚合物 (第 207 頁) 分子是複雜的：它係由小的分子 (部分) 結合成一個整體，而整個分子的結構可從部分的結合識別出。（名詞爲 complexity、complex、動詞爲 complex）

簡單的 (形)　描述非由部分構成的任何事物，它不複雜 (↑)，易於理解。例如簡單離子 (即錯離子) 異簡單離子的比較。（名詞爲 simplicity）

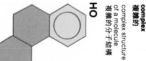
complex 複雜的
complex structure of a molecule
複雜的分子結構

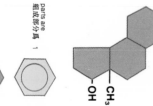
parts are
組成部分為

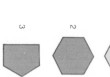

simple 簡單的
a simple molecule
簡單的分子

**uniform** 均勻的
uniform colour 顏色均勻

**non-uniform** 不均勻
non-uniform colour 顏色不均勻

**uniform** (*adj*) describes a constant (p.106) value of a quantity, or quality, which is spread over space or time, e.g. (a) a uniform colour is a constant colour spread over a surface; (b) a uniform acceleration is a constant acceleration spread over a period of time. To contrast *uniform* and *constant*: there is a *constant* concentration of carbon dioxide in the air (it is 0.03%); there is a *uniform* concentration of carbon dioxide in the atmosphere (the same concentration spread over the surface of the Earth). **uniformity** (*n*).

**steady** (*adj*) describes a value of a quantity which is kept constant (p.106) although there is a tendency (p.216) for the value to vary, e.g. to maintain a steady vacuum in vacuum distillation (the pressure tends to rise).

**standard¹** (*adj*) (1) describes the value of a quantity which is accepted by everyone as a constant (p.106) value, e.g. (a) standard atmospheric pressure is accepted as 760 mm of mercury; (b) standard electrode potentials (p.130) are those defined (↑) under known conditions of concentration and temperature. (2) describes an instrument (p.23) or a piece of apparatus (p.23) having a stated value of a quantity, or an accurate scale for a quantity, e.g. a standard graduated flask with an accurate volume at a given temperature. **standard** (*n*), **standardization** (*n*), **standardize** (*v*).

**standard²** (*n*) an accepted value of a quantity given in a definition, e.g. the standard for length is the metre, defined in S.I. units (p.212).

**normal** (*adj*) describes conditions, particularly room conditions, which are between the expected limits of variation, e.g. in northern countries, the normal room temperature is accepted to be 18°C; the temperature varies daily above or below this value, but on most days the temperature will not be too different from this value. **normality** (*n*).

**abnormal** (*adj*) describes conditions which are greatly different from normal (↑) conditions, e.g. in northern countries, a room temperature of 45°C would be considered abnormal. **abnormality** (*n*).

均勻的(形) 描述在空間或時間上延續的一個恒定不變(第 106 頁)的量或質的值。例如:(a) 均勻的顏色是塗在一個表面上均勻不變的顏色;(b) 均勻加速是加速度是延續一段時間保持不變。uniform(均勻的)與 constant(不變的)之比較:空氣中二氧化碳的濃度是不變的(爲 0.03%);大氣中二氧化碳的濃度是均勻的(在地球表面上分佈的濃度相同)。(名詞爲 uniformity)

穩定的(形) 描述一個量的值,盡管存在着使數值變化的趨勢(第 216 頁),但仍保持該量的值不變(第 106 頁)。例如在真空蒸餾中(有壓力升高的傾向)保持穩定的真空度。

標準的(形) (1) 描述一個量之值獲人們公認是一個不變的(第 106 頁)值。例如:人們認爲標準大氣壓是 760 mm 水柱高;(b) 標準電極電位(第 130 頁)是在已知濃度和溫度條件下定義的(↑)的電極電位;(2) 描述一件儀表(第 23 頁)或一件器件(第 23 頁)有一個聲稱的量值,或有一個量的準確標度。例如一個標準量瓶在給定溫度下具有準確的體積。(名 詞 爲 standard,standardize(動詞爲 standardization,動詞爲 standardize)

標準(名) 以一個定義規定的一個量的接受值。例如按國際單位(第 212 頁)制規定,長度的標準是米。

正常的(形) 描述條件,尤其是室內條件在期望的偏差限度之內。例如北方國家的人們認爲標準室溫是 18°C;每日的溫度在比值上、下波動,但往往大多數日子,溫度與此值相差不大大。(名詞爲 normality)

反常的,不正常的(形) 描述條件與正常(↑)條件大不相同。例如北方國家的人們認爲室溫45°C是反常的。(名詞爲 abnormality)

**exception** (n) anything left out of a statement or description, e.g. the chlorides of the alkaline earth metals (p.117) are not hydrolyzed (p.66) in solution with the exception of beryllium and magnesium chlorides (these two chlorides are hydrolyzed). **except** (v).

**exclude** (v) to keep out, or to put out, e.g. (a) in the preparation of iron (II) sulphate crystals, atmospheric oxygen is excluded to prevent oxidation of the salt (the air is kept out of the apparatus); (b) in the reduction of copper (II) oxide by hydrogen, air is excluded by the hydrogen before heating takes place (hydrogen pushes the air out). **exclusion** (n). **exclusive** (adj).

**excess** (n) an amount that is more than the amount needed, e.g. if 100cm³ of hydrochloric acid is the exact amount of acid needed to dissolve 5 g of zinc, then 150 cm³ of the acid is added to the zinc to make sure all the metal is dissolved. Excess acid has been added to the zinc, and there is an excess of 50 cm³ of hydrochloric acid. A volume of 150 cm³ of hydrochloric acid is 50 cm³ *in excess* of the 100 cm³ of acid needed to dissolve 5 g of zinc. **excessive** (adj).

**intense** (adj) describes a high degree of a quantity, e.g. (a) an intense heat is at a very high temperature. (b) an intense radiation is a very powerful radiation. **intensity** (n).

**converse** (n) an equal and opposite action, or an equal and opposite statement, e.g. (a) condensation is the converse of evaporation, the two actions are equal and opposite; (b) the statement *'ideal gases obey Boyle's law'* has as its converse 'non-ideal gases do not obey Boyle's law'. **converse** (adj).

**introduce** (v) to put a solid, a liquid, or a gas into a vessel (p.25) when skill is needed in the technique (p.43), e.g. to introduce a small quantity of a volatile liquid into a mercury barometer tube when measuring vapour pressure. **introduction** (n).

**insert** (v) to put an object in a fixed position, e.g. a thermometer in a cork.

---

除外：例外(名) 陳述或敍述中不考慮的任何東西。例如鹼土金屬(第 117 頁)的氯化物在溶液中不發生水解(第 66 頁)，但鈹化鎂、氯化鎂除外(這兩種氯化物可水解)。(動詞爲 except)

排除(動) 不使進入、或除去。例如：(a) 製備硫酸亞鐵(II)晶體時必須除去大氣中的氧以防止鹽發生氧化(不讓空氣進入儀器內)；(b)用氫還原氧化銅(II)時，用氫排除空氣之後再進行加熱(氫將空氣推出)。(名詞爲 exclusion，形容詞爲 exclusive)

過剩量(名) 多於所需量的一個數量。例如，如果溶解 5 g 鋅所需要的正確量爲 100cm³ 鹽酸，那麼再溶解 5g 鋅時加入 150cm³ 則鹽酸全部金屬都溶解。鋅中加入了過量的酸，酸的過剩量爲 50cm³。由於溶解 5 g 鋅只需用 100cm³ 鹽酸，因此 150cm³ 體積的鹽酸過剩量爲 50cm³。(形容詞爲 excessive)

劇烈的(形) 描述一個程度(第 227 頁)高的量。例如：(a) 酷熱是指溫度極高；(b)強輻射是指很有力的輻射。(名詞爲 intensity)

逆：逆敍(名) 指大小相等而方向相反之作用，或意思相同相反之陳述。例如：(a) 冷凝爲蒸發之逆過程，兩者的作用相同而方向相反；(b)"理想氣體服從波義耳定律"這條陳述的逆敍爲"非理想氣體不服從波義耳定律"。(形容詞爲 converse)

引入：導入(動) 當技術(第 43 頁)上需要技能時，將一種固體、液體或氣體放入一個器皿(第 25 頁)中。例如在測量蒸汽壓時將少量揮發性液體引入水銀氣壓計的導管中。(名詞爲 introduction)

插入(動) 將一個物件放在一個固定的位置。例如將溫度計插入一隻軟木塞中。

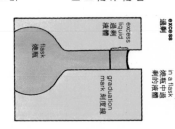

**excess** 過剩

excess liquid 過剩液體

flask 燒瓶

excess liquid in a flask 燒瓶中過剩的液體

graduation mark 刻度線

**abundant** (*adj*) describes something occurring in great quantity, especially spread over a large area, but which is not necessarily going to be used, e.g. coal is abundant in Europe.

**plentiful** (*adj*) describes something which is in great quantity and is available (p.85) for use, e.g. (a) petrol is plentiful in Arabia (there is a great quantity of it and it is put to use), (b) coal burned in a plentiful supply of air forms carbon dioxide and water. *Contrast limited* (↓).

**adequate** (*adj*) describes an amount that is equal to the amount needed for a particular purpose. The actual amount may not be stated, but is assumed (p.222) to be known, e.g. add an adequate amount of alkali to make the mixture alkaline (the amount is not known). **adequacy** (*n*).

**sufficient** (*adj*) describes an amount that is equal to the amount needed for a particular purpose, and is used with the name of the material or substance. To contrast *sufficient* and *adequate*: there is *sufficient* acid to neutralize the alkali; there is an *adequate* amount of acid to neutralize the alkali. **sufficiency** (*n*), **suffice** (*v*).

**insufficient** (*adj*) describes a quantity that is not sufficient (↑).

**inadequate** (*adj*) describes an amount that is not adequate, e.g. (a) the amount of acid was inadequate to neutralize the alkali; (b) the voltage of the electrical supply was inadequate to carry out the electrolysis; (c) his knowledge of mathematics was inadequate for his work in chemistry. **inadequacy** (*n*).

**limited²** (*adj*) (1) describes something which is only available (p.85) for use in an inadequate (↑) amount. It is the opposite of plentiful (↑), e.g. coal burned in a limited supply of air forms carbon monoxide and water. (2) describes the application of theories and laws and the use of instruments, devices and apparatus when there are limits, e.g. (a) the law of constant composition is limited to stoichiometrical compounds (p.82); (b) ammeters have a limited range of measurement.

100 cm³ alkali
neutralizes 100cm³ acid
100 cm³ 鹼中和100 cm³ 的酸

alkali 鹼

**sufficient**
足夠的

sufficient
alkali 足夠量的鹼

alkali
more than
100 cm³
鹼的量多於
100 cm³

neutralize
中和

acid 酸
100 cm³

alkali 鹼
100 cm³

adequate
amount
of alkali
適當量
的鹼

**adequate**
適當的

**豐富的**（形）描述某些東西大量存在，尤其是分佈在一個大面積上，但這些東西未必立即使用。例如在歐洲有豐富的煤。

**大量的；許多的**（形）描述某些東西的存量很巨大並可資利用（第85頁）。例如：(a) 在阿拉伯有大量的石油（有大量的石油可供利用）；(b) 煤在有大量空氣供應下燃燒生成二氧化碳和水。"有限的"（↓）作對比。

**適當的**（形）描述一個量與特定作用途所需之量相等。實際的量也許無指出，但可假定（第222頁）此量的是已知的。例如加入適當量的鹼以製備混合鹼的鹼。（名詞為 adequacy）

**足夠的；充足的**（形）描述一個量與特定用途所需之量相等，並和材料或物質的名稱連在一起使用。sufficient（足夠的）與 adequate（適當的）之比較：有足夠的酸中和鹼；有適當量的酸中和鹼。（名詞為 sufficiency，動詞為 suffice）

**不足夠的**（形）描述一個量是不足的（↑）。

**不適當的**（形）描述一個量不是適當（↑）的，或形容某物是需要的而不是適當的。例如：(a) 對中和鹼來說，酸的量不適當；(b) 電源之電壓不適合於進行電解；(c) 他的數學知識對他的化學工作不相適應。（名詞為 inadequacy）

**有限的**（形）(1) 描述某物只可以以不適當的（↑）量供利用（第85頁）。其意義與大量的（↑）相反。例如，煤在有限的空氣供應下燃燒時生成一氧化碳和水。(2) 描述理論和定律的應用，以及裝置和工具的使用而存在限制。例如：(a) 定組成比律只限用於化學計算量的化合物（第82頁）；(b) 安培計的測量範圍有限。

**deficient** (*adj*) describes a material, object, or an idea that lacks a part or is lacking in quantity in a part of it, e.g. (a) if rubber is deficient in its sulphur content it will be too soft for many purposes; (b) the laboratory is deficient in fume cupboards. **deficiency** (*n*), **deficit** (*n*).

**supplementary** (*adj*) describes an amount which is in addition to a previous (p.220) amount and is needed to complete or to improve a reaction or process. (2) describes an angle, which with another angle adds up to 180°, e.g. an angle of 124° is the supplementary angle of 56°. **supplement** (*n*), **supplement** (*v*).

**apply** (*v*) (1) to cause a force or a potential to act at a point or place, e.g. (a) to apply pressure to a gas in a vessel using a column of mercury; (b) to apply a voltage of 12 volts to the electrodes in a voltameter. (2) to use a theory or a law, e.g. Boyle's law is applied to all gases, but is only valid (↓) when applied to ideal gases (p. 107). **application** (*n*).

**valid** (*adj*) describes a statement (p.222) or experiment that is accurate and is in agreement with scientific experience, e.g. the gas laws are valid for predictions of volume changes at low pressures.

**relation** (*n*) a connection: between quantities that can vary; between cause and effect (p.214); or between objects, e.g. (a) there is a relation between the mass of a metal deposited during electrolysis and the quantity of electric charge passed through the electrolyte; (b) there is a spatial (p.211) relation between the atoms in a molecule. **relate** (*v*), **related** (*adj*).

**relative** (*adj*) describes the relation (↑) between one physical property of a substance and the same physical property of a standard (p.229) substance, e.g. (a) relative vapour density is the ratio of the density of a gas divided by the density of hydrogen; (b) relative molecular mass is the ratio of the mass of one molecule of a substance divided by 1/12 of the mass of one atom of carbon-12.

缺欠的：不足的（形）　描述一種材料、物體 或一個概念缺少了一部分或者在其一部分中 的重量不足。例如：(a) 如果橡膠的硫含量不 足，則對許多用途都會顯得太軟；(b) 這 間實驗室缺乏通風櫥。（名詞為 deficiency，deficit）

補充的：補角的（形）　(1) 描述一種數量， 前者（第 220 頁）的一個數量，而此量對於完成 或改善一個反應過程或是必要的。(2) 描述一個角和另一個角共為 180°，例 如，一個 124° 的角，其補角為 56°。（名詞為 supplement，動詞為 supplement）

施加：應用（動）　(1) 引致一個力或潛力對一點 或一個位置起作用。例如：(a) 利用一支水 銀柱對容器中的氣體施加壓力；(b) 給電壓 計的電極施加 12 伏電壓。(2) 利用一種理論 或一條定律。例如，將波義耳定律應用於一 切氣體，但只在應用於理想氣體（第 107 頁） 時才是有效的(↓)。（名詞為 application）

有效的　描述一條陳述（第 222 頁）或一項實驗不 但準確而且與科學實驗一致。例如氣體定律 對預測氣體在低壓時的體積變化是有效的。

關係（名）　指可變化的量之間、原因與結果（第 214 頁）之間　或物物體之間的關係。例如： (a) 電解過程中所沉積的金屬質量和流過電 解質的電荷量之間存在着某種關係；(b) 在 一個分子中各原子之間的空間（第 211 頁）關 係。（動詞為 relate，形容詞為 related）

相對的：相關的（形）　描述一種物質的一項物理 性質和一種標準（第 229 頁）物質的同一項物 理性質之間的關係(↑)。例如：(a) 相對蒸汽 密度是氣體密度除以氫氣密度的比值；(b) 相對分子質量是物質的一個分子的質量除以 一個碳 -12 原子質量之 1/12 的比值。

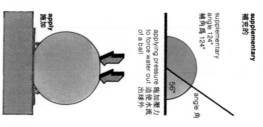

**supplementary**
補充的

supplementary
angle 124°
補角為 124°

56°

angle 角

applying pressure 施加壓力
to force water out 迫使水流
of a ball 出球外

**apply**
施加

各自的：個別的（形）描述存在兩個或以上的物體時，有關物體與其敘述的次序。例如氧、溴和銅各自的物態各自氣體、液體和固體。（形容詞爲 respectively）

**respective** (*adj*) describes the order relating objects to their descriptions when there are more than two such objects, e.g. the respective states of matter of oxygen, bromine and copper are gas, liquid and solid. **respectively** (*adv*).

對應（動）在某些部分、功能、結構或狀況方面是相似（↓）而不是相同。例如：(a) 醇分子中的羥基對應於水分子中的羥基；(b) 硝酸鹽（第52頁）離子（第123頁）的結構和碳酸根離子（第49頁）相對應。（名詞爲 correspondence）

**correspond** (*v*) to be similar (↓) in some part, function, structure or situation, without being identical, e.g. (a) the hydroxyl group in an alcohol corresponds to the hydroxyl group in water; (b) the nitrate (p.52) ion (p.123) and the carbonate ion (p.49) correspond in structure. **correspondence** (*n*).

類似的（形）如果兩件東西有許多相似的性質、品質或特徵，但是有可資辨別（第224頁）的性質、結構、品質或特徵，那麼此兩件東西是類似的。例如過渡金屬元素（第121頁）有類似的特徵，而且知每個金屬都可與其他的相辨別。（名詞爲 similarity）

**similar** (*adj*) two things are similar if they have many like properties, structures, qualities or characteristics, but have properties, structures, qualities or characteristics that distinguish (p.224) them. For example the characteristics of the transitional metals (p.121) are similar, yet each metal can be distinguished from the others. **similarity** (*n*).

等同的：全同的（形）形容兩件東西有同數目之性質和特徵，而且這些性質和特徵所佔據（↓）空間辨別之外，不可能用其他方法辨別。

**identical** (*adj*) describes two things which have the same number of properties and characteristics and these properties and characteristics are exactly the same. It is not possible to distinguish between identical objects except by the space they occupy (↓).

直接的（形）(1) 描述一個量的增加與另一個相關量的增加的一種關係；(2) 描述一種推斷（第43頁），從已知一個子集的性質而推斷整個集個集具有相同之性質。

**direct** (*adj*) (1) describes a relation in which an increase in one quantity is related to the increase in a related quantity. (2) describes an inference (p.43) in which the properties of a subset are known and the inference is that the properties of the whole set will be the same.

相反的：逆的（形）(1) 描述一種關係，其中一個量的增加與另一個量的減少有關係；(2) 描述一種推斷（第43頁），從已知一個整集的性質而推斷子集具有相同之性質。

**inverse** (*adj*) (1) describes a relation in which an increase in one quantity is related to a decrease in another quantity. (2) describes an inference (p.43) in which the properties of a whole set are known and the inference is that the properties of a subset will be the same.

佔有（動）存在於一個特定位置中而使另一個物體不能存在於此位置中。例如：(a) 晶體（第91頁）中的離子有晶格（第92頁）中之位置；(b) 鹼金屬（第117頁）類佔有週期系（第119頁）的I族。（名詞爲 occupation、occupant，形容詞爲 occupied）

**occupy** (*v*) to be in a particular place so that another object cannot be in that place. e.g. (a) the ions in a crystal (p.91) occupy positions in the lattice (p.92); (b) the alkali metals (p.117) occupy places in group I of the periodic system (p.119). **occupation** (*n*), **occupant** (*n*), **occupied** (*adj*).

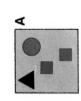

A

**correspond 對應**
the black triangles are in corresponding positions in A, B and C
黑色三角形處於A、B、C的對應位置

B

**similar 相似的**
A, B are similar
A和B相似

C

**identical 等同的**
A, C are identical
A和C是同樣的

**occupy 佔據**
the green colour occupies the same position in A, B, C
綠色佔據A、B、C的相同位置

# International System of Units (SI)
# 國際單位制

**PREFIXES FOR SI UNITS 國際單位制用的詞頭**

| MULTIPLE 倍數 (或分數) | FIGURE 數字 | PREFIX 詞頭 | SYMBOL 符號 |
|---|---|---|---|
| $10^{12}$ | 1 000 000 000 000 | tera 太 | T |
| $10^{9}$ | 1 000 000 000 | giga 吉 | G |
| $10^{6}$ | 1 000 000 | mega 兆；百萬 | M |
| $10^{3}$ | 1 000 | kilo 千 | k |
| $10^{-3}$ | 0.001 | milli 毫 | m |
| $10^{-6}$ | 0.000 001 | micro 微 | μ |
| $10^{-9}$ | 0.000 000 001 | nano 納 (毫微) | n |
| $10^{-12}$ | 0.000 000 000 001 | pico 皮 (微微) | p |

## BASIC UNITS

Système International d'Unités (SI) is based on seven basic units; the seventh, the candela, is not included in the list below, as it is not used in chemistry.

**metre** (*unit of length*)  *symbol:* m
 Defined from a wavelength in the spectrum of krypton.

**kilogram** (*unit of mass*)  *symbol:* kg
 The mass of the international prototype kept at Sèvres.

**second** (*unit of time*)  *symbol:* s
 Defined from a frequency in the spectrum of the caesium-133 atom.

**kelvin** (*unit of temperature*)  *symbol:* K
 The fraction 1/273.16 of the temperature of the triple point of water.

**mole** (*unit of amount*)  *symbol:* mol
 The amount of substance which contains as many elementary units as there are atoms in 0.012 kilogram of carbon-12. The elementary unit must be specified and may be an atom, a molecule, an ion, an electron, a radical, etc.

**ampere** (*unit of electric current*)  *symbol:* A
 Defined from a newton and a metre.

## 基本單位

國際單位制 (SI) 以七個基本單位作為基礎；第七個單位坎德拉並非化學用之單位，不在此列出。

**米** (長度單位)  符號：m
 由氫光譜中的一個波長下定義。

**千克** (質量單位)  符號：kg
 保存於法國塞弗爾市的國際市克原型 (原器) 之質量。

**秒** (時間單位)  符號：s
 由銫-133 原子的光譜中的一個頻率下定義。

**開爾文** (溫度單位)  符號：K
 水三相點溫度的 $\frac{1}{273.16}$。

**摩爾** (物質量的單位)  符號：mol
 含有基本單元數等於 0.012 千克碳-12 原子之數目的物質的量。基本單元須指明，可以是原子、分子、離子、電子或原子團等。

**安培** (電流單位)  符號：A
 由牛頓和米下定義。

## DERIVED UNITS 導出單位

| QUANTITY. 量 | SYMBOL FOR QUANTITY 量的符號 | UNIT 單位 | SYMBOL FOR UNIT 單位符號 | BRIEF DEFINITION 簡單定義 |
|---|---|---|---|---|
| velocity 速度 | v | m/s 米／秒 | — | displacement/time 位移／時間 |
| acceleration 加速度 | a | m/s² 米／秒² | — | velocity/time 速度／時間 |
| force 力 | F | newton 牛頓 | N | mass × acceleration 質量 × 加速度 |
| energy 能 | E | joule 焦耳 | J | force × distance 力 × 距離 |
| pressure 壓強：壓力 | P | pascal 帕斯卡 | Pa | force/unit area 力／單位面積 |
| density 密度 | ρ | kg/m³ 千克／米³ | — | kilogram/cubic metre 千克／立方米 |
| frequency 頻率 | f | hertz 赫茲 | Hz | number of oscillations/time 振盪次數／時間 |
| concentration 濃度 | — | mol/dm³ 摩爾／分米³ | M | moles/cubic decimetre 摩爾數／立方分米 |
| electric charge 電荷 | Q | coulomb 庫侖 | C | amperes × time 安培 × 時間 |
| electric potential 電位 | V | volt 伏特 | V | joule/coulomb 焦耳／庫侖 |
| electric resistance 電阻 | R | ohm 歐姆 | Ω | volt/amperes 伏特／安培 |
| volume 體積：容積 | V | dm³ 分米³ | — | cubic decimetre 立方分米 |

# Symbols Used in Chemistry
# 化學中使用的符號

| LETTER/SYMBOL 字母／符號 | QUANTITY, OBJECT OR OPERATOR | |
|---|---|---|
| A | mass number. $A_r$ relative atomic mass | 質量數。$A_r$ 相對原子質量 |
| E | energy, electromotive force | 能量；電動勢 |
| e | electron. $_{-1}^{0}e$ charge and mass of electron | 電子。$_{-1}^{0}e$ 電子的電荷和質量 |
| F | Faraday constant | 法拉第常數 |
| f | frequency | 頻率 |
| H | heat of reaction | 反應熱 |
| I | electric current | 電流 |
| k | a constant | 常數 |
| L | Avogadro constant | 亞伏加德羅常數 |
| M | concentration in moles per cubic decimetre; $M_r$ relative molecular mass | 每立方分米的摩爾數目表示的濃度 $M_r$ 相對分子質量 |
| m | mass | 質量 |
| N | number of molecules | 分子數 |
| N | neutron number | 中子數 |
| n | any number; mole fraction; number of moles | 任何數；摩爾分數；摩爾數目 |
| n | a neutron. $_{0}^{1}n$ charge and mass of neutron | 中子。$_{0}^{1}n$ 中子的電荷和質量 |
| p | pressure | 壓力 |
| p | a proton. $_{1}^{1}p$ charge and mass of proton | 質子。$_{1}^{1}p$ 質子的電荷和質量 |
| Q | quantity of electric charge | 電荷量 |
| R | molar gas constant, resistance | 摩爾氣體常數；阻力；電阻 |
| r | gas constant, radius | 氣體常數；半徑 |
| T | thermodynamic temperature (measured in kelvin) | 熱力學溫度（以開爾文計算） |
| t | time. $t_{\frac{1}{2}}$ half life | 時間。$t_{\frac{1}{2}}$ 半衰期 |
| V | volume, electric potential difference; $V_m$ molar volume | 體積；電位差。$V_m$ 摩爾體積 |
| Z | atomic number | 原子序（數） |
| △ | a change, e.g. $\triangle H$ change in heat | 一種變化，例如 $\triangle H$ 為熱量之變化 |
| θ | temperature difference, temperature (Celsius scale) | 溫差：溫度（攝氏溫標） |
| ρ | density | 密度 |

# Important constands
# 重要常數

s.t.p. = standard temperature and pressure which is 1.00 atmospheres or 760 mm of mercury or 101 kPa and 273 K or 0°C.

Temperature of the triple point of water is 273.16K.

Absolute zero of temperature is 0 K or − 273°C.

Volume of 1 mole of gas at s.t.p. is 22.4 dm³ (molar volume).

Molar gas constant, 8.314 J K⁻¹ mol⁻¹.

Avogadro constant, L = 6.02 × 10²³ mol⁻¹.

Faraday constant, F, 9.65 × 10⁴ C mol⁻¹.

Mass of electron, 9.11 × 10⁻³¹ kg.

Ratio of masses, proton/electron, 1840.

Ratio of masses, neutron/electron, 1840.

Charge on an electron, 1.6 × 10⁻¹⁹ C.

Ionic product of water $K_w$ = 1.008 × 10⁻¹⁴ mol² dm⁻⁶ (298 K or 25°C).

1 calorie = 4.18 joules.

1 electron-volt (1 eV) = 1.6 × 10⁻¹⁹ J.

標準溫壓 = 標準溫度和壓力，以 1.00 大氣壓（或 760 毫米汞柱高或 101 kPa）和 273 K（或 0°C）表示。

水的三個相點之溫度為 273.16 K

溫度絕對零度為 0 K 或 −273°C

在標準溫壓下，1 摩爾氣體之體積為 22.4 升（摩爾體積）

摩爾氣體常數為 8.314 焦耳・開氏度⁻¹・摩爾⁻¹

亞伏加德羅常數，F = 9.65 × 10⁴ 庫侖・摩爾⁻¹

法拉第常數，L = 6.02 × 10²³ 焦耳・摩爾⁻¹

電子質量為 9.11 × 10⁻³¹ 千克

質量比：質子／電子，1840

質量比：中子／電子，1840

一個電子之電荷為 1.6 × 10⁻¹⁹ 庫侖

水的離子積 $K_w$ = 1.008 × 10⁻¹⁴ 摩爾²・分米⁻⁶（298 k 或 25°C）

1 卡 = 4.18 焦耳

1 電子伏（1 eV）= 1.6 × 10⁻¹⁹ 焦耳

# Common alloys 普通合金

| NAME OF ALLOY 合金名稱 | APPROXIMATE COMPOSITION 近似成分 | USES 用途 |
| --- | --- | --- |
| brass 黃銅 | zinc 鋅 35%－10%, copper 銅 65%－90% | decorative metal work 裝飾性金屬工件 |
| bronze 青銅 | | |
| —common 普通 | zinc 鋅 2%, tin 錫 6%, copper 銅 92% | machinery, decorative work 機械、裝飾性工件 |
| —aluminium 鋁 | aluminium 鋁 10%, copper 銅 90% | machinery castings 機器鑄件 |
| —coinage 鑄幣合金 | zinc 鋅 1%, tin 錫 4%, copper 銅 95% | coins 硬幣 |
| dentist's amalgam 牙醫用汞合金 | copper 銅 30%, mercury 汞 70% | dental fillings 補牙填料 |
| duralumin 硬鋁合金 | magnesium 鎂 0.5%, manganese 錳 0.5%, copper 銅 5%, aluminium 鋁 94% | framework of aeroplanes 飛機的架構 |
| gold 金合金 | | |
| —coinage 金幣 | copper 銅 10%, gold 金 90% | coins 金幣 |
| —dental 牙料用 | silver 銀 28%－14%, copper 銅 14%－28%, gold 金 58% | dental fillings 補牙填料 |
| lead, battery plate 蓄電池極板鉛合金 | antimony 銻 6%, lead 鉛 94% | accumulators 蓄電池 |
| manganin 錳銅合金 | nickel 鎳 1.5%, manganese 錳 16%, copper 銅 82.5% | resistance wire 電阻絲 |
| nichrome 鎳鉻合金 | chromium 鉻 20%, nickel 鎳 80% | heating elements, resistance wire 加熱元件、電阻絲 |
| pewter 白鑞(錫鉛合金) | lead 鉛 20%, tin 錫 80% | utensils 家用器皿 |
| silver 銀合金 | | |
| —coinage 鑄幣 | copper 銅 10%, silver 銀 90% | coins 銀幣 |
| solder 焊料 | tin 錫 50%, lead 鉛 50% | joining iron objects 鉾合鐵製物件 |
| steel 鋼合金 | | |
| —stainless 不銹鋼 | nickel 鎳 8%－20%, Chromium 鉻 10%－20%, iron 鐵 80%－60% | utensils 家用器皿 |
| —armour 鐵甲 | nickel 鎳 1%－4%, chromium 鉻 0.5%－2%, iron 鐵 98%－95% | armour plating 鐵甲板 |
| —tool 工具 | chromium 鉻 2%－4%, molybdenum 鉬 6%－7%, iron 鐵 95%－90% | tools 各種工具 |

# Common abbreviation in Chemistry
## 化學上常用的縮寫詞

| abbr. | full | 中文 |
|---|---|---|
| abs. | absolute | 絕對的 |
| anhyd. | anhydrous | 無水的 |
| approx. | approximately | 近似地 |
| aq. | aqueous | 水的；含水的 |
| b.p. | boiling point | 沸點 |
| conc. | concentrated | 濃縮的 |
| concn. | concentration | 濃度 |
| const. | constant | 常數 |
| crit. | critical | 臨界的 |
| cryst. | crystalline | 結晶的 |
| d. | decomposed | 分解的 |
| decomp. | decomposition | 分解 |
| dil. | dilute | 稀釋 |
| dist. | distilled | 蒸餾的 |
| e.g. | for example | 例如 |
| e.m.f. | electromotive force | 電動勢 |
| eqn. | equation | 反應式 |
| expt. | experiment | 實驗 |
| fig. | figure (diagram) | 插圖（圖）；數字（值） |
| f.p. | freezing point | 冰點 |

| hyd. | hydrated | 水合的 |
|---|---|---|
| i.e. | that is | 亦即；即是 |
| insol. | insoluble | 不溶的 |
| liq. | liquid | 液體 |
| max. | maximum | 最大值（的）；最高值（的） |
| min. | minimum | 最小值（的）；最低值（的） |
| m.p. | melting point | 熔點 |
| p.d. | potential difference | 位差；電位差 |
| ppt. | precipitate | 沉澱 |
| r.a.m. | relative atomic mass | 相對原子質量 |
| r.m.m. | relative molecular mass | 相對分子質量 |
| sol. | soluble | 溶解的 |
| soln. | solution | 溶液 |
| sp. | specific | 特殊的；比的；明確的 |
| s.t.p. | standard temperature and pressure | 標準溫度和壓力 |
| temp. | temperature | 溫度 |
| vac. | vacuum | 真空 |
| v.d. | relative vapour density | 相對蒸汽密度 |
| wt. | weight | 重量 |
| ° | degree (celsius) | 度（攝氏） |

# Understanding scientific words
## 理解意義的科學用詞彙

New words can be made by adding **prefixes** or **suffixes** to a shorter word. Prefixes are put at the front of the shorter word and suffixes are put at the back of the word. Words can also be broken into parts, each of which can have a meaning, but cannot be used alone.

(i) correct → incorrect
    correct → correctness
    correct → incorrectness

(ii) **isomorphism** is broken into iso-morph-ism
    **iso-** is a prefix which means 'identical in structure'
    **morph** is a word part which means 'form or shape'
    **-ism** is a suffix which means 'a condition'

Hence *isomorphism* means the condition of having identical forms or shapes; it describes the condition of two crystalline substances.

Prefixes describing numbers or quantities are taken from Greek or Latin words. The following table shows the common prefixes from these two languages. Prefixes, suffixes, and word parts are listed alphabetically in separate sections after the table.

一個簡短的單詞添加語頭或詞尾構成新詞。詞頭置於該短詞之前，詞尾置於其後。單詞也可斷分成各有其意義但不能單獨使用的幾部分。

(i) correct —— incorrect （加詞頭）
    correct —— correctness （加詞尾）
    correct —— incorrectness （加詞頭和詞尾）

(ii) **isomorphism** 可以斷分成 iso-morph-ism
    **iso-** 爲詞頭，義爲 "結構相同"
    **morph-** 爲詞部，義爲 "形式或形狀"
    **-ism** 爲詞尾，義爲 "某種狀態"

因此，isomorphism 義爲具有相同形式或形狀之狀態；它描述兩種晶物質之狀態。

表示數字或數量的詞頭都源於希臘或拉丁詞。下表列出源於這兩種語言的常用詞頭、詞尾之後按字母順序分段列出各詞頭、詞尾和構詞成分。

| | GREEK PREFIX 希臘詞頭 | LATIN PREFIX 拉丁詞頭 | PREFIX 詞頭 | MEANING 意義 | 詞源 |
|---|---|---|---|---|---|
| 1 一 | mono- | uni- | hemi- | half 半 | Gr 希臘 |
| 2 二 | di- | bi- | semi- | half 半 | L 拉丁 |
| 3 三 | tri- | ter- | poly- | many 多；聚 | Gr 希臘 |
| 4 四 | tetra- | quad- | multi- | many 多 | L 拉丁 |
| 5 五 | penta- | quinq- | omni- | all 全；總 | L 拉丁 |
| 6 六 | hexa- | sex- | dupli- | twice 兩倍；兩次 | L 拉丁 |
| 7 七 | hepta- | sept- | tripli- | three times 三倍；三次 | Gr 希臘 |
| 8 八 | octo- | oct- | hypo- | less, under 次；較少；在下 | Gr 希臘 |
| 9 九 | nona- | novem- | hyper- | more, over 超；過；高；在上 | Gr 希臘 |
| 10 十；分 | deca- | deci- | sub- | under 亞；次；在下；較低 | L 拉丁 |
| 100 百；厘 | hecta- | centi- | super- | over 過；超；高於 | L 拉丁 |
| 1000 千；毫 | kilo- | milli- | iso- | same, equal, identical 相同、相等、同一 | Gr 希臘 |

## PREFIXES 詞頭

**a-** without, lacking, lacking in, e.g. *amorphous*, being without shape; *asymmetrical*, without symmetry, or lacking in symmetry.

無：非；缺：缺對稱的。例如 *amorphous* 無定形的；*asymmetrical* 不對稱的。

**allo-** different, or different kinds, e.g. *allotropy*, the existence of an element in two or more different forms.

異：異類。例如 *allotropy* 同素異形現象，指一種元素存在兩種或以上的不同形狀。

**amphi-** on both sides, e.g. *amphoteric*, having the nature of both an acid and a base.

兩：雙：兼：在兩側。例如 *amphoteric* 兩性的。

**an-** the same prefix as **a-**, used in front of words beginning with a vowel, or the letter *h*, e.g. *anisotropic*, not having the same properties in all directions; *anhydrous*, being without, or lacking, water, in a crystal.

同詞頭與 **a-** 加在以元音或字母 *h* 為首之詞前。同詞頭與 **a-**。例如 *an-isotropic* 各向異性的；*anhydrous*（指晶體中）無水或缺水的。

**anti-** opposite in direction, e.g. *anticatalyst*, a catalyst which slows down a chemical reaction, i.e. works in the opposite direction to a catalyst.

方向或或位置異相反。例如 *anticatalyst* 反催化劑，指使催化劑減慢之反方向催化作用的催化劑。

**auto-** caused by itself, e.g. *autoxidation*, reaction of a substance with atmospheric oxygen at room temperature; the substance oxidizes itself; *autocatalysis*, a chemical reaction in which the products act as catalysts for the reaction.

自：自動。例如 *autoxidation* 自氧化，物質在室溫下自與大氣的氧反應；*autocatalysis* 自催化，產物起催化劑作用的化學反應。

**cis-** on the same side, e.g. *cis-compound*, an isomer in which two like groups are on the same side of the double bond in the compound. See *trans-*.

在同側（順式）。例如 *cis-* compound 順式化合物，指分子中兩個相同基團在雙鍵同一側的同分異構物。見 *trans-*。

**co-** acting together, with, e.g. *cohesion*, the force holding two or more objects together.

共：共同：和。例如 *cohesion* 內聚力，指保持兩個或多個物體的力。

**counter-** acting against, acting in the opposite direction, e.g. *counteract*, to act against, such as a mild alkali counteracts the effect of acid on skin; *counterclockwise*, turning in the opposite direction to the hands of a clock.

對抗：對逆：反方向作用。例如 *counteract* 抵消：反作用，例如弱鹼抵銷皮膚的影響；*counterclockwise* 逆時針的，指與時針方向相反的轉動。

**de-** opposite action, e.g. *decompression*, the lessening of a pressure, it is the opposite action to compression; *deactivate*, to make less active, it is the opposite of activate.

相反的動作。例如 *decompression* 減壓，與壓縮作用相反：*deactivate* 鈍化，與活化相反。

**dia-** through, across, e.g. *diameter*, the line going across a circle, through the centre.

通過：橫過。例如 *diameter* 直徑，為通過圓心之橫過一段直線。

**dis-** opposite action, e.g. *discharge*, to take an electric charge away from a charged body, the opposite of charge; *disconnect*, to break, or open, a connection, the opposite of connect.

反：非：對反。例如 *discharge* 放電，排除帶電體之電荷並為相反之反；*disconnect* 分離：分開，使連接斷開或脫開：連接之反。

**equi-** having the same number, equal, e.g. *equimolecular*, having the same number of molecules; *equilibrium*, the condition of two rates of reaction being equal and opposite, so that there is no further change in a reversible reaction.

同數：相等。例如 *equimolecular* 等分子數目的；*equilibrium* 平衡，兩個可逆反應速度相等的再發生變化之情況。

**im-** the opposite, not. (Used with words beginning with b, m, p.) For example, *imperfect*, not perfect, the opposite of perfect; *impermeable*, not permeable.

不：非：無等否定含義（置非以 b、m、p 起首之詞前）。例如 *imperfect* 不完美的，完美之反；*impermeable* 不通過的。

**in-** the opposite, not. (Used with all words other than those beginning with b, m, p.) For example, *inactive*, the opposite of active; *inadequate*, not adequate.

不：非：無等否定含義（置非以 b、m、p 起首的一切詞之前）。例如 *inactive* 不活潑的：活潑之反，*inadequate* 不適當的。

**infra-** below, e.g. *inframolecular*, having a size smaller than a molecule, so the size is below molecular size.

在下：較低。例如 *infra-molecular* 亞分子的，即尺寸小於一個分子的。

**inter-** between, among, e.g. *interface*, a common surface between two liquids or two solids; *interstice*, a narrow space between two solid objects.
之間；之中。例如 *interface* 界面，為兩液體或兩固體間的公共面；*interstice* 間隙，為兩固體之間的狹窄空間。

**macro-** great, large, e.g. *macromolecule*, a large molecule composed of many smaller molecules, as in a polymer.
大量的；巨大的。例如 *macromolecule* 大分子，由許多較小分子組成的，如聚合物。

**micro-** small, especially if too small to be seen by the human eye alone, e.g. *microbalance*, a balance used for measuring masses of less than 1 mg; *microanalysis*, analysis using very small amounts of substances.
微小的。尤指微小得不能讓單單人眼觀察。例如 *microbalance* 微量天平，測量少於一毫克重量之天平；*microanalysis* 微量分析，以極少量物質作分析的方法。

**non-** not, e.g. *non-electrolyte*, a substance which is not an electrolyte; *non-ferrous*, any metal other than iron.
非；不；無。例如 *non-*electrolyte 非電解質；*non-*ferrous 非鐵的：(任何)非鐵金屬。

**ortho-** straight, right-angled, upright, e.g. *orthogonal*, with parts at right-angles; *orthorhombic*, a crystal system with three unequal axes at right-angles.
直；直角的；直立的。例如 *orthogonal*(各部分成)直角的；*ortho-*rhombic 斜方晶的，指三個不等軸互成直角之晶系。

**pan-** all, complete, every, e.g. *panchromatic*, covering all wavelengths of light in the spectrum.
總；全。全色。例如 *panchromatic* 全色的，包括光譜中的全部光波長。

**para-** at the side of, by, e.g. *paracasein*, an insoluble form of casein, formed when soluble casein coagulates.
側；旁。例如 *paracasein* 副酪蛋白，由可溶酪蛋白凝固所成之不溶解形酪蛋。

**pseudo-** has the same appearance, but is false, e.g. *pseudoalum*, a substance which has the appearance of an alum, but is not an alum.
貌同質假的。例如 *pseudoalum* 假礬，一種具外觀，卻非礬的物質。

**re-** again, e.g. *reactivate*, to make something activated again; *recrystallize*, to crystallize again.
再；重。例如 *reactivate* (使某物)再活化；*recrystallize* 再結晶。

**syn, sym-** joined together, united, e.g. *synthesis*, combining elements or compounds to make new compounds.
連；合。例如 *synthesis* 合成，指元素或化合物結合形成新的化合物。

**trans-** across, on the opposite side of, e.g. *trans-compound*, an isomer in which two like groups are on opposite sides of the double bond in the compound. See *cis-*.
橫貫；在相反兩側(反式)。例如 *trans-compound* 反式化合物，化合物分子中兩相同基團在雙鍵兩側之同方異構物。見 *cis-*。

**ultra-** beyond, e.g. *ultrafilter*, a filter which has holes so small it filters out colloids; it thus has uses beyond those of the ordinary filter.
超。例如 *ultrafilter* 超濾器，能濾出膠質的多微孔濾器，其用處為普通濾器所不及。

**un-** not, the opposite, e.g. *unsaturated*, means not saturated; *unstable*, means not stable; *unpaired*, means not in a pair, and so by itself.
不；反等否定意義。例如 *unsaturated* 不飽和；*unstable* 不穩定；*unpaired* 不成對的，即獨自的。

## SUFFIXES 詞尾

**-able** forms an adjective which shows an action can possibly take place, e.g. *changeable*, something which can change; *transformable*, something which it is possible to transform.
構成形容詞，表示一種作用有可能發生。例如 *changeable*(某物) 可以改變；*transformable*(某物) 可以轉變。

**-al** of, or to do with; forms a general adjective, e.g. *experimental*, of, or to do with, experiment; *fractional*, of, or to do with, fractions; *thermal*, of, or to do with, heat.
構成普通形容詞，表示"……的"，與實驗有關的。例如 *experimental* 或用分數詞來表示"關於……的"；*fractional* 分數的或用分數詞表示；*thermal* 熱力的。

**-ed** forms the past participle of a verb, can be used as an adjective; it shows an action under the control of an experimenter, e.g. *varied*, describes a quantity changed by an experimenter; *dehydrated*, describes a substance from which water has been removed under the control of an observer.
構成動詞的過去分詞，作形容詞。表示一個量受實驗者改變作用。例如 *varied* 已改變的，形容一種物質已在觀察控制下除去所含的水分。*dehydrated* 已脫水的，形容一種物質已在觀察控制下除去所含的水分。

**-er (-or)** forms a noun from a verb and describes an agent, e.g. *mixer*, a device which mixes; *desiccator*, a device that desiccates; *generator*, a device that generates a gas.
從動詞構成名詞，描述一種工具。例如 *mixer* 混合器；*desiccator* 乾燥器；*generator* 氣體發生器。

**-gram**
forms a noun describing a record which is written or drawn, e.g. chromatogram, the recorded result from an experiment on chromatography; telegraph.

構成名詞，記述寫成繪成下之記錄。例如 chromatogram 色譜：色層法實驗之記錄；telegram 電報，由電報機所錄之訊息。

**-graph**
forms a noun describing an instrument or device that records variation in a quantity, or other information, e.g. thermograph, a kind of thermometer which records changes of temperature over a period of time; telegraph, a device which records information in words.

構成名詞，描述記錄量變或其他信息用之儀器或裝置。例如 thermograph 溫度記錄器，能錄一段時間內溫度變化之溫度計；telegraph 電報機，用字記錄信息的裝置。

**-ic**
forms a general adjective, e.g. basic, of, or to do with, a base; cyclic, of, or to do with, a cycle; ionic, of, or to do with, ions.

構成一般形容詞，表示"屬……的"，"關於……的"。例如 basic 鹼性的；cyclic 循環的；ionic 離子的。

**-ify**
forms a verb which is causative in action, e.g. purify, to cause to become pure; solidify, to cause to become solid.

構成動詞的現代分詞，作形容詞，表示非受實驗者控制化。例如 purify 淨化；solidify 固化。

**-ing**
forms the present participle of a verb, can be used as an adjective, it shows an action not under the control of an experimenter, e.g. fluctuating, describes a quantity varying above and below an average value, which cannot be controlled by an observer; disintegrating, describes a radioactive substance undergoing disintegration, as the process cannot be controlled by an observer.

構成動詞的現代分詞，作形容詞，表示非受實驗者控制化，且不受觀察者控制之某量值上、下變化。例如 fluctuating 波動的；disintegrating 形容放射性物質經受蛻變，不受觀察者所控制之過程。

**-ity**
forms a noun of a state or quality, e.g. purity, the quality or state of being pure; acidity, the quality of being acid.

構成名詞，表示一種狀態或性質。例如 purity 純度，純淨，指純的質或狀態；acidity 酸性，指是酸的質。

**-ive**
forms an adjective by replacing -ion in nouns; the adjective describes an agent producing the effect described by the noun, e.g. inhibitive, describes an agent causing inhibition; oxidation → oxidative, describes a process causing oxidation; explosion → explosive, describes an agent causing an explosion.

構成形容詞，取代名詞中的 -ion 構成形容詞；形容詞產生該名詞所描述的效應之作用物。例如 inhibition —— inhibitive 抑制的的作用物；oxidation —— oxidative 氧化的；explosion —— explosive 爆炸的，描述引致氧化之作用物；描述引致爆炸之作用物。

**-ize**
forms a verb which is causative in the formation of something, e.g. ionize, to cause ions to be formed; polymerize, to cause polymers to be formed.

構成動詞，表示形成某物的成因。例如 ionize 離子化；polymerize 聚合。

**-lysis**
forms a noun describing the action of breaking down into simpler parts, e.g. hydrolysis, the decomposition of a compound by the action of water; electrolysis, the decomposition of a substance by an electric current.

構成名詞，表示分裂成為較簡單部分的作用。例如 hydrolysis 水解；electrolysis 電解。

**-meter**
forms a noun describing an instrument which measures quantitatively, e.g. thermometer, an instrument which measures temperature accurately; voltmeter, an instrument which measured electric potential in volts.

構成名詞，描述定量用的一種儀器。例如 thermometer（準確測溫度之）溫度計；voltmeter（以伏特為單位測量電位之）伏特計。

**-metry**
forms a noun describing a particular science of accurate measurement, e.g. thermometry, the science of measuring temperature; hydrometry, the science of measuring the density of liquids.

構成名詞，描述準確測量的專門學科。例如 thermometry 測溫學；hydrometry 液體密度測量學。

**-ness**
forms an abstract noun of state or quality, e.g. sweetness, the quality of being sweet; softness, the quality of being soft.

構成表示"狀態或性質"之抽象名詞。例如 sweetness 甜性；softness 軟性。

**-ous**
forms an adjective showing possession, or describing a state, e.g. anhydrous, being in the state of not possessing water, homologous, in the state of being a homologue; homogenous, in the state of having the same properties throughout a substance.

構成表示"具有"，或表示一種狀態。例如 anhydrous 無水的，不含水之狀態；homologous 同系的，例如 homogenous 均勻的，指全部物質都具相同性質之狀態。

**-philic**
forms an adjective describing a liking for something, e.g. protophilic, describes a substance which accepts protons.
構成形容詞，描述親某物。例如 protophilic 親質子的，描述接受質子的物質。

**-phobic**
forms an adjective describing a dislike for something, e.g. lyophobic, describes a colloid which does not go readily into solution.
構成形容詞，描述疏某物。例如 lyophobic 疏液的，描述不易加入溶液中之膠體。

**-scope**
forms a noun describing an instrument which measures qualitatively, e.g. spectroscope, an instrument by which spectra can be observed qualitatively; hygroscope, an instrument which measures qualitatively the humidity of the atmosphere.
構成名詞，描述定性測定之儀器。例如 spectroscope 分光鏡，係可定性觀察光譜之儀器；hygroscope，係可定性量測大氣濕度之儀器。

**-scopy**
forms a noun describing the use of instruments for observation in science, e.g. microscopy, the use of microscopes for scientific observation.
構成名詞，描述將儀器用於科學觀察。例如 microscopy 顯微鏡檢查法，係指顯微鏡用於科學觀察。

**-stat**
forms a noun describing a device which keeps a quantity constant, e.g. hydrostat, a device which keeps water in a boiler at a constant level; thermostat, a device which keeps a liquid, or an object, at constant temperature.
構成名詞，描述使量恆定之裝置。例如 hydrostat，（使鍋爐水位不變的）恆溫器；thermostat（使液體或物體維持恆溫的）恆溫器。

**-tion**
forms an abstract noun. With -ation, it forms a noun of action, e.g. pollution, the result of polluting; concentration, the degree to which a solution is concentrated; distillation, the noun of action from distil; precipitation, the noun of action from precipitate.
構成抽象名詞。用 -ation 時構成表示作用的名詞。例如 pollution 污染，作用；concentration 濃度；distillation 蒸餾作用；precipitation 沉澱作用。

## WORD PARTS 詞部

**aqua**
water, to do with water, e.g. aqueous, a solution containing water; aquaion, an ion with molecules of water associated with it.
水，與水有關。例如 aqueous 水的，含水之溶液；aquaion 水合離子，係篩含有水分子的離子。

**chrom**
colour, to do with colour, e.g. panchromatic, all the colours, and hence all the wavelengths of the visible spectrum; chromatography, the analysis of complex substances in which a coloured record of the analysis is produced.
顏色，與顏色有關。例如 panchromatic 全色的，即可見光譜之全部波長；chromatography 色譜分析法，產生有顏色記錄的分析法。

**gen**
to produce, e.g. homogenize, to make a mixture of solid and liquid substances into a viscous liquid of the same texture throughout; generate, to produce energy or a flow of gas.
引起，產生。例如 homogenize 均化，使固體和液體混合物地成為相同的黏稠液體；generate 產生能量或氣流。

**hydr**
water or liquids, e.g. dehydrate, to remove water; anhydrous, describes a substance without water.
水或液體。例如 dehydrate 脫水，除水；anhydrous 無水的，描述無水的物質。

**hygro**
damp or humid, e.g. hygroscopic, attracting water from the atmosphere to become damp; hygrometer, an instrument that measures the relative humidity of the atmosphere.
濕或潮濕。例如 hygroscopic 吸濕的；hygrometer（量度大氣相對濕度的）濕度計。

**morph**
shape or form, e.g. amorphous, describes a substance which is without a crystalline form; polymorphism, existing in different forms.
形狀或形式。例如 amorphous 無定形的，描述非晶體態的物質；polymorphism 多形晶性，同質多晶形現象。

**photo**
light, e.g. photolysis, decomposition caused by light; photohalide, any halide which is decomposed by light.
光。例如 photolysis 光分解作用；photohalide 能為光分解的感光性鹵化物。

**pneumo**
air or gas, e.g. pneumatic trough, a trough for the collection of gases.
空氣或氣體。例如 pneumatic trough 集氣槽。

**pyro**
great heat, e.g. pyrolysis, decomposition caused by heating; pyrometer, a special kind of thermometer for measuring very high temperatures.
高溫。例如 pyrolysis 熱解，加熱引致的分解；pyrometer 高溫計，係量度極高溫度用之特種溫度計。

**therm**
heat, e.g. thermostable, stable when heated; thermal, of, or to do with heat; thermometer, an instrument for the quantitative measure of temperature.
熱。例如 thermostable 耐熱的，描述加熱時保持穩定的；thermal 熱的，和熱有關的；thermometer 溫度計，係定量量度溫度之儀器。

Bilingual Edition Publisher : Willie Shen
雙語版出版人：沈維賢

Author : Arthur Godman
原著者： 亞瑟・戈德曼

Managing Editor : Aman Chiu
策劃編輯： 趙嘉文

Translator: Chan Kai Kan
翻譯： 陳繼勤

Reviser : Wong Yee Chu
審訂： 黃宜鑄

Editor : Chan Kai Kan
編輯： 陳繼勤

transition elements 過渡元素

| period 週期 | group I I族 | group II II族 | transition elements 過渡元素 | | | | | group III 族 |
|---|---|---|---|---|---|---|---|---|
| 1 | 1<br>Hydrogen 氫<br>H<br>1.01 | | | | | | | |
| 2 | 3<br>Lithium 鋰<br>Li<br>6.94 | 4<br>Beryllium 鈹<br>Be<br>9.01 | | | | | | 5<br>Boron<br>B<br>10.8 |
| 3 | 11<br>Sodium 鈉<br>Na<br>22.99 | 12<br>Magnesium 鎂<br>Mg<br>24.31 | | | | | | 13<br>Aluminium<br>Al<br>26.9 |
| 4 | 19<br>Potassium 鉀<br>K<br>39.10 | 20<br>Calcium 鈣<br>Ca<br>40.08 | 21<br>Scandium 鈧<br>Sc<br>44.96 | 22<br>Titanium 鈦<br>Ti<br>47.90 | 23<br>Vanadium 釩<br>V<br>50.94 | 24<br>Chromium 鉻<br>Cr<br>52.00 | 25<br>Manganese<br>Mn<br>54.9 | 31<br>Gallium<br>Ga<br>69.7 |
| 5 | 37<br>Rubidium 銣<br>O<br>85.47 | 38<br>Strontium 鍶<br>Sr<br>87.62 | 39<br>Yttrium 釔<br>Y<br>88.91 | 40<br>Zirconium 鋯<br>Zr<br>91.22 | 41<br>Niobium 鈮<br>Nb<br>92.91 | 42<br>Molybdenum 鉬<br>Mo<br>95.94 | 43<br>Technetium<br>T<br>98.9 | 49<br>Indium<br>In<br>114.8 |
| 6 | 55<br>Cesium 銫<br>Cs<br>132.91 | 56<br>Barium 鋇<br>Ba<br>137.33 | 57 ●<br>Lanthanum 鑭<br>La<br>138.91 | 72<br>Hafnium 鉿<br>Hf<br>178.49 | 73<br>Tantalum 鉭<br>Ta<br>180.95 | 74<br>Tungsten 鎢<br>W<br>183.85 | 75<br>Rhenium<br>Re<br>186. | 81<br>Thallium<br>Tl<br>204. |
| 7 | 87<br>Francium 鈁<br>Fr（鈁）<br>(223) | 88<br>Radium 鐳<br>Ra<br>226.03 | 89 ●●<br>Actinium 錒<br>Ac<br>227.03 | 104<br>Rutherfordium 鑪<br>Rf<br>(261) | 105<br>Hahnium 𨧀<br>Hn<br>(260) | 106<br>(263) | | |